Chemical Kinetics and Reaction Mechanisms

Chemical Kinetics and Reaction Mechanisms

Dr. Kamlesh Bilgaiyan

ARCLER PRESS

www.arclerpress.com

Chemical Kinetics and Reaction Mechanisms

Dr. Kamlesh Bilgaiyan

Arcler Press
4164 Lakeshore Road
Burlington, ON L7L 1A4
Canada
www.arclerpress.com
Email: orders@arclereducation.com

ISBN: 978-1-77956-959-2 (Hardcover)

ABOUT THE AUTHOR

Dr. Kamlesh Bilgaiyan (Ph D & NET) a well known widely acclaimed assistant professor of Uttar Pradesh top rated Bundelkhand University, Jhansi (NAAC A+ Accredited University). Besides 10 U K design patents and two published patent, several publications. He has contributed pool of research by guiding two PhDs and more than 27 M Phil dissertations. He has published more than 20 papers in international and national reputed journals. Dr Bilgaiyan is having vast administrative experience of more than 20 years in various administrative positions. Presently he has working as Coordinator of B Sc, institute of Basic Science. He is also subject experts for Board of studies of various universities. Presently working on kinetics, material science and physicochemical properties of various plant extracts. Having more than 23 Years research experience.

TABLE OF CONTENTS

List of Figures .... *xi*

Preface .... *xvii*

Chapter 1 Introduction to Chemical Kinetics 1

Introduction 1

1.1. Fundamental Concepts 7

1.2. Rate Laws 12

1.3. Reaction Mechanisms 16

1.4. Temperature and Kinetics 20

1.5. Catalysts in Chemical Reactions 25

1.6. Experimental Techniques in Kinetics 29

Summary 34

References 36

Chapter 2 Rate of Reaction and Order 41

Introduction 41

2.1. Factors Affecting Reaction Rate 46

2.2. Measuring Reaction Rates 52

2.3. Reaction Order Explained 57

2.4. Determining Reaction Order 62

2.5. Applications of Reaction Rate and Order 67

2.6. Case Studies in Reaction Rates 71

Summary 75

References 77

Chapter 3 Temperature Dependence of Reaction Rates 83

Introduction 83

3.1. Theoretical Background 87

3.2. Kinetic Molecular Theory 93
3.3. Experimental Evidence 98
3.4. Arrhenius Equation 102
3.5. Catalysts and Temperature 107
3.6. Real-World Applications 111
Summary 115
References 118

Chapter 4 Complex Reactions 123
Introduction 123
4.1. Types of Complex Reactions 128
4.2. Mechanisms of Complex Reactions 132
4.3. Factors Influencing Complex Reactions 136
4.4. Applications of Complex Reactions 140
4.5. Case Studies of Complex Reactions 146
4.6. Challenges in Studying Complex Reactions 151
Summary 155
References 158

Chapter 5 Polymerization Kinetics 163
Introduction 163
5.1. Types of Polymerizations 167
5.2. Mechanisms of Polymerization 172
5.3. Factors Affecting Polymerization Rates 177
5.4. Kinetic Models in Polymerization 181
5.5. Experimental Techniques for Studying Polymerization 185
5.6. Applications of Polymerization Kinetics 190
Summary 196
References 198

Chapter 6 Catalysis 207
Introduction 207
6.1. Types of Catalysts 212
6.2. Mechanisms of Catalysis 218
6.3. Factors Affecting Catalytic Activity 223
6.4. Kinetics of Catalytic Reactions 227

6.5. Applications of Catalysis 232
6.6. Challenges in Catalysis 237
Summary 241
References 244

Chapter 7 Fast Reactions 249
Introduction 249
7.1. The Basics of Chemical Kinetics 253
7.2. Types of Fast Reactions 258
7.3. The Role of Temperature 263
7.4. The Influence of Concentration 268
7.5. Techniques for Studying Fast Reactions 273
7.6. Applications of Fast Reaction Kinetics 278
Summary 282
References 285

Chapter 8 Reaction Dynamics 289
Introduction 289
8.1. Fundamental Principles of Reaction Dynamics 294
8.2. Types of Chemical Reactions 298
8.3. Theoretical Frameworks 303
8.4. Kinetics and Thermodynamics 308
8.5. Experimental Techniques in Reaction Dynamics 313
8.6. Applications of Reaction Dynamics 317
Summary 322
References 325

Chapter 9 Photochemical Kinetics 331
Introduction 331
9.1. Fundamental Principles of Photochemistry 336
9.2. Mechanisms of Photochemical Reactions 342
9.3. Factors Influencing Photochemical Kinetics 347
9.4. Experimental Techniques in Photochemical Kinetics 352
9.5. Applications of Photochemical Kinetics 357
9.6. Challenges and Limitations in the Field 362

Summary 366
References 368

Chapter 10 Surface Reactions and Adsorption Kinetics 371
Introduction 371
10.1. Fundamentals of Surface Chemistry 374
10.2. Adsorption Phenomena 380
10.3. Kinetic Models of Adsorption 385
10.4. Experimental Techniques for Studying Adsorption 390
10.5. Factors Influencing Adsorption Kinetics 395
10.6. Applications of Surface Reactions and Adsorption Kinetics 400
Summary 406
References 408

Index 413

LIST OF FIGURES

Figure 1.1. Chemical kinetics
Figure 1.2. Understanding chemical kinetics
Figure 1.3. Importance of chemical reaction
Figure 1.4. Svante Arrhenius
Figure 1.5. Fundamental concepts of chemical kinetics
Figure 1.6. Rate of reaction
Figure 1.7. Factors affecting reaction rate in chemical kinetics
Figure 1.8. The role of concentration in chemical kinetics
Figure 1.9. Zero-order, first-order, and second-order reactions
Figure 1.10. Arrhenius equation
Figure 1.11. Temperature dependence of the rate constant
Figure 1.12. Catalysis
Figure 1.13. Role of catalysts in chemical reactions
Figure 1.14. Zeolites among the most widely used catalysts in the petrochemical industry
Figure 1.15. How to measure the rate of a chemical reaction
Figure 2.1. Rate of reaction
Figure 2.2. Concept of reaction rate
Figure 2.3. Reaction order
Figure 2.4. Factors affecting rate of reaction
Figure 2.5. Reaction rate tends to increase with concentration phenomenon explained by collision theory
Figure 2.6. Effect of temperature on the kinetic energy distribution of molecules in a sample
Figure 2.7. Presence of a catalyst increasing the rate of a reaction by lowering its activation energy
Figure 2.8. In a reaction that produces a gas, the volume of the gas produced can be measured using the gas-syringe method

Figure 2.9. General spectrophotometer diagram

Figure 2.10. Two different ways to measure the volume of a gas that is produced

Figure 2.11. Chemical reaction orders

Figure 2.12. Order of reaction

Figure 2.13. Zero-order reactions

Figure 2.14. Graphical representation of a first-order reaction

Figure 2.15. Methods of determining reaction order

Figure 2.16. Integrated rate laws

Figure 2.17. Initial rates method

Figure 3.1. Temperature dependence of reaction rates. (A) Reaction rates calculated according to the Arrhenius formula. The blue curve represents an enzymatic reaction (E_a = 40 kJ/mol [9.5 kcal/mol]), whereas the green curve denotes the uncatalyzed counterpart with higher activation energy (E_a = 85 kJ/mol [20.3 kcal/mol]). The corresponding Q_{10} values are shown for a temperature shift between 20 and 30°C. (B) Temperature dependence of an imaginary enzymatic reaction rate based on Peterson et al. that takes enzyme denaturation into account, with a melting temperature of 45°C, in equilibrium condition. The Q_{10} values indicate a gradual decrease with temperature: Q_{10} (10/20°C) = 1.81, Q_{10} (20/30°C) = 1.65, and Q_{10} (30/40°C) = 1.27

Figure 3.2. Definition of reaction rates

Figure 3.3. Temperature influence on rate of reaction

Figure 3.4. Collision theory of chemical reactions

Figure 3.5. Activation energy

Figure 3.6. Influence of temperature on particle motion and thermal expansion

Figure 3.7. Kinetic molecular theory

Figure 3.8. Kinetic theory of gases

Figure 3.9. A distribution of molecular kinetic energies as a function of temperature. The blue curve is for a low temperature, whereas the red curve is for a high temperature

Figure 3.10. Understanding Arrhenius equation

Figure 3.11. Graph of Arrhenius equation

Figure 3.12. The addition of a catalyst to a reaction lowers the activation energy, increasing the rate of the reaction. The activation energy of the uncatalyzed reaction is shown by E_a, while the catalyzed reaction is shown by E'_a. The heat of reaction (ΔH) is unchanged by the presence of the catalyst

Figure 4.1. A complex reaction mechanism involving energy barriers and transition states

Figure 4.2. A complex chemical reaction passes through different transition states and adopts different completion routes before yielding the products

Figure 4.3. Importance of chemical processes

Figure 4.4. Reversible reactions

Figure 4.5. Potential-energy curve. The activation energy represents the minimum amount of energy required to transform reactants into products in a chemical reaction. The value of the activation energy is equivalent to the difference in potential energy between particles in an intermediate configuration (known as the transition state or activated complex) and particles of reactants in their initial state. The activation energy thus can be visualized as a barrier that must be overcome by reactants before products can be formed

Figure 4.6. Industrial processes

Figure 4.7. Biological systems

Figure 4.8. Environmental chemistry

Figure 4.9. Haber process

Figure 4.10. Schematic of photosynthesis in plants. The carbohydrates produced are stored in or used by the plant

Figure 4.11. Combustion reaction

Figure 5.1. Polymerization kinetics

Figure 5.2. Polymerization

Figure 5.3. Types of polymerizations

Figure 5.4. Addition polymerization

Figure 5.5. Condensation polymerization

Figure 5.6. Mechanisms of polymerization

Figure 5.7. Steps of chain-growth polymerization of ethylene to form polyethylene

Figure 5.8. Exploring differential scanning calorimetry (DSC)

Figure 5.9. Gel permeation chromatography (GPC)

Figure 5.10. Nuclear magnetic resonance (NMR) spectroscopy

Figure 5.11. Material science

Figure 5.12. Pharmaceutical development

Figure 5.13. Environmental problems caused by polymers

Figure 6.1. Catalysis

Figure 6.2. How does a catalyst work

Figure 6.3. Chemical reactions

Figure 6.4. Common catalyst types with emphasis on heterogeneous and homogeneous catalysts

Figure 6.5. Homogeneous catalysts

Figure 6.6. Heterogeneous catalysts

Figure 6.7. Three-dimensional structure of an enzyme. Biocatalysis utilizes these biological macromolecules to catalyze small-molecule transformations

Figure 6.8. Transition state theory model of catalysis. X‡ (A) is the transition state for a reaction in solution. X‡ (B) is the transition state for the enzyme-mediated reaction

Figure 6.9. Diagram of a catalytic reaction, showing the energy level as a function of the reaction coordinate. For a catalyzed reaction, the activation energy is lower

Figure 6.10. Activation energy

Figure 6.11. Rate law

Figure 6.12. Order of reaction

Figure 6.13. Effect of catalysts

Figure 6.14. Various applications of catalysis in the modern world

Figure 6.15. Process flow chart of biodiesel production from pretreated *WCO*. Legend: M-1, M-2: mixer; H-1: heater; R-1: esterification reactor; T-1: methanol rectification tower; H-2: cooler; B-1: liquid–liquid separator; T-2: biodiesel rectification tower

Figure 6.16. Environmental applications

Figure 6.17. Pharmaceutical synthesis

Figure 7.1. The kinetics of chemical reactions

Figure 7.2. Basic concepts in chemical reactions

Figure 7.3. Understanding the rate of reaction

Figure 7.4. Factors that affect reaction rate

Figure 7.5. Elementary reactions

Figure 7.6. Catalytic reactions

Figure 7.7. Measure the change in temperature during chemical reactions

Figure 7.8. Influence of temperature on reaction rates

Figure 7.9. The Arrhenius equation

Figure 7.10. Reaction profile for an elementary step over an activated barrier of height E_a

Figure 7.11. Effect of concentration

Figure 7.12. A continuous flow fast kinetic system

Figure 7.13. Diagram of a stopped-flow instrument

Figure 7.14. A flash-photolysis relaxation experiment

Figure 7.15. Understanding the environmental chemistry

Figure 8.1. Understanding reaction dynamics: Exploring the pathways of chemical reactions

Figure 8.2. Formulas for rate of reaction

Figure 8.3. Factors affecting rates of chemical reaction

Figure 8.4. Types of chemical reactions

Figure 8.5. Synthesis reaction

Figure 8.6. Decomposition reaction

Figure 8.7. Redox reaction

Figure 8.8. Transition state theory

Figure 8.9. Reaction coordinate diagram for a typical zeolite-catalyzed, aqueous-phase reaction. Initially, the reactant (*R*) is solvated by the liquid phase outside the porous catalyst, and the proton in the catalyst is not hydrated (a). The hydration of the proton by water forms the specific acid (b), which is followed by the adsorption of *R* in the catalyst pore (c). The hydrogen bonding or protonation of *R* in the catalyst pore is accompanied by a partial or complete de-solvation of the proton (d), which is followed by the acid-catalyzed chemical transformation by going through a transition state (*TS*). Water molecules in brown and red are involved in the specific acid solvation sphere and in the *TS*, respectively. The solid curved lines represent the zeolite pore walls. The enthalpy changes (ΔH), apparent activation energy ($E_{a,\ app}$), and intrinsic activation energy ($E_{a,\ int}$) are dependent on interactions with the solvent environment

Figure 8.10. Reaction energy diagram showing kinetic and thermodynamic relationships

Figure 8.11. Gibbs free energy

Figure 8.12. An enzyme or catalyst lowers the activation energy

Figure 8.13. Industrial applications in chemical synthesis

Figure 8.14. Basics of enzyme kinetics graphs

Figure 9.1. Photochemical reaction

Figure 9.2. Fundamentals of photochemistry

Figure 9.3. Interaction of light with matter

Figure 9.4. Absorption spectra and energy levels

Figure 9.5. Quantum yield and radiative and non-radiative decay processes

Figure 9.6. Type I and type II mechanisms of photochemical reactions

Figure 9.7. Potential curves of ground and electronic excited states of a photochemical reaction. (*R*) is the ground-state reactant; (**R*) is the excited-state reactant; (*I*) is the ground-state intermediate; (**I*) is the excited-state intermediate; (**P*) is the excited-state product; and (*P*) is the final product of the overall process

Figure 9.8. Reaction intermediate

Figure 9.9. Scheme of the trivial mechanism of energy transfer
Figure 9.10. Catalyst selection and wavelength of light
Figure 9.11. Solvent effects on S_N1 and S_N2 reactions
Figure 9.12. Types of spectroscopic techniques
Figure 9.13. Solar energy conversion
Figure 9.14. Phototherapy treatment on hands
Figure 10.1. Surface chemistry
Figure 10.2. Single crystalline form of solid insulin
Figure 10.3. Schematic representation of the reaction steps in a general reaction in heterogeneous catalysis and the three key approaches our group adapted (green) in building a microkinetic model, including solving the issues of reaction barriers in the adsorption/desorption steps and coverage effects in surface reaction steps. Some microkinetic results from different approaches are listed
Figure 10.4. Adsorption
Figure 10.5. Types of adsorptions
Figure 10.6. Factors affecting adsorption
Figure 10.7. Adsorption isotherms
Figure 10.8. Adsorption kinetic models
Figure 10.9. Langmuir adsorption isotherm
Figure 10.10. Freundlich isotherm
Figure 10.11. Schematic representation for the general procedure of adsorption study
Figure 10.12. Relationship between adsorption and temperature
Figure 10.13. Variation in the adsorption of nitrogen gas on charcoal with varying pressure and temperature
Figure 10.14. Surface area and porosity
Figure 10.15. The significance of heterogeneous catalysis in industrial processes
Figure 10.16. Material science and nanotechnology

PREFACE

Chemical change is the dynamic pulse of chemistry. Reactions flash in microseconds, others move slowly over months, and most wait patiently for subtle adjustments in temperature, solvent, or surface. Thermodynamics tells us what is possible; kinetics tells us what actually happens and how quickly. By measurements of rates, by design of conditions, and by stepwise pathway suggestions, we connect observations to mechanisms that explain product distributions, selectivity, and yields. Kinetics is a control means as well: it instructs one on how to slow down breakdown, speed up synthesis, or shift the direction of a series of reactions toward a desired outcome in laboratories, reactors, living cells, and the atmosphere.

Kinetics provides a lingua franca linking molecular movement and macroscopic transformation. It couples differential equations and spectroscopic signatures, potential energy landscapes and rate constants, elementary steps and rate laws in the flask. The instruments are conceptual and pragmatic: design an experiment, set down a rate expression, fit parameters, test a mechanism against data, and cycle back until evidence and reason converge. This book is concerned with kinetics as a way of thinking and not with a list of formulas. The aim is to build judgement—identifying patterns in data, selecting the appropriate approximations to use, and knowing why one mechanism is sensible and another is not.

Chemical Kinetics and Reaction Mechanisms is designed to allow students to move from fundamental tools to advanced topics of relevance to industry and research.

Chapter 1 introduces the scope and objective of chemical kinetics, sets up basic definitions (rate, rate constant, elementary step, molecularity), and defines the boundary between kinetics and thermodynamics. It presents differential and integrated rate laws conceptually, explains how mechanisms yield rate expressions, and pre-empts the steady-state and pre-equilibrium approximations.

Chapter 2 deals with reaction rate and order. It derives the equations and diagnostics for zero-, first-, and second-order processes; treats mixed and fractional orders; and illustrates how stoichiometry and mechanism dominate observed order. The students learn initial-rate method, isolation method, and integrated-rate plotting, and half-life analyses and pseudo-first-order strategies. Pragmatic advice is given for choosing concentration ranges, detecting side reactions, and avoiding common misfits that come about by treating curved data as linear.

Chapter 3 outlines temperature dependence of rate. The Arrhenius equation is derived and explained, including mention of activation energy, frequency factor, and curvature in Arrhenius plots. Transition state (Eyring) theory is introduced to compare rate constants with activation enthalpy and activation entropy as a thermodynamic window on the activated complex. The chapter also describes pressure and solvent effects, corrections for tunneling, and isokinetic relations, and demonstrates that minor temperature errors are responsible for major activation parameter uncertainties.

Chapter 4 addresses complex reactions that cannot be accounted for by simple one-step scheme. Parallel, consecutive, and reversible systems are discussed, and so are autocatalysis, inhibition, and chain processes. Mechanism building is the chapter's principal topic: write out elementary steps, apply steady-state or pre-equilibrium approximation where necessary, and compare model predictions with data. Students learn techniques for unscrambling a rate-controlling step, for diagnosing branching in chain processes, and for discriminating between competing paths by selective probes or inhibitors.

Chapter 5 develops polymerization kinetics. Step-growth and chain-growth processes are compared, including radical, ionic, and coordination mechanisms. Initiation, propagation, termination, and transfer are addressed quantitatively in terms of conversion, number- and weight-average degrees of polymerization, and molecular-weight distributions. More advanced concepts like gelation, living polymerization, and copolymerization reactivity ratios are presented with helpful examples correlating synthesis control with kinetic parameters.

Chapter 6 presents catalysis in homogeneous, heterogeneous, and enzymic systems. Langmuir–Hinshelwood and Eley–Rideal models are treated for surface reaction, along with microkinetic schemes involving adsorption, surface reaction, and desorption. In solution, the catalytic cycle is written and investigated to yield turnover frequency, selectivity, and ligand or promoter concentration dependence. Enzyme kinetics is treated using the Michaelis–Menten formalism and its modifications like inhibition and allosteric effects. The chapter ends with brief information regarding catalyst poisoning, deactivation, and mass-transfer masking.

Chapter 7 discusses rapid reactions that are close to detection and mixing limits. Methods are stopped-flow, rapid-scan spectroscopy, temperature-jump and pressure-jump relaxation, and flash photolysis. Diffusion-limited rates and the Debye–Smoluchowski model are outlined, and guidelines for recognizing when diffusion rather than chemical conversion controls the rate observed are given. Pragmatic issues—dead time, instrument response, and global fitting—are addressed so that readers can realize how to extract rapid-kinetics experiments with confidence.

Chapter 8 addresses reaction dynamics, connecting microscopic motion to rate constants. Potential energy surfaces, reaction coordinates, and the transition state definition are qualitatively introduced. The chapter elaborates on trajectory calculations, steric effects, energy disposal, and mode selectivity. RRKM theory is treated for unimolecular reactions, as well as falloff behavior and pressure dependence. Nonadiabatic transitions and surface crossings are treated to illustrate how electronic structure influences observed kinetics.

Chapter 9 discusses photochemical kinetics. Quantitative techniques include light absorption, the Beer–Lambert law, and quantum yield. Excited-state processes such as internal conversion, intersystem crossing, fluorescence, phosphorescence, and photodissociation are placed into perspective with the Jablonski diagram. Quenching, energy transfer, and photosensitization kinetic schemes are developed, including Stern–Volmer analysis and time-resolved measurements. Applications extend from synthesis to materials modification and sunlight-induced environmental transformations.

Chapter 10 covers surface reaction and adsorption kinetics. Adsorption isotherms (Langmuir, Temkin, Freundlich) are built, then used to develop rate equations for reactions that progress via intermediates adsorbed on the surface. Site balance, coverage, and competitive adsorption are emphasized. Students learn to identify intrinsic kinetics from transport rates, diagnose diffusion or heat effects in reactors, and use microkinetic modeling to bridge elementary steps with measurable rates and selectivity.

This book is designed to serve upper-level undergraduate and graduate courses in chemistry and chemical engineering, as well as researchers in materials, catalysis, and environmental chemistry, and other related disciplines who need a clear introduction to kinetic analysis. Instructors will find an organization suitable for one-semester courses on kinetics or two-module presentations that tie together rate laws and mechanisms with catalysis or reaction dynamics. Practitioners will be shown pictures of real decisions: how to choose a rate model, what to quantify next, how to test whether a parameter is identifiable, and when a less complex mechanism explains the evidence as well as a more complex one. The goal is consistent competence—movement from foundations to master practice—so that readers can design experiments, model data, and make compelling arguments for mechanisms that stand up.

CHAPTER 1

INTRODUCTION TO CHEMICAL KINETICS

Contents

Introduction.......... 1
1.1. Fundamental Concepts.......... 7
1.2. Rate Laws.......... 12
1.3. Reaction Mechanisms.......... 16
1.4. Temperature and Kinetics.......... 20
1.5. Catalysts in Chemical Reactions.......... 25
1.6. Experimental Techniques in Kinetics.......... 29
Summary.......... 34
References.......... 36

INTRODUCTION

Chemical kinetics is the study of chemical reaction rates and factors that influence the rates of chemical reactions. By examining reaction mechanisms, kinetic experiments, and mathematical models, chemists can quantify the effect of reactant concentration, temperature, and catalysts on reaction rates. The understanding provided allows for product formation over time to be predicted, which is useful in many applications including industrial manufacturing, drug discovery, and environmental science. These findings not only contribute to theory but also are vital in practical applications where reaction control maximizes product yield and reduces waste. Reaction kinetics mathematical modeling is central to describing the relationship between reactants and products. This involves differential equations derived from rate laws that specify how the reactant concentration falls as products rise with time. Here, model inadequacies frequently happen when theoretical models fail to coincide with experimental outcomes.

Figure 1.1. *Chemical kinetics.*

Source:https://cards.algoreducation.com/_next/image?url=https%3A%2F%2Ffiles.algoreducation.com%2Fproduction-ts%2F__S3__202614ba-f7a2-4435-88c7-9a4b374a0f73&w=1920&q=75.

Through probabilistic model discussions, in such a situation, it is required to incorporate consideration of uncertainty since their neglect generates discrepancies that detract from the credibility of a model. The inclusion of a stochastic inadequacy operator, suggested in recent studies, can thereby render kinetic models more solid by addressing sources of uncertainty and with regard to significant physical constraints, e.g., conservation laws. Also, the employment of high-level statistical methodology, e.g., hierarchical Bayesian inference, plays a central role in rendering kinetic models more accurate. This approach enables model hyperparameter and parameter tuning against record data, enabling scientists to learn about the operation of chemical reactions. For instance, in the case of dealing with complex reaction mechanisms like hydrogen combustion, incorporating these probabilistic models is able to reveal the effects of variations in conditions on reaction pathways. With the development of the discipline, theoretical prediction and experimental testing come closer to each other, accumulating a common knowledge in reaction kinetics that can guide technological innovations and solve common scientific issues. An in-depth introduction to chemical kinetics thus ultimately focuses on the mutual interaction among theoretical thinking and applied science in different areas of science (Figure 1.1).

Definition of Chemical Kinetics

Chemical kinetics is an important branch of physical chemistry that describes the rates of chemical reactions and the mechanisms of their course. An examination of reaction rates helps chemists gain necessary information regarding the dynamics of molecule-to-molecule interactions, making chemical processes more accessible to understand. The field not only quantifies the rate at which the reactions occur, but it also investigates the conditions such as temperature, concentration, and the presence of catalysts that influence these rates. These issues with the rate of reactions are important, particularly in advanced systems wherein multiple simultaneous reactions may occur. By offering an outlet for the study of such responses on a quantitative basis, chemical kinetics enables applications as varied as industrial production to environmental science, showing its applicability to both theoretical and practical settings. The domain of chemical kinetics also encompasses many models that provide a comprehensive description of reaction mechanisms and their foundation. Such models range from elementary, simple reactions to more complex mechanisms involving multiple steps and intermediate species. One of the key areas is the study of reaction orders, which establish the relationship between rate of reaction and concentration of reactants. This is central to determining the rate laws, quantitatively explaining the relationship between rate of reaction and concentration. The discipline has been extended to cover the topic of kinetic control and selectivity, so that reactions can be classified based on their thermodynamic as well as kinetic properties. These classifications are used to determine the most favorable paths for chemical conversions and, consequently, foster investigation and industrial processes (Andreasen et al., 2011). Moreover, in the dynamic world of chemical kinetics, the inclusion of probabilistic models has transformed traditional paradigms to accept uncertainties of reaction mechanisms and parameters. This is particularly relevant to systems with many variables, for which uncertainties like measurement uncertainty and the indeterminacy of initial conditions can lead to inadequate models. Consequently, novel methods that incorporate stochastic terms, including the inadequacy operator, have emerged to better model these uncertainties. This not only respects fundamental physical laws but allows reaction models to be tuned using advanced statistical methods like Bayesian inference. By using these methods to model provided reaction mechanisms, such as hydrogen combustion, researchers are in a position to expand their understanding of kinetics, thanks to which more accurate predictions of the behavior of reactions and increased control over chemical processes are possible (Morrison et al., 2018) (Figure 1.2).

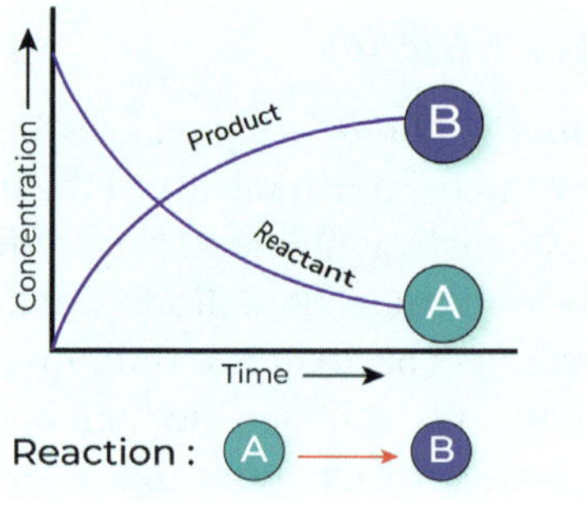

Figure 1.2. *Understanding chemical kinetics.*

Source: https://media.geeksforgeeks.org/wp-content/uploads/20240219183653/Chemical-Kinetics.png.

Importance in Chemical Reactions

The importance of acquiring knowledge about chemical reactions within the context of chemical kinetics cannot be overstated because it pertains to our ability to predict and manage chemical behavior. Chemical kinetics examines reaction rates as well as their factors so that scientists can construct experiments that yield information on reaction mechanisms. Through the identification of these underlying mechanisms, scientists can design conditions for maximum effectiveness, such as in the industrial environment, where reaction efficiency is paramount. For instance, the introduction of computational tools like KiNetX has transformed the field by providing computer-based mechanisms that analyze intricate chemical reaction networks and efficiently identify kinetically important species. This capability accelerates the identification of feasible reaction pathways, a significant step in the identification of chemical behaviors and their applications in predictive modeling and reaction design (Proppe et al., 2018). Reduction of complex mechanisms to skeletal forms is another significant aspect of chemical kinetics that marks its significance in chemical reactions. By focusing on key species and dampening irrelevant reactions, researchers can create truncated models that remain predictively accurate but easier to analyze. This can be particularly well demonstrated in the example of butanol isomers, where extensive kinetic modeling was used to generate skeletal mechanisms from detailed models. The process of reduction used techniques such as directed relation graphs and sensitivity analyses, thereby considering the intricacies of pressure-dependent reactions and furthering our understanding of combustion processes. The subsequent skeletal mechanisms well simulated global combustion behaviors, evidencing the efficiency of these reduced models in producing successful simulations while greatly simplifying the systems under consideration (Brady et al., 2017). Furthermore, chemical kinetics is of greatest importance in relating conditions of a reaction to outcome, and this is critical for theoretical and applied chemistry. Optimization of reactions for specific purposes is made possible by chemists from the ability to study reaction parameters like temperature, pressure, and concentration. In applications in real life, such as simulation of engine of biofuels like butanol, understanding the kinetics of each fuel plays a role in improving performance and efficiency. The butanol isomers versus gasoline surrogates contrast illustrates the competitive nature of fuels and the need for advanced kinetic models that are equipped to simulate performance accurately. By integrating chemical kinetics with experimental design and analysis, researchers are able to further advance the leading-edge development of renewable energy solutions as well as enhance the efficiency of chemical processes in numerous applications (Brady et al., 2017) (Figure 1.3).

Figure 1.3. *Importance of chemical reaction.*

Source: https://cdn.britannica.com/51/196751-131-CED1DE53/exposure-science-laboratory-test-tubes-chemical-reaction.jpg.

Historical Background and Development

The development of chemical kinetics as a distinct field of study emerged from a rich historical context deeply intertwined with the evolution of chemistry itself. In the late 19th century, scientists began to question not only the what of chemical reactions but also the how and why underlying these processes. The pioneering work of researchers like Svante Arrhenius, who devised the theory of activation energy in 1889, was a turning point in this area. His studies laid the basis of reaction rate understanding based on factors like temperature and concentration. This change in thinking signaled the move away from qualitative accounts of reactions themselves, and indicated more quantitative analytical treatments, which eventually became a requirement before physical chemistry emerged as an officially recognized field of study and chemical kinetics as an academic subject (Gooday et al., 2004).

Figure 1.4. *Svante Arrhenius.*

Source: https://upload.wikimedia.org/wikipedia/commons/6/6c/Arrhenius2.jpg.

In the 20th century, continued advances in sophisticated experimental methods and the development of theoretical frameworks fueled the evolution of chemical kinetics. Of note is that the Arrhenius equation provision provided a mathematical model for the description of the rate of reactions with respect to temperature, enhancing predictability throughout the field. The groundbreaking contribution influenced various fields of chemical study and established the precedent for stringent scientific practice, leading to codified practices in studying reaction mechanism and kinetics. Besides, researchers like Michael Polanyi and John Langmuir also achieved much by combining concepts from thermodynamics into kinetics, which laid a more comprehensive understanding of reaction dynamics. All these advances formed the basis upon which further research on catalysis and enzyme kinetics was explored, and which increased the functional application of chemical kinetics in other areas of science (Cynthia et al., 1989). Development of chemical kinetics over the centuries reveals the interaction between empirical observation and formulation at a theoretical level, leading to breakthroughs with long-term effects on scientific exploration as much as on industrial processes. With scientists continuing to explore the nuances of rates and mechanisms of reactions, the foundational principles established in the 19th and 20th centuries remain at the heart of chemistry today. For instance, the understanding gained through reaction rate history has important implications in the area of pharmacology, where drug interaction kinetics knowledge can inform treatment. And with the advance of computational chemistry, simulation of complex reaction pathways has been made possible, rendering chemical kinetics predictively more effective and demonstrating the ongoing relevance of its earlier development (Gooday et al., 2004) (Figure 1.4).

1.1. FUNDAMENTAL CONCEPTS

An understanding of reaction rates and conditions governing them is required in order to grasp the underlying principles of chemical kinetics. Foremost among these conditions are temperature, concentration, and catalysts that together decide the rate at which reactants are transformed into products. The Arrhenius equation is a key theory of chemical kinetics that describes the relationship between reaction rate and temperature and activation energy, demonstrating how raising temperatures can have a significant role in increasing reaction rates significantly by providing energy for reaction barriers. Concentration is also a vital factor since higher concentration of reactants is likely to lead to more molecule collisions, hence more reaction rates. Catalysts, which reduce the activation energy but are not involved in the reaction, provide additional efficiency to chemical processes. Such concepts are thus central in predicting and controlling rates of chemical reactions in the lab and industry. A major aspect of chemical kinetics is the investigation of reaction mechanisms, ranging from the step-by-step process of elementary reactions to the ultimate conversion from reactants to products. Precise understanding of these mechanisms is necessary since they inform us about how reactions proceed on the molecular level and thus enable scientists to identify intermediate species and transition states. Kinetic experiments are often employed to find rate laws typical of a reaction, which inform us about how the rate depends on the concentration of reactants. Moreover, knowledge about reaction mechanisms enables chemists to employ techniques such as the concept of dominance in network modeling. Specifically, this approach can condense complex kinetic models into the most important steps of a kinetic series, thus making it simpler to analyze complicated biochemical networks in systems such as cell signaling and metabolism, ultimately benefiting our general understanding of chemical behavior within different environments. The significance of simplifying models in chemical kinetics is increased by advances in computational biology and the intersection of kinetic modeling and network theory. Techniques at multiple scales allow scientists to handle complexity in biochemical networks by simplifying them based on time scale distinctions and relative importance of various components. With the aid of tropical geometry principles, model reduction techniques can be improved to allow theoretical investigation of large reaction networks that were initially difficult to study. Reduction not only gives insight into kinetics but also enables practical applications like drug design and metabolic engineering. Concepts borrowed from chemical kinetics therefore extend far beyond the mere determination of simple rates and inform more general scientific inquiry and applications that encompass environmental remediation and the dynamics of complex biochemical processes, highlighting the universal applicability of these underlying principles (Gorban et al., 2012; Egli et al., 2020) (Figure 1.5).

Figure 1.5. *Fundamental concepts of chemical kinetics.*

Source: https://learnbin.net/wp-content/uploads/2025/07/Concepts-of-Chemical-Kinetics.jpg.

1.1.1. Reaction Rate and Its Measurement

The reaction rate is a fundamental measure of the rate at which reactants are being converted into products during a chemical reaction. It is determined by the rate of change in concentration of a reactant or a product per unit of time, and it can be affected by a number of factors like temperature, concentration, and the employment of catalysts. The rate of reaction can be determined from a variety of techniques such as spectrophotometry and chromatography that measure concentration changes with time. Utilization of changes in temperature and concentration control also provides data for reaction kinetics, which assists in formulating rate equations. Understanding and quantifying reaction rates is critical to making predictions about how reactions will be able to proceed under various conditions, so that chemists can design and optimize chemical processes more effectively in the laboratory and industry. In chemical kinetics, how temperature affects reaction rates is of critical importance because it can help to understand the energy barriers for which a reaction must overcome to be able to occur. An increase in temperature usually leads to an increase in the rate of reactions, according to the principles outlined in the Arrhenius equation, illustrating how the temperature dependency of the rate constant is obtained. Using this model, chemists are able to predict reaction rates when thermal conditions are varied. Furthermore, temperature effects may be investigated using techniques such as isoconversional procedures, where kinetic parameters are calculated upon investigating heating rate dependency of reaction rates. Such procedures, as described by Čechová et al. (2012), enable the explanation of polymer and composite material stability and degradation mechanisms and the complex interconnection between temperature and reaction kinetics under actual circumstances. Measurement of reaction rates also includes the characterization of complex processes of protein folding and other biochemistry reactions. For example, bacteriorhodopsin folding has seen demonstrations of the potential of kinetic modeling and φ-value analysis in characterizing the folding transitions of membrane proteins. By quantifying the reaction rates of various mutants, researchers can infer a more elaborate understanding of the structural dynamics of folding, at the heart of the understanding of biological functions and the impact of mutations. The findings of such research indicate that the transition state is not rigidly structured but is an ensemble of conformations that is dynamic (Bowie et al., 2014). This level of intensity reflects the necessity of precise measurement and interpretation of reaction rates in biochemical systems, ultimately contributing to the broader discussion of chemical kinetics (Figure 1.6).

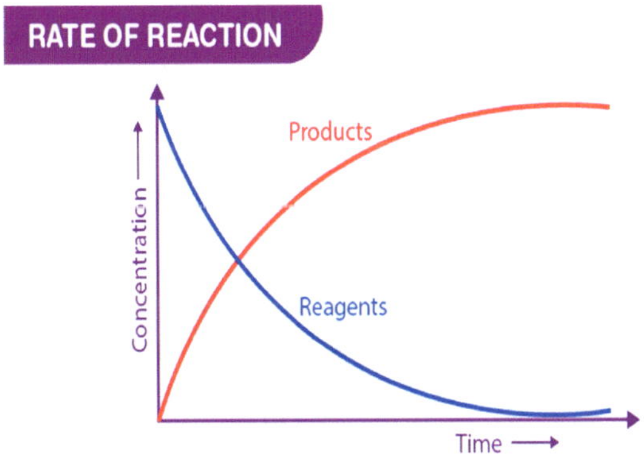

Figure 1.6. *Rate of reaction.*

Source: https://cdn1.byjus.com/wp-content/uploads/2019/02/Rate-of-Reaction.png.

1.1.2. Factors Affecting Reaction Rates

The rate at which a chemical reaction occurs is dictated by several important factors that individually hold some level of significance in determining the rate of process. Concentration of reactants is among these factors; increased concentration will typically equate to an accelerated rate of reaction. This is due to the presence of greater numbers of reactant molecules that are available to collide, thus increasing the likelihood of successful chemical interaction. Also, temperature is another important parameter; higher temperatures give molecules additional kinetic energy, that is, stronger and more frequent collisions. As a result, the reaction will proceed at a higher rate. In light of all these factors, it is evident that balanced conditions have to be attained in order to optimize reaction conditions for many chemical operations and thereby increase efficiency and yield. Apart from concentration and temperature, the action of catalysts can also greatly influence rates of reaction.

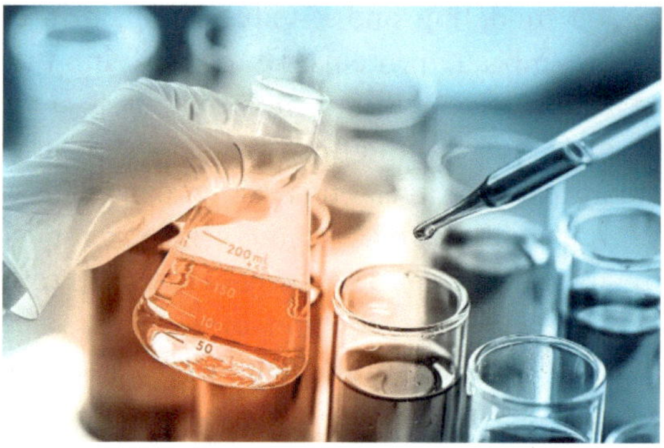

Figure 1.7. *Factors affecting reaction rate in chemical kinetics.*

Source: https://cdn-gddfb.nitrocdn.com/XMOCPBDkJQcfbMeyrNhpgTZaRuIae-dyd/assets/images/optimized/rev-d57283f/www.atascientific.com.au/wp-content/uploads/2010/10/shutterstock_236654359.jpg.

Catalysts are substances that speed up the rate of a reaction without being permanently changed. They work by having a lower activation energy for the reaction route that they provide, so more molecules are involved in the reaction at any given time. This is a principle applicable to many industrial processes where the efficiency of reaction is maximized, for example, in chemical and pharmaceutical production. The application of catalysts to allow concurrent processes, albeit with difficulty, is as noted in the leaching of complex sulfide ores, according to Chou et al. (1987). What this underlines is the need to appreciate not just how catalysts influence reaction rates but also the conditions which may influence their efficiency. Moreover, the physical state of reactants also plays an important role in the rate determination of a reaction. Reactions between gaseous, liquid, and solid reactants are of variable rates because their rates are affected by surface area and molecular motion. For instance, solids in powdered form react faster than larger solid pieces because the higher surface area offers greater opportunities for collision. Liquid and gas reactants are likely to exhibit higher rates of reaction than solids because the latter cannot flow and mix easily. Moreover, the effect of molecular orientation when colliding should not be ruled out; only then do reactions occur when reactants collide with appropriate orientation and energy. Identification of these factors is important in chemical kinetics since they provide means of acquiring maximum reaction optimization and efficient process design in scientific and industrial processes, such as those shown in gene regulatory networks (Alberts et al., 2007) (Figure 1.7).

1.1.3. The Role of Concentration in Kinetics

In chemical kinetics, the reactant concentration is a determining factor in reaction rates. On the basis of collision theory principles, increased concentrations raise the rate of molecular collisions, hence the chance of successful collisions that result in product formation. This connection is typically quantified in terms of rate laws, which specify how the rate of a chemical reaction correlates with concentrations of the reactants. To illustrate, in bimolecular reactions, the rate of reaction is proportional to the product of both reactant concentrations and indicates that concentration plays a central role in determining the rate at which reactants produce products. Recognition of such dependence is essential to conceptual models as well as instrumental use in chemical engineering and kinetics research. Moreover, the principle of the rate constant, which constitutes the essence of reaction kinetics research, is revealed to unfold intricate dependences on concentration under different conditions. For example, a novel method based on holographic interferometry has played a major role in measuring surface reaction rate constants, and the proof that the dissolution rate of minerals such as gypsum can be significantly less than is commonly assumed when due consideration is given to concentration effects and hydrodynamics. This emphasizes the complexity of concentration effects in reactions, particularly in heterogeneous systems where mass transport limitations may limit predicted reaction rates. Therefore, concentration together with other kinetic values needs to be quantified in building reliable models that can reliably represent chemical reaction kinetics. Intensity dynamics also reflect how concentration influences kinetic activity in varying conditions. In situations where fluids of varying chemical compositions come into contact with one another, velocity gradients and ensuing shear stress will have a tendency to dominate local rates of reaction. As defined in current research, competition between shear flow and chemical reactivity is a function of the Péclet and Damköhler numbers, which are essential for the understanding of mass transport and reaction kinetics. In a regime of extreme stretching, where forces associated with advection dominate, competition between concentration and kinetic terms is realized, which gives rise to localized hotspots of reactivity that do not adhere to classical kinetic predictions. This multifaceted relationship showcases the multifaceted concentrating role in kinetic systems, underscoring its all-importance in theoretical and applied chemistry (Bandopadhyay et al., 2016; Colombani et al., 2008) (Figure 1.8).

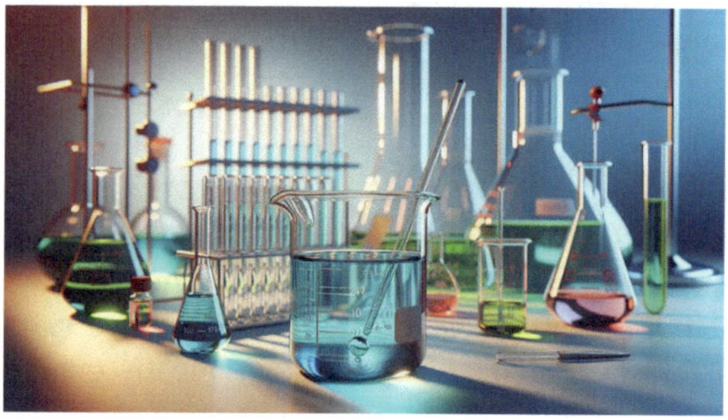

Figure 1.8. *The role of concentration in chemical kinetics.*

Source:https://cards.algoreducation.com/_next/image?url=https%3A%2F%2Ffiles.algoreducation.com%2Fproduction-ts%2F__S3__49afe419-6aed-4134-8e5a-ba6b16c68e25&w=1920&q=75.

1.2. RATE LAWS

The concept of rate laws is a central component in chemical kinetics, which exerts vast influence on our understanding of reaction mechanisms. Rate laws quantitatively measure the relationship between the rate of a chemical reaction and the concentration of the reacting compounds. Empirical observation leads to such laws, which take the generalized form rate = $k[A]^m[B]^n$, where k is the rate constant and m and n are the orders of reactants A and B, respectively. Chemists can find values of m and n through systematic experiments, which provide information regarding the reaction mechanism and inter-reactant interaction. This basic analysis not only enables one to predict rates of reaction but also serves as the foundation for further theoretical developments in chemical kinetics, thereby enhancing our comprehension of the dynamical properties of chemical reactions. The resolution to the determination of rate laws stems from the fact that reaction mechanisms are not a simplification but an involved system prescribing the complex pathways through which reactants are transformed into products. The generalization of rate laws to multicomponent reactions is an essential consideration, especially considering the interactions within such systems. The interactions in such systems may produce nonlinear effects against which traditional modeling needs to be more responsive. According to recent studies, chemical kinetics principles encompass basic tools for process modeling across many scientific fields. For instance, invoking kinetic theory concepts enables chemists to write nonlinear multicomponent diffusion equations that model the development of mixtures with intense component interaction. This is an example of the necessity to invoke subtle concepts borrowed from chemical kinetics to define physical phenomena correctly, thereby expanding the scope of studies in chemical kinetics and related sciences (Gorban et al., 2011). The incorporation of kinetic laws in multicomponent systems poses key questions regarding thermodynamic limitations. More particularly, the maintenance of the dissipation inequalities is of vital importance for determining the applicability of the kinetic models employed. The more complex the reaction is, the more oversimplified approaches of former kinetic theories have to evolve in order to cope with refined behavior endowed by reactants and products. For instance, the development of a generalized Mass Action Law subject to thermodynamic restrictions highlights the delicate balance required in chemical kinetics. Such developments indicate that as scientists extend the frontiers of traditional rate laws into nonlinear domains, sound methods will need to be established for ensuring compliance with fundamental physical laws. Accordingly, the exploration of rate laws becomes a matter of more than mere calculation to entail a broader investigation of theoretical constructs underlying chemical change and thereby solidify the critical significance of chemical kinetics in shedding light upon real phenomenon (Pinto et al., 2017).

1.2.1. Definition and Significance of Rate Laws

The rate law theory is a unifying concept to the field of chemical kinetics, providing an elementary framework for understanding how a chemical reaction rate varies with a multitude of factors. Rate laws are a quantitative representation of how reaction rate varies with reactant concentration, and this allows chemists to forecast how changes in concentration affect reaction rate. Through the determination of the rate law for a reaction, scientists can also make assumptions regarding the reaction mechanism, giving insights into the stepwise character of the processes that govern the conversion of reactants to products. This is particularly applicable to multi-step complex reactions, where determining the rate-determining step can give significant insight into the mechanism of the overall reaction. The significance of rate laws extends beyond mere prediction of reactions as they play a vital role in metabolic modeling and interpretation of biological processes. For instance, metabolic networks can be transformed into kinetic models by sampling reaction elasticities with appreciable enhancement in predictability of metabolic behavior under various conditions. The availability of software such as MATLAB for use in elasticity sampling allows models to be constructed easily, which can be made reversible rate laws—a feature that is crucial when metabolite reactions are influenced by multiscalar thermodynamic forces, regulation, and flux distributions. Moreover, such dynamic models enable researchers to analyze the effect of enzyme modifications and interactions on metabolic pathways, thereby enhancing our knowledge of biochemical systems in health and disease. In addition to their usefulness in biochemistry, rate laws also find more general use in environmental science and in pharmacology, where reaction rate is key. In pharmacokinetics, for example, rate laws explain drug metabolism in the body, with important insight into dose-response relationships. By monitoring the change in drug metabolism rate with concentration, researchers are able to customize therapeutic regimens to the individual patient. Furthermore, principles that are borrowed from rate laws direct experimental investigation towards the measurement of the effect of various factors on reaction kinetics, thereby providing a credible basis for scientific research in a broad variety of disciplines. In this regard, the study of rate laws emphasizes the interdependence of chemical kinetics with practical applications in natural sciences as well as applied sciences.

1.2.2. Zero-Order, First-Order, and Second-Order Reactions

Chemical kinetics has various reaction orders, including zero-order, first-order, and second-order reactions. Zero-order reactions have a constant reaction rate, independent of the concentration of reactants. This occurrence typically occurs where a catalyst is saturated, and the reaction takes place at a top speed regardless of the availability of a substrate. The reaction rate thus depends on temperature and the effectiveness of catalysts. Understanding zero-order kinetics is significant in situations such as drug metabolism and reactions at the surface, in which concentration neither influences nor changes the rate of reaction, thus emphasizing the significance of outside influences more than the extent of the substrate for defining kinetics (Haseltine et al., 2007). The reverse is the case with first-order reactions, where the rate of a single reactant's concentration is linear. This linear relationship implies that as the concentration reduces, the rate of reaction declines proportionally. Most common biochemical reactions, i.e., breakdown of radioactive isotopes and certain enzyme-catalyzed ones, are representatives of first-order kinetics. Mathematical expression of first-order reactions allows simple integration, resulting in an explicit concentration-time relationship. This inherent relationship is found to be very helpful in predicting reaction behavior and engineering efficient processes in a number of fields, from drugs to the environment (Gorban et al., 2011). Second-order reactions entailed more complex interaction as it is proportional to the concentration of two reactants' product or the square of a single reactant's concentration. Second-order reaction is commonly entailed in processes with bimolecular interactions such as the concepts of collision theory for gaseous reactants. Complexity of second-order kinetics entails precise experimental determinations to properly interpret concentration dependencies. Because second-order reactions are likely to dominate highly reactive systems, understanding of them plays an important role in the modeling of reaction mechanisms and product yield calculation under different conditions. Proficiency in the kinetics of zero-order, first-order, and second-order reactions, as a whole, provides an integral platform for growth of knowledge in the field of chemical kinetics (Haseltine et al., 2007) (Figure 1.9).

Zero-order	$[A]_t = -kt + [A]_0$
First-order	$\ln[A]_t = -kt + \ln[A]_0$
Second-order	$\frac{1}{[A]_t} = kt + \frac{1}{[A]_0}$

Figure 1.9. *Zero-order, first-order, and second-order reactions.*

Source: https://www.jove.com/files/media/science-education/science-education-thumbs/11377.jpg.

1.2.3. Determining Rate Constants

Measurement of rate constants is a central aspect of chemical kinetics, with implications not only for our understanding of reaction mechanisms but also for applications in most practical fields. Rate constants inform us as to how quickly a reaction will proceed under certain conditions, and they are necessary for predicting reaction behavior under conditions of actual use. A variety of experimental techniques, including spectroscopic techniques and calorimetry, are employed to obtain the constants with accuracy. For instance, the Initial Growth Rates SSTD NMR technique is a sensitive tool for the determination of kinetic parameters of complex chemical systems, especially for recalcitrant small organic and organometallic molecules (Angulo et al., 2015). These state-of-the-art methods allow chemists to obtain more precise measurements of the rate constants and thus create more accurate predictive models of chemical reactions. Moreover, the study of the kinetics of a reaction is more than just calculating rate constants; it is to explore the dynamics and the transition states of reactions. A good example would be in helical membrane protein research, where researchers have employed φ-value analysis to uncover the folding kinetics of bacteriorhodopsin. By building a complete kinetic model, they could assess folding rates and thereby obtain useful thermodynamic data regarding the role of retinal hydrolysis in folding (Bowie et al., 2014). Such an analysis helps to suggest that rate constants are not mere figures but significant parameters that bear witness to the intrinsic properties of the chemical systems being studied. In short, determination of the rate constant is at the heart of chemical kinetics, bridging theoretical and practical. As evident from advances in methods such as the Initial Growth Rates SSTD NMR method and φ-value analysis, researchers continually further improve their methods for better approximating underlying kinetics of reactions. These methods not only enable us to comprehend complex systems more fully but also hold out the promise of innovation in fields as disparate as materials science and biochemical research. Further investigation of the rate constants will assuredly continue to unravel the secrets of chemical reactivity, contributing further to the development of new technologies and therapeutic treatments.

1.3. REACTION MECHANISMS

Comprehension of reaction mechanisms is core to chemical kinetics because it provides a step-by-step description of the processes in a chemical reaction. A reaction mechanism is a description of how reactants are transformed into products in a sequence of elementary steps, which are primarily through bond making and bond breaking. This knowledge is not only vital for reaction rate prediction but also for describing intricate reaction pathways. For example, in combustion mechanism studies, researchers have revealed a simplified scheme of reactions that can adequately describe fuel behavior. By sensitivity analysis, the model based on 40 reacting species and 118 elementary reactions was reduced to emphasize merely 16 species and 21 elemental reactions, emphasizing the necessity of identifying important reactive components in a mechanism (Acosta et al., 2025). Mechanistic research enhances predictive capability across a variety of scenarios, such as environmental applications. A kinetic experiment of reactions that occur in electrokinetic treatments, for instance, points out the necessity of incorporating real-time reaction processes to model transport phenomena with accuracy. To that extent, the dynamics of a series of reactions, such as aqueous species complexation and precipitation, must be linked with transport mechanisms such as electromigration. The predictive power of such models is greatly improved when kinetic rates are included in order to better simulate the complex chemical interactions of the treated system. A comparison between models based on thermodynamic equilibrium and kinetic rate laws indicated the latter to be a major improver of predictive accuracy and to uncover the decisive role played by reaction mechanisms in practical applications (Ceccarini et al., 2017). Moreover, a full insight into reaction mechanisms aids in the development of new chemical processes and enhancement of known ones. In catalysis, understanding of the reaction path can enhance catalyst design, thus the selectivity and efficiency of the reaction. This is critical in industrial chemistry, where the time of reaction and yield of the product are of prime concern. By learning about reaction rate mechanisms, chemists can maximize conditions favoring desirable pathways without a heavy by-product burden. Reaction mechanism development typically leads to the derivation of rate laws that can be experimentally verified and applied to the scaled-up process. Thus, reaction mechanisms not only lead us to better appreciate chemical kinetics but also enable us to advance technology and sustainability efforts.

1.3.1. Definition of Reaction Mechanisms

In chemical kinetics, the mechanism of reaction is a step-by-step sequence of elementary steps that lead to the conversion of reactants to products. The sequence is responsible for outlining how reactions occur at a molecular level and provides an insight into the energetics and kinetics involved. An elementary step within a mechanism is a specific molecular interaction under given kinetic laws, often expressed by stoichiometric algebra. The importance of reaction mechanisms extends far beyond chemistry; they are fundamental principles in many different fields of science, from particle physics to sociology. Chemists can then predict reaction rates and conditions based on these mechanisms, making it possible to control product formation as well as the optimization of reaction routes toward desirable objectives (Gorban et al., 2011). The construction of reaction mechanisms typically involves subtle insight into the interaction between molecular species that react. As noted in recent advancements, numerous aspects of a reaction may have intricate behavior that necessitates nonlinear diffusion equations for multicomponent systems. Through the modification of chemical kinetics concepts, researchers have developed methods to simulate such interactions more accurately. The use of reaction mechanisms in a thermodynamic framework enhances not only the ability to model such processes but also ensures adherence to thermodynamic requirements, such as energy conservation and entropy laws. The approach has been successful in addressing nonlinear diffusion and multicomponent reaction problems and thus emphasizes the interdisciplinary character of reaction mechanism study (Gorban et al., 2011). Moreover, the complexity of reaction mechanisms is often depicted through graphical plots, where each species and reaction is in a bipartite graph. This method allows for the study of system behavior and the identification of important structural features that are able to provoke unique behavior, such as multistationarity in biochemical networks. The use of degree theory and graph-theoretic methodology provides a context for analyzing systematically how specific structural properties of the mechanism determine observable characteristics. Such analysis is pivotal in describing how complex biochemical networks, such as cellular signaling networks, can exhibit multiple stable states. This information is not just relevant to chemical kinetics but also to constructing more effective biochemical networks that can respond to various cellular conditions.

1.3.2. Elementary Steps and Intermediates

The study of elementary steps and intermediates is at the heart of the determination of chemical reaction mechanisms, especially in chemical kinetics. By elementary steps, reference is to the discrete steps of a complex reaction mechanism, with each step referring to an individual molecular event. The discrete steps combine to create the overall reaction pathway, and their study reveals the dynamics of the reaction. Intermediates, which are short-lived species produced by these elementary steps, are of prime importance to the progression towards end products. The kinetic activity of a reaction can generally be accounted for by examination of the rate constants of each elementary step to achieve a more profound understanding of reaction rates and mechanism. In addition, deconstruction of these steps exposes the impact of numerous parameters like temperature and concentration on reaction kinetics, where the complex interrelationship between molecular interactions and measurable macroscopic properties emerges. The idea of kinetics is not limited to straightforward reactions but includes complicated scenarios where numerous pathways and intermediates compete. Within reaction networks that are complicated, the degree of kinetic rate control becomes a central framework for considering the relationship between thermodynamic and kinetic factors. With the application of this rigorous analytical tool, researchers are able to establish the salient components of the kinetic model and separate the contribution made by each elementary step to the reaction rate. For instance, networks that consist of many intermediates can exhibit behaviors that are rather dissimilar from those that would have been predicted based on more simple models. By employing Campbell's principles, scientists can stepwise analyze the influence that mechanisms and conditions of reactions have on the formation and stability of intermediates in such networks (Andreasen et al., 2011). This enhances the ability of kinetic models to predict and allows for the optimization of reaction conditions in industry as well as academia. In addition, as biochemical networks grow more intricate, parametric uncertainty and combinatorial explosion become worse problems. These networks often involve numerous well-separated time scales and thus an approach that can simplify analysis without compromising accuracy is required. The concept of dominance among model components enables this simplification to be accomplished so that big and complicated networks can be approximated to more manageable forms based on their relative effect on dynamics. These strategies, such as quasi-steady-state and quasi-equilibrium approximations, offer practical approaches to the assessment of reaction networks while addressing the intricate interaction between intermediates and elementary steps (Gorban et al., 2012). This hierarchical organization is what facilitates step-by-step unscrambling of dynamics in complex systems and thereby advances both theory and applications across a variety of fields, from biochemistry to materials science.

1.3.3. The Relationship Between Mechanisms and Rate Laws

The relationship between rate laws and mechanism is at the heart of chemical kinetics. A rate law provides a description of a reaction's velocity as a function of reactant concentration, typically involving rate constants and orders of reaction. Each elementary step in a reaction mechanism provides information on the manner in which reactants are transformed into products, and these will often be a specific rate law corresponding to their molecularity. For instance, in a simple elementary reaction, its rate law can often be predicted directly from its stoichiometry. Complex mechanisms involving multiple steps, however, may introduce inconsistencies and necessitate careful scrutiny to connect the resulting overall rate law with its mechanistic origin. To cast more insight on this connection, the specific rate laws may often suggest the dominant mechanism that governs a reaction. For example, analysis of temperature-time plots derived from techniques such as constant rate thermal analysis (CRTA) can allow one to identify certain kinetic models governing the involved degradative processes based upon characteristic shapes. It has been observed by experiments that polymers like polytetrafluoroethylene (PTFE) and polyethylene (PE) have varying degradation mechanisms such as phase boundary or random scission that strongly affect their individual rate laws. By close examination of these curves, the mechanistic pathway consistent with seen kinetic behavior can be determined: a critical finding as it enables chemists to make predictions regarding the effects of changes in conditions on reaction rates and mechanisms, reaffirming the relationship between kinetics and mechanisms (Luque et al., 2011). Moreover, the dependence of rate laws on reaction mechanisms reveals no mystery that the significance of comprehensive kinetic studies in the prediction and control of chemical behavior cannot be underscored. Experimental evidence-based rate laws not only measure reaction speeds but also provide insight into the underlying dynamics controlling chemical transformation. The variety in kinetic models again demonstrates that one rate law is incapable of characterizing complex reactions. For example, with a number of various mechanisms for degradation in operation in different materials, kinetics can typically rely on empirical investigation supplemented by theory models to reasonably accurately predict reaction behavior. The study of such relationships is important to applications ranging from optimization in industrial processes to material development, emphasizing the significance of the understanding of the connection between mechanisms and rate laws in the broader scheme of chemical kinetics (Pinto et al., 2017).

1.4. TEMPERATURE AND KINETICS

Temperature effect on the rate of chemical reactions is one of the major concepts in chemical kinetics. When temperature increases, kinetic energy of the molecules that are reactants also raises, and therefore more and more energetic collisions take place. The aforementioned principle plays a critical role in chemical reactions such as selective oxidation of ethylene, for which various temperatures have a significant impact on reaction rates and product yields. Higher temperatures could increase reaction rate to some extent, but the extent of influence on selectivity towards valuable products, for example, ethylene oxide relative to complete combustion products, is worth ascertaining. Experimental studies in different reactor systems demonstrate that temperature could affect the adsorption properties on catalyst surfaces, ultimately changing the kinetics and mechanism of reaction routes (Borman et al., 1996). Thus, temperature regulation not only accelerates reactions but also becomes a primary factor deciding the efficacy of catalytic reactions. In addition to kinetic effects, temperature also largely decides the state and sensitivity of reaction mixtures. For example, during metallic nanoparticle synthesis in microemulsions, temperature variations decide nanoparticle size and quality. The rate of reaction is temperature-sensitive according to the Arrhenius equation, and hence the synthesis reaction is significantly influenced by the reaction rates. If the temperature is increased, the diffusion of reactants in surfactant layers is faster, thus raising the rate of the reaction. The product morphology thus forms larger nanoparticles, showing temperature dependency, kinetics, and the morphology of the product. Such an observation brings to light the fact that temperature variations influence not only the speeds of reactions but also the resulting physical properties of products (Ritacco et al., 2018). Therefore, temperature becomes a fundamental parameter in reaction condition optimization in different chemical processes. Temperature is also instrumental in determining the chemical reaction's thermodynamic equilibrium. With a rise in temperature in exothermic reactions, there is a potential displacement of the position of equilibrium, as per Le Chatelier's principle, toward the reactants over the products. But with rise in temperature in endothermic reactions, there is a potential increase in product formation and overall reaction feasibility. The interplay between kinetic and thermodynamic forces offers a fascinating area of investigation in chemical kinetics. Being able to discern the influence of temperature on reaction kinetics and equilibrium is critical, especially in complicated systems with numerous competing reactions. Optimizing temperature with the other reaction conditions to achieve maximum rates and selectivity is the challenge. This fine balance is critically important in industrial settings, where optimum results must be achieved with maximum efficiency if chemical production processes are to be economically viable (Borman et al., 1996; Ritacco et al., 2018).

1.4.1. The Arrhenius Equation

The Arrhenius equation is at the heart of chemical kinetics research because it describes quantitatively the dependence of reaction rates on temperature. The equation specifically illustrates how the rate constant (k) increases with a rise in temperature, demonstrating the exponential relationship between temperature and activation energy for chemical reactions to take place. Mathematically stated as $k = Ae^{-Ea/\mathrm{RT}}$, the equation encompasses the frequency factor (A), activation energy (E_a), universal gas constant (R), and temperature (T) in Kelvin. The equation highlights the major role of thermal energy in facilitating molecular collisions, leading to successful reactions. This equation is applied by chemists to forecast reaction behavior at varying temperatures, consequently enabling optimization of reaction conditions in the laboratory and industry. In practical applications, the Arrhenius equation can be combined with other kinetic models to better understand governing mechanisms of the reaction rate. The classic first-order kinetic model, for instance, is commonly applied in the investigation of catalytic activity degradation in enzymatic browning reactions. The process has been investigated more recently, with specific attention paid to the temperature-dependent behavior of polyphenol oxidase (PPO) in fruits like Golden apples. With the application of both the old and new regression techniques, parameters can be obtained that not only characterize the reaction at specific temperatures but also assist with a more integrated kinetic model that better fits the experimental data gathered. Models that involve temperature dependencies as part of this scenario can enhance the precision in estimates, which can end up leading to improved control and understanding of enzymatic processes (Machado et al., 2005). Moreover, consequences of the Arrhenius equations extend beyond standard chemical reactions into numerous other disciplines, including astrophysics. In brown dwarf and luminous extrasolar planet atmospheric chemistry, kinetic-based models depict the dynamics between rivaling molecular abundances under evolving atmospheric conditions. Temperature, gravity, and vertical mixing parameters interact, thus controlling the formation of predominant molecules such as methane and carbon monoxide. The articulation of these relations via kinetics-directed modeling allows scientists to forecast chemical behavior in conditions vastly different from the usually confined realm of the laboratory. The Arrhenius equation is thus a staple tool, bridging theoretical comprehension and practical applications within both chemical and physical fields, such as important atmospheric phenomena on other bodies (Marley et al., 2014) (Figure 1.10).

$$k = Ae^{-Ea/RT}$$

Figure 1.10. *Arrhenius equation.*

Source:https://cdn1.byjus.com/wp-content/uploads/2015/12/Arrhenius-Equation-1.png.

1.4.2. Effect of Temperature on Reaction Rates

Temperature plays a key role in regulating rates of reaction of chemical reactions, essentially controlling how reactants collide and interact in a setting. As temperature increases, kinetic energy is greater, and particles move more aggressively with more rapid collisions. This increase in molecular motion not only raises the number of collisions between reactants but also the likelihood that such collisions will result in an effective reaction. According to the Arrhenius equation, which defines the rate of chemical reactions, a rise in temperature is linked with a decrease in the activation energy barrier reactants have to overcome in order to form products. Therefore, understanding the temperature-reaction rate relationship is central to predicting reaction behavior and optimizing reaction conditions in the vast majority of practical applications. For a study of how temperature affects individual reactions, many models and empirical relationships have been used to characterize these dynamic interactions. For instance, for the selective oxidation of ethene, reaction rates are strongly influenced by temperature and the presence of reaction product. The studies reveal that carbon dioxide enhances selectivity towards the target ethene oxide at high temperatures, a testament to the complex interaction between concentration of reactants, temperature, and behavior of reaction product. The significance of temperature is further realized when contrasting other reactor designs and conditions, such as for a cooled tubular reactor versus internal recycle reactors. In such conditions, the variation in reaction kinetics between them can lead to varying rate expressions, thereby establishing how complexities in temperature affect the net rates of reaction in chemical systems (Borman et al., 1996). Moreover, the relationship between temperature-reaction kinetics is not limited to ordinary chemical processes; it also holds for photochemical processes in various atmospheres, including those of exoplanets. For instance, a study of the photochemistry of 51 Eridani b shows that photochemistry of sulfur is strongly temperature-dependent with the product of elemental sulfur being present at various thermal points. Such temperature-related behavior is capable of producing varying physical states of sulfur ranging from gaseous to condensed states based on the environment's effective temperature. This serves to illustrate the wider implications of temperature on reaction rates; whereby even small changes can significantly alter the direction of reactions and their products. Therefore, the investigation of both classical kinetics and reaction performances under various conditions is critical in the appreciation of the extensive application of temperature in chemical kinetics (Marley et al., 2016) (Figure 1.11).

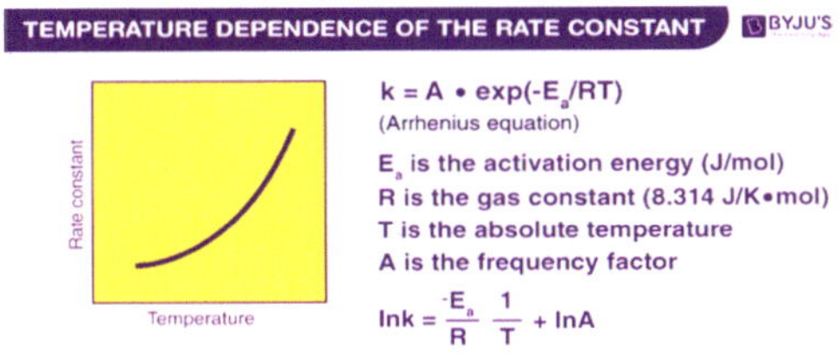

Figure 1.11. *Temperature dependence of the rate constant.*

Source: https://cdn1.byjus.com/wp-content/uploads/2017/10/Temperature-Dependence.png.

1.4.3. Catalysis and Its Temperature Dependence

The catalysis phenomenon is inextricably linked to the temperature-dependent nature of chemical kinetics due to the fact that reaction rates significantly depend on thermal conditions. An increase in temperature normally increases the kinetic energy of the reactants and thus causes molecules to collide with more force and more frequency. This is aptly demonstrated by the Arrhenius equation that posits an exponential correlation between the rate constant increase and the increase in temperature. Temperature variation in catalytic reactions can lead to alternative reaction pathways and mechanisms. For example, biphasic hydroformylation of 1-octene with a rhodium sulfoxantphos catalyst has been observed where process conditions such as temperature influenced the initial rate and selectivity of the reaction significantly. Thermal variations facilitated the synthesis of a linear aldehyde product with a remarkable n:iso ratio of 30, which underscored the significance of temperature conditions in catalytic selectivity and efficiency (Delmas et al., 2011).

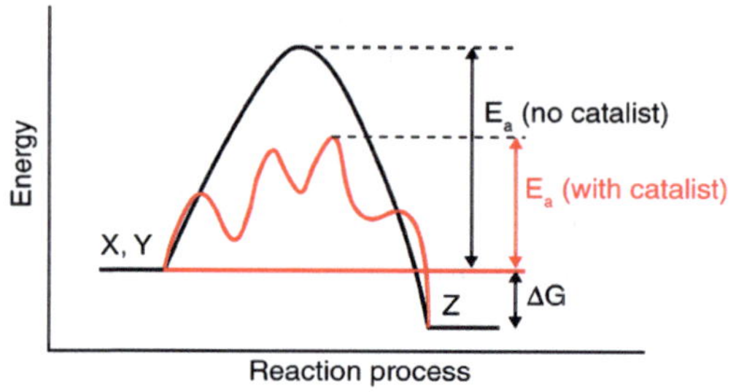

Figure 1.12. *Catalysis.*

Source: https://cdn1.byjus.com/wp-content/uploads/2017/08/Catalysis-How-does-a-catalyst-work-1.png.

Temperature not only speeds up kinetic processes but also influences the reactivity and stability of catalytic materials. Catalytic reactions that occur on electrode surfaces during reaction, for instance, can be subject to noticeable temperature dependencies. As temperature alters the encounter between reactants and catalysts, it ultimately influences activation barriers that determine reaction pathways. Recent work has discovered that temperature-driven modification can affect transition state dynamics, revealing complexities in charge-trapping intermediate behavior on semiconductor surfaces. For water oxidation reactions, possessing reaction-dependent time constants and temperature-dependent varying kinetic effects emphasizes the critical interplay between thermal conditions and catalytic performance. These findings expand our understanding of how temperature affects not only the efficiency of the reaction but also how the processes that regulate product creation under catalysis (Aschaffenburg et al., 2019). Catalytic optimization further relies on a delicate understanding of how temperature alters various operating parameters in heterogeneous and homogeneous environments. For example, within gas-phase catalytic reactions, reactant solubility may be a temperature-dependent function, affecting available reaction concentrations. Additionally, change of temperature may lead to various competitive pathways towards product formation, thus making the temperature dependence of catalysis complex. This dynamic interplay emphasizes how chemists must monitor thermal conditions closely for desired reaction outcomes. For the hydroformylation reaction catalyzed by rhodium, adjustment of temperature allowed large variations in rates and selectivity, which emphasizes the possibility of the optimization of the efficacy of catalysts with little undesirable byproduct. Lastly, understanding the intricate relationship between temperature and catalysis gives rise to the optimized catalytic systems for it to pave the way for improvement of chemical kinetics (Delmas et al., 2011) (Figure 1.12).

1.5. CATALYSTS IN CHEMICAL REACTIONS

The role of catalysts in reactions is the core area of the study of chemical kinetics, as these have significant impacts on the rate at which a reaction will advance without being consumed during the reaction process. Catalysts operate by introducing a new reaction path with lower activation energy, enabling reactant interactions that otherwise would occur more slowly or not at all. Their ability to speed up reaction rates is at the core of industrial processes and biological processes. For instance, enzymes act as biochemical catalysts, making biochemical reactions occur at a higher rate needed for life. The role of a catalyst in speeding up a reaction is not necessarily a matter of top speed but rather how to do optimization of conditions for the reaction to be finished and with minimum energy usage and waste of resources in chemical processes. Thermodynamics of catalytic reactions is yet another essential area connected to their role in chemical kinetics. Knowledge of energy transformations during catalytic reactions enables improved reaction behavior predictions under specific conditions. Thermodynamic viability of an enzyme-catalyzed reaction, say, emphasizes the sensitive balance between reaction kinetics and energy conservation as biochemical principles reveal. This is explained in part by the coarse-graining process outlined in the literature, which gives stoichiometries and fluxes of reaction in a thermodynamically consistent manner. These observations allow scientists to construct more precise models of catalytic action and its implications for larger sets of biochemical reactions and observe how catalysts influence rates and comply with the principles governing thermodynamic efficiencies, as outlined by Esposito et al. (2018). Besides, the study of catalysts is more than the mere acceleration in reaction rate; it involves the delicate aspect of how the various parameters influence them. The surface area, the composition, and the structure of the catalysts can significantly influence the catalytic efficiency of the catalysts. Transition metals, for instance, are normally catalysts due to their unique electronic properties and ability to transfer electrons in chemical reactions. Such factors directly relate to rates of molecular interaction, lending support to claims put forth within the domain of chemical kinetics as conceptualized by Pinto et al. (2017). The study of these factors enables researchers to tune catalysts for the enhancement of preferred reactions better, with implications toward sustainable policies and innovative technologies in chemical synthesis. The current research on catalyst dynamics is a leading example of the new face of chemical kinetics, where catalytic mechanisms dictate the trends of the future work and application (Figure 1.13).

Figure 1.13. *Role of catalysts in chemical reactions.*

Source: https://www.sciencing.com/sciencing/different-types-catalysts-6824467/51ad5471459a4c35b2cd34b4c43f102e.jpg.

1.5.1. Definition and Types of Catalysts

Catalysts are vital in chemical kinetics since they enhance the rates of chemical reactions without undergoing permanent changes. Catalysts enable reactants to change into products at faster rates by providing a different reaction pathway that involves less activation energy. There are two broad categories of catalysts: homogeneous and heterogeneous catalysts. Homogeneous catalysts are in the same phase as the reactants, usually solution, so that intimate molecular interactions may occur. Heterogeneous catalysts are in another phase, e.g., solid catalysts that facilitate liquid or gas reactions, where surface interactions govern the reaction process. Such distinction is important to understand real-world applications and helps chemists to select the proper catalyst on the basis of certain requirements of a system or reaction. Among the variety of catalyst types, zeolites became a favorite due to their unique pore structures and ability to catalyze numerous catalytic reactions, particularly within the petrochemical industry. Zeolite catalysis has the advantage of shape-selectivity in that some molecule sizes are able to fit inside and react within their pores. Not only does this improve reaction selectivity, but it also improves the yield of products as well. For instance, zeolites are commonly applied in methanol-to-hydrocarbons conversion, showing their catalytic efficiency in complicated reactions. Recent studies utilizing advanced molecular dynamics techniques investigate the kinetic pathways of zeolite-catalyzed reactions, revealing the effects of topology, acidity, and temperature on catalysis. As such, understanding zeolite catalysis properties and processes adds depth to the investigation of catalytic reactions in chemical kinetics (Bailleul et al., 2016). Additionally, advancements in catalytic materials, such as graphene-type materials, are being investigated for their potential uses in electrochemistry and catalytic reactions. These emerging materials possess distinct advantages, such as enhanced conductivity and the ability to form active sites using non-metallic heteroatom incorporation. Research has proven through studies that graphene catalysts can be utilized in oxygen reduction reactions, a step towards fuel cell technology. The opportunity to optimize these nanostructured catalysts for sustained performance is evidence of the face of catalysis in contemporary chemistry. As research continues, these advanced materials are likely to redefine catalytic reactions and enhance reaction performance critical in energy storage and conversion, echoing the accelerating advances in the area of chemical kinetics (Alammari et al., 2018).

1.5.2. Mechanism of Catalytic Action

The catalytic process is a fundamental concept in chemical kinetics, giving insight into how the catalyst accelerates reactions while undergoing no irreversible change. Catalysts work by offering an alternative pathway for the reactants of lower activation energy. This reduction in energy requirement enhances reaction rate, a crucial aspect in industrial and biological processes. Notably, enzyme catalysis in biological systems demonstrates this mechanism, whereby enzyme binding of substrates results in an enzyme-substrate complex. This complex stabilizes the transition state and allows for substrate-to-product conversion with greater efficiency. The significance of understanding these mechanisms is not only to enhance reaction rates but also to allow the development of more effective catalysts for specific chemical reactions, thus solving issues in such varied areas as medicines, energy production, and environmental studies. On the molecular scale, the catalytic process tends to perform some or all of the following: adsorption, reaction, and desorption. Initially, reactant molecules adhere to the surface of the catalysts to form an intermediate complex. This phase is important because it changes the electronic surroundings of the reactants, thereby facilitating their transformation into products by reduced energy barriers.

The role played by the catalyst at this point cannot be overrated; by providing a surface where the reaction takes place, catalysts not only accelerate the reaction but also modify the mechanism through which reactants are transformed. In heterogeneous catalysis, for example, this type of interaction is crucial as it often makes possible the performance of reactions under milder conditions than would otherwise be the norm in homogeneous systems. Understanding the fine detail of molecular interaction implicated in these phases enhances our range of options for designing catalytic processes to achieve optimally desired outcomes. New knowledge in the study of catalytic mechanisms has brought insights into transient species with a potential to significantly influence reaction courses. Quasi-steady-state approximation (QSSA) is among the analytical techniques used to approximate reaction mechanisms involving large concentrations of small but significant species that play key roles in transition states of reactions. By singular perturbation analysis, researchers can decouple reaction networks, effectively removing reactive intermediate species of high activity from the controlling equations of reactions, thus generating more tractable models. This approach has been found to be useful in various contexts, including classical systems such as Michaelis-Menten kinetics. With our growing maturity in the mechanistic basis for catalysis with techniques such as those described by Haseltine et al. (2007), we are better placed to optimize catalytic efficiency and address some of the most vexing problems in chemical kinetics and beyond and to make significant contributions to basic chemistry as well as applied sciences.

1.5.3. Industrial Applications of Catalysts

The industrial catalytic applications are central to enhancing the efficiency and sustainability of chemical reactions. The impact of catalysts on reaction rates is very significant while they are not consumed in the process, making them crucial in petrochemicals, drug production, and the manufacture of biodiesel. They enable reactions that would otherwise proceed too slowly or require enormous energy input. One of the uses is in the production of fatty acid methyl esters (FAME) from waste cooking oil, a function achieved using transesterification. The use depicts how catalyst choice and new equipment can optimize production processes to promote environmental sustainability as well as meet energy demands (Aubin et al., 2013). Also, reaction intensification with catalysts leads to effective utilization of resources, decreasing waste while the production is enhanced. Heterogeneous catalysts are an option of choice in the synthesis of biodiesel because they can be separated and reused. Recent kinetic studies on liquid-phase transesterification indicate that reaction kinetics depend on the choice of catalysts. For example, commercially available ion-exchange resins in gel-type and macroporous types possess special performance characteristics. Experiments have demonstrated that gel-type resins, such as Lewatit K1221, exhibit significantly higher catalytic activity compared to their macroporous counterparts, even when in the same acid site concentrations. The reason behind this disparity may be due to the swelling properties of the resins, in which case active sites are increasingly exposed during ethyl acetate transesterification with methanol, thereby influencing the kinetics of the reaction (Clercq et al., 2012). These results highlight the importance of catalyst design tailored to specific reactions for industrial applications. The understanding established from research into catalyst efficiency and reaction kinetics has significant implications for achieving maximum industrial process improvement. Enhanced understanding of reaction variables such as temperature, molar ratios, and catalyst morphology provides a plan for yield and processing time improvements. Integrated reaction-separation continuous flow reactors have been very effective when paired with the right catalyst, as seen in FAME manufacturing applications. The technology is not limited to biodiesel but to other chemical conversion reactions in various industries. As companies compete to satisfy the increasing demand for regulatory compliance and cleaner production by consumers, the function of catalysts will continue to be paramount to advancing chemical engineering and sustainable industrial processes (Aubin et al., 2013) (Figure 1.14).

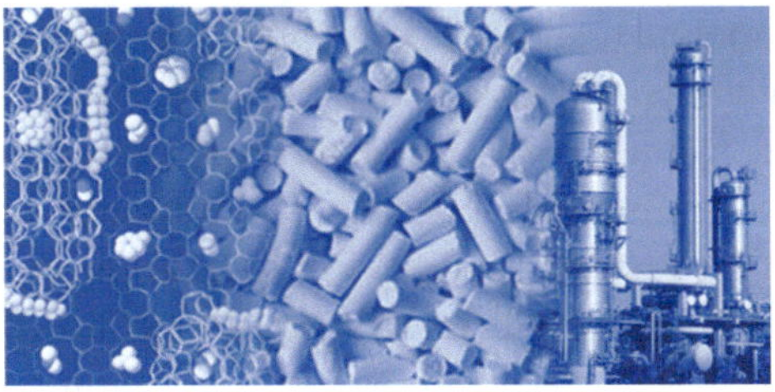

Figure 1.14. *Zeolites among the most widely used catalysts in the petrochemical industry.*

Source: https://ethz.ch/en/news-and-events/eth-news/news/2014/05/better-catalysts-for-the-petrochemical-industry/_jcr_content/news_content/fullwidthimage/image.imageformat.930.1672512403.jpg.

1.6. EXPERIMENTAL TECHNIQUES IN KINETICS

Experimental techniques in kinetics constitute an examination of great relevance in demystifying reaction mechanisms and chemical process design. A prominent technique among those that can be mentioned is spectrophotometry, which allows for the observation of concentration time-dependent changes via measurements of absorbance at selected wavelengths. Spectrophotometry is particularly convenient in quantifying rates of reactions for colored products or reactants, hence providing real-time insight into progress in the reaction. Furthermore, several other techniques such as high-performance liquid chromatography (HPLC) and nuclear magnetic resonance (NMR) spectroscopy can also yield complementary data which casts further insight into reaction pathways and kinetics. Each technique has to be critically evaluated in terms of the specific reactions of interest, with the guarantee of interpretations being grounded on solid data collection strategies.

Another experimental kinetics area that holds considerable interest involves examining reaction behavior at the quantum level. Recent work emphasizes the significance of quantum entanglement in ultracold reactions, suggesting that the phenomenon has radically shifted reaction dynamics towards achieving dynamical equilibrium. Techniques employed in experimental exploration of ultracold regimes often involve injecting advanced cooling methods and trap apparatuses for the containment and manipulation of reactants. Of possible interest may be that the chaotic character of behavior in these systems suggests that classical kinetic theories do not accurately describe underlying dynamics. Accordingly, physicists and chemists alike are compelled to embrace a more interdisciplinary approach wherein ideas of quantum mechanics find direct application in experimental technique, ultimately enhancing our comprehension of intricate reaction systems (Becker et al., 2015). In the pursuit of chemical kinetics knowledge, researchers are exploring computational approaches alongside classical experimental approaches. Computer calculations have the potential of performing important functions in determining reaction pathways and potential energy surfaces that will facilitate the design of optimum experimental conditions. Even in the study of complicated systems with multiplicity of reactions occurring concurrently, calculations can identify the system's response to varying parameters, e.g., temperatures and concentrations, of the kinetic outcomes. Computational predictions can guide experimental procedures such that selective experiments can differentiate basic reaction mechanisms. Coupled with empirical approaches, this twin strategy can yield a more complete understanding of kinetics and reveal new opportunities for applications across fields from pharmaceuticals to materials science (Pinto et al., 2017).

1.6.1. Methods for Measuring Reaction Rates

A basic aspect of chemical kinetics is the determination of the rate of chemical reactions, or the reaction rates. Various ways exist to make such determinations, each having its own merits and demerits based on the nature of the reaction as well as the conditions under which the reaction occurs. One such technique is the monitoring of the variation of reactant or product concentrations with time by spectroscopic or titrimetric determinations. These methods allow for measurement of the concentration in real time and provide valuable data to be utilized in reaction rate determination. Additionally, an examination of the kinetic data acquired can be utilized in formulating rate laws, which correlate the rate of reaction to the concentration of reactants and thus determine the reaction mechanism.

Figure 1.15. *How to measure the rate of a chemical reaction.*

Source:https://postapplescientific.com/wp-content/uploads/2023/07/PostAppleScientific-244587-Rate-Chemical-Reaction-blogbanner1.jpg.

Another significant method in the calculation of reaction rates is through manometric methods, particularly for gas-phase reactions. The technique involves the measurement of pressure variation in a closed system upon the course of a reaction, from which reaction rates can be determined based on the ideal gas law. For instance, in olefin hydroformylation, pressures of gases of hydrogen and carbon monoxide can be recorded as they vary with time in order to gather information regarding the kinetic nature of the reaction. For biphasic hydroformylation with a rhodium catalyst in an ionic liquid, for example, pressure drop techniques were employed to measure solubilities and dynamics of gas-phase reactants to enable precise insight into the effects of varying pressures on reactions rate and order on a component-by-component basis in each system (Delmas et al., 2011). In addition to these methods, the use of temperature experiments and activation energy measurements is also crucial in reaction kinetics evaluation. By conducting reactions at various temperatures and applying the Arrhenius equation, researchers can determine the relationship between the rate of reaction and temperature changes. This not only helps in the estimation of the activation energies but also provides information on the mechanism of the reaction and the transition states. Through the combination of various kinetic measurement methods, such as spectroscopic analyses, manometric methods, and temperature dependency methods, detailed information on the reaction rates is available. Collectively, these approaches contribute significantly to chemical kinetics advances, which advance the conditions of reactions for industrial applications and also theoretical ideas in chemistry (Jacobs et al., 1960) (Figure 1.15).

1.6.2. Use of Spectroscopy in Kinetics

Spectroscopy is one of the main tools used in the study of chemical kinetics, providing chemists with valuable data regarding reaction dynamics that cannot be obtained by traditional methods. Using a variety of spectroscopic techniques, scientists are able to monitor the concentration of reactants, intermediates, and products in real time. With this ability, one can examine reaction pathways and mechanisms in a detailed manner, gaining an insight into complex reactions. For example, time-resolved frequency comb spectroscopy was employed to elucidate the kinetics of the reaction between hydroxyl radicals and carbon monoxide. These techniques allow for the measurement of absolute yields and branching ratios of chemical species, bridging the gap between theoretical prediction and actual kinetic behavior, as reported in transient intermediate work such as cis- and trans-HOCO and reaction dynamics of isotopic analogues such as deuteroxyl radicals, as provided by Bjork et al. (2017). Equally important, spectroscopic techniques allow it to be straightforward to explore energy landscapes in reactions of biological macromolecules. Single-molecule mechanical unfolding experiments illustrate how spectroscopy is capable of quantifying the roughness of energy landscapes without placing restrictive assumptions upon reaction coordinates. Temperature variation in experiments enables researchers to decode fine-grained information about free energy barriers and transition states. Such data provides possibilities for insight into how biomolecules respond under different conditions, shedding light on molecular stability and reaction dynamics. The sophistication of such analysis relates to the importance of integrating experimental data with robust theoretical models to ensure accurate interpretation of the energy landscape and kinetic behavior, as in protein-protein interactions and thermal stability highlighted by Hyeon et al. (2006). The function of spectroscopies is not merely the observation without the aid of theory; it assists in kinetic modeling of a reaction as well. Spectral data of high resolution can be applied in kinetic models to make rate constant predictions and identification of reaction pathways more valid. The transition states in reactions with varying force or loading rate, for example, will critically determine kinetic behavior, thus requiring a nuanced understanding of how spectroscopic information can be utilized to direct reaction coordinate selection. The challenge is to make sure that spectroscopic techniques are adapted appropriately to the specific systems under study, especially for soft or plastic biomolecules, where the location of the transition state can change substantially. Overall, spectroscopy does not simply provide additional insights into chemical kinetics but also provokes methodology development to be in a position to cope with the complexity that pervades systems that are chemically and biochemically heterogeneous.

1.6.3. Computational Methods in Studying Kinetics

Utilization of computational tools in the field of chemical kinetics has revolutionized the methodology of research in reaction dynamics and mechanistic pathways. A major breakthrough in this field is the utilization of genome-scale kinetic models. Genome-scale kinetic models are constructed from high-throughput data sets that provide detailed information on biological networks. The platform used in building these models utilizes statistical evaluation and reaction network reconstruction, and this enables scientists to gain insight into complex kinetic behaviors. By using methods that decompose the Jacobian matrix into its chemical, kinetic, and thermodynamic components, researchers are able to achieve insight into the intrinsic interaction guiding the rate of reactions. This mathematical foundation not only assists in model building but also serves to highlight the inherent character of complexity in big biochemical systems, providing a pathway for further study of dynamic models in kinetic studies (Jamshidi et al., 2008). Computational methods are also more and more used to address the stochastic nature of chemical kinetics, particularly in systems with prevailing low numbers of molecules such as those found in genetic control networks. Stochastic models are being increasingly popular because they quantify the noise and heterogeneity of biological pathways in the best possible manner. Researchers employ parallel computing techniques to ensure these simulations are more efficient, thus enabling researchers to study complex cellular processes on a large scale. Parallelism, as achieved by OpenMP, has been found to be quite effective in reducing the computation time drastically without compromising the accuracy of stochastic simulations. These advancements mirror the continuous advancement of computational techniques in chemical kinetics, bridging the gap between empirical data and theoretical prediction and providing beneficial insights into the prognostic potential of kinetic models (Burrage et al., 2005). In addition, application of computational techniques to the investigation of kinetics has wide-ranging implications in drug discovery and bioengineering. The ability to mimic reaction kinetics and predict results using in silico methods facilitates the design of effective therapeutic compounds. As scientists realize the potential of computational tools, they are able to simulate the interactions between drugs and biological targets and thus make pharmaceutical design and optimization more rational. This computational integration also allows enzyme kinetics and metabolic pathways to be explored to an unprecedented degree of detail. Computational insight into the kinetic parameters puts biochemists in command of biochemical reactions, with implications for industrial processes and the clinic. Ultimately, these approaches are refining our capability to dissect the complexity of chemical kinetics, advancing various scientific disciplines.

SUMMARY

In summarizing an investigation into chemical kinetics, it is clear that the complex nature of chemical reactions is controlled by a host of variables such as temperature, concentration, and the influence of catalysts. Insight into these dynamics is essential in order to predict reaction rates and mechanisms. As we have discussed in the course of this chapter, sophisticated models and theoretical constructs enable one to understand such processes, successfully modeling the complexities inherent in reaction systems. Such models are critical for both research in academia and real-world applications, since they provide the fundamental knowledge required for disciplines such as materials science through to biochemistry. In addition, major advances in analytical methods have made it possible to gain a more profound insight into kinetics research. For instance, the newer techniques like the self-consistent atmospheric retrieval framework SCARLET, presented by Benneke et al. (2015), are reflective of how the understanding based on chemical kinetics is not confined to conventional environments but carries its relevance forward to astrophysical environments as well. This interdisciplinary utilization of kinetics reflects the value of kinetics in diverse scientific investigations, depicting an overarching story of chemical processes that include both terrestrial and extraterrestrial ones. Finally, the impact of chemical kinetics is wide-ranging and significant, reaching beyond theoretical chemistry to affect actual endeavors like drug discovery and environmental preservation. Understanding the assembly properties of viral proteins, as highlighted by Bakker et al. (2013), underscores the significance of kinetic principles in biology, demonstrating how chemical kinetics can guide novel therapeutic strategies. As research continues to advance, ongoing pursuit of reaction kinetics will provide critical information that fuels scientific and technological innovation, ultimately toward a greater knowledge of the microscopic and macroscopic world.

Chemical kinetics examines the rate of chemical reactions and the factors that govern them, providing vital information on reaction mechanism and dynamics. The most important in this field is how the concentration, temperature, and catalysts regulate the rate of reaction. For instance, increasing the concentration of reactants will generally raise the rate of reaction because there are more molecules bumping into one another at a higher frequency. Moreover, temperature is important; higher temperatures will increase kinetic energy, resulting in more energetic collisions and the potential to overcome the activation energy barrier. These concepts are essential for predicting the behavior of chemical reactions under various conditions, therefore being the basis for applications ranging from industrial synthesis to environmental chemistry. In addition to the critical factors that govern reaction rates, the antimicrobial activity of microemulsions presents fascinating examples of how kinetics can influence practical application. Research has indicated that oil-in-water microemulsions show very disparate antimicrobial activity based on their position in the stability region of their pseudo-ternary phase system. Experiments with other microorganisms, like *Staphylococcus aureus* and *Candida albicans*, indicate that these microemulsions are very active, particularly when positioned near the middle of their stability zone. The kinetic data obtained in such research highlight not only the effectiveness of microemulsions as antimicrobial agents but also the nuanced interplay between stability and concentration in kinetic paradigms, demarcating the broader potential of kinetic theory to real-world applications (Al-Adham et al., 2012). Further, the study of secondary organic aerosols (SOA) depicts another facet of chemical kinetics dealing with atmospheric chemistry and climate interaction. Experiments on the water-soluble organic portion of SOA formed through alkene ozonolysis illustrate critical characteristics in cloud formation and aerosol behavior. With the application of Köhler Theory Analysis, researchers obtained outstanding parameters, such as surface tension depression and effective molar mass, that function to control CCN activity. The implications of these findings underscore the reality that the kinetics of CCN activation are not just dictated by the chemistry of the aerosols but also by their physical attributes. Thus, understanding these kinetics is paramount for simulating atmospheric processes and the prediction of climate forcing, thereby reaffirming the significance of reaction dynamics in addressing environmental challenges (Asa-Awuku et al., 2010).

REFERENCES

1. Acosta, W., Kundu, K., & Lee, C. M. (2025). Jet-A reaction mechanism study for combustion application. https://core.ac.uk/download/pdf/42816564.pdf (accessed on 10th September 2025).
2. Al-Adham, I. S. I., Al-Nawajeh, A., Collier, P. J., & Khalil, G. (2012). The antimicrobial activity of oil-in-water microemulsions is predicted by their position within the microemulsion stability zone. https://core.ac.uk/download/228177199.pdf (accessed on 10th September 2025).
3. Alammari, F., Allen, M., Cai, Y., Chen, X., Conner, J., Farooque, A., Gong, Y., et al. (2018). Graphene-based nanostructures in electrocatalytic oxygen reduction. http://arxiv.org/abs/1805.03149 (accessed on 10th September 2025).
4. Alberts, B., Phillips, A., Blossey, R., Bundschuh, R., Cherry, J. L., De Jong, H., Fall, C. P., et al. (2007). Compositionality, stochasticity and cooperativity in dynamic models of gene regulation. http://arxiv.org/abs/q-bio/0702016 (accessed on 10th September 2025).
5. Andreasen, A., & Stegelmann, C. (2011). Finding the key transition states and intermediates controlling net reaction rates and selectivity. https://core.ac.uk/download/pdf/290114.pdf (accessed on 10th September 2025).
6. Angulo, J., Bain, A. D., Bellon, L., Brown, L. J., Castañar, L., et al. (2015). Kinetics of intramolecular chemical exchange by initial growth rates of spin saturation transfer difference experiments (SSTD NMR). https://core.ac.uk/download/41989369.pdf (accessed on 10th September 2025).
7. Asa-Awuku, A., Flagan, R. C., Gao, S., & Nenes, A. (2010). Water-soluble SOA from alkene ozonolysis: Composition and droplet activation kinetics inferences from analysis of CCN activity. https://core.ac.uk/download/4884188.pdf (accessed on 10th September 2025).
8. Aschaffenburg, D. J., Chen, X., & Cuk, T. (2019). Selecting between two transition states by which water oxidation intermediates on an oxide surface decay. http://arxiv.org/abs/1903.01675 (accessed on 10th September 2025).
9. Aubin, J., Mazubert, A., & Poux, M. (2013). Intensified processes for FAME production from waste cooking oil: A technological review. https://core.ac.uk/download/18540047.pdf (accessed on 10th September 2025).
10. Bailleul, S., De Wispelaere, K., & Van Speybroeck, V. (2016). Towards molecular control of elementary reactions in zeolite catalysis by advanced molecular simulations mimicking operating conditions. https://core.ac.uk/download/55855326.pdf (accessed on 10th September 2025).

11. Bakker, S. E., Barker, A. M., Borodavka, A., Dykeman, E., et al. (2013). Packaging signals in single-stranded RNA viruses: Nature's alternative to a purely electrostatic assembly mechanism. https://core.ac.uk/download/16460597.pdf (accessed on 10th September 2025).
12. Bandopadhyay, A., Le Borgne, T., Dentz, M., Méheust, Y., et al. (2016). Enhanced reaction kinetics and reactive mixing scale dynamics in mixing fronts under shear flow for arbitrary Damköhler numbers. http://arxiv.org/abs/1604.08092 (accessed on 10th September 2025).
13. Becker, D., Bény, C., Osborne, T. J., Ospelkaus, S., et al. (2015). Ultracold chemistry and its reaction kinetics. https://core.ac.uk/download/52919848.pdf (accessed on 10th September 2025).
14. Benneke, B. (2015). Strict upper limits on the carbon-to-oxygen ratios of eight hot Jupiters from self-consistent atmospheric retrieval. https://core.ac.uk/download/216243642.pdf (accessed on 10th September 2025).
15. Bjork, B. J., Bui, T. Q., Changala, P. B., Nguyen, T. L., et al. (2017). Direct measurements of DOCO isomers in the kinetics of OD+CO. https://core.ac.uk/download/211372083.pdf (accessed on 10th September 2025).
16. Borman, P. C., Schouten, E. P. S., et al. (1996). Influence of reaction products on the selective oxidation of ethene. https://core.ac.uk/download/pdf/11452680.pdf (accessed on 10th September 2025).
17. Bowie, J. U., Park, C., Schlebach, J. P., Woodall, B. T., et al. (2014). Bacteriorhodopsin folds through a poorly organized transition state. https://core.ac.uk/download/323079571.pdf (accessed on 10th September 2025).
18. Brady, K. B., Hui, X., Niemeyer, K. E., Sung, C. J., et al. (2017). Reduced chemistry for butanol isomers at engine-relevant conditions. http://arxiv.org/abs/1706.02043 (accessed on 10th September 2025).
19. Branco Pinto, D., Constales, D., Marin, G., Yablonsky, G. S., et al. (2017). Kinetic intersections as a source of information on rate coefficients. https://core.ac.uk/download/91288153.pdf (accessed on 10th September 2025).
20. Burrage, K., & Tian, T. (2005). Parallel implementation of stochastic simulation for large-scale cellular processes. https://core.ac.uk/download/81556.pdf (accessed on 10th September 2025).
21. Ceccarini, A., Gomez-Lahoz, C., Iannelli, R., Masi, S., et al. (2017). Influence of chemical reaction kinetics on electrokinetic remediation modelling results. https://core.ac.uk/download/214838833.pdf (accessed on 10th September 2025).

22. Čechová, E. (2012). Thermooxidative stability of PMMA composites. https://core.ac.uk/download/30296004.pdf (accessed on 10th September 2025).
23. Chou, K. T. (1987). Oxygen application to chloride leaching of complex sulfide ores. https://core.ac.uk/download/162573786.pdf (accessed on 10th September 2025).
24. Colombani, J. (2008). Measurement of the pure dissolution rate constant of a mineral in water. http://arxiv.org/abs/0911.5054 (accessed on 10th September 2025).
25. Conradi, C., & Mincheva, M. (2014). Graph-theoretic analysis of multistationarity using degree theory. http://arxiv.org/abs/1411.2896 (accessed on 10th September 2025).
26. Criado Luque, J. M., Perejón Pazo, A., Pérez Maqueda, L. A., Sánchez Jiménez, M. A., et al. (2011). Constant rate thermal analysis for thermal stability studies of polymers. https://core.ac.uk/download/187420555.pdf (accessed on 10th September 2025).
27. De Clercq, J., Thybaut, J., & Van de Steene, E. (2012). Experimental kinetic study of transesterification of ethyl acetate with methanol catalyzed by gel and macroporous acidic ion exchange resins. https://core.ac.uk/download/55729062.pdf (accessed on 10th September 2025).
28. Delmas, H., Deshpande, R. M., Julcour-Lebigue, C., Kelkar, M., et al. (2011). Kinetics of hydroformylation of 1-octene in ionic liquid–organic biphasic media using rhodium sulfoxantphos catalyst. https://core.ac.uk/download/12043525.pdf (accessed on 10th September 2025).
29. Egli, M., Faybishenko, B., Ghanbarian, B., Hunt, A. G., et al. (2020). Predicting water cycle characteristics from percolation theory and observational data. https://core.ac.uk/download/305124798.pdf (accessed on 10th September 2025).
30. Eisenberg, B. (1999). *From Structure to Function in Open Ionic Channels*. http://arxiv.org/abs/1011.2939 (accessed on 10th September 2025).
31. Esposito, M., Rao, R., & Wachtel, A. (2018). Thermodynamically consistent coarse graining of biocatalysts beyond Michaelis–Menten. https://core.ac.uk/download/154761372.pdf (accessed on 10th September 2025).
32. Gooday, G. (2004). Cry "Good for history, Cambridge and Saint George"? Essay review of Mary Jo Nye (Ed.); *The Cambridge History of Science*, Vol. 5. *The Modern Physical and Mathematical Sciences*. https://core.ac.uk/download/52713.pdf (accessed on 10th September 2025).

33. Gorban, A. N., Barenblatt, G. I., Bertini, L., Blesgen, T., Boudart, M., Briggs, G. A., et al. (2011). Quasichemical models of multicomponent nonlinear diffusion. http://arxiv.org/abs/1012.2908 (accessed on 10th September 2025).

34. Gorban, A. N., Noel, V., Radulescu, O., Zinovyev, A., et al. (2012). Reduction of dynamical biochemical reaction networks in computational biology. https://core.ac.uk/download/pdf/82880909.pdf (accessed on 10th September 2025).

35. Haseltine, E. L., Mastny, E. A., & Rawlings, J. B. (2007). Two classes of quasi-steady-state model reductions for stochastic kinetics. https://core.ac.uk/download/4876324.pdf (accessed on 10th September 2025).

36. Hyeon, C., Thirumalai, D., Derenyi, I., Evans, E., Fisher, T. E., Hummer, G., Lide, D. R., et al. (2006). Measuring the energy landscape roughness and the transition state location of biomolecules using single molecule mechanical unfolding experiments. http://arxiv.org/abs/cond-mat/0612433 (accessed on 10th September 2025).

37. Jacobs, T. A. (1960). Shock-tube measurements of the homogeneous rate of decomposition on NH_3 in NH_3–Ar mixtures. https://core.ac.uk/download/4887824.pdf (accessed on 10th September 2025).

38. Jamshidi, N., & Palsson, B. Ø. (2008). Formulating genome-scale kinetic models in the post-genome era. https://core.ac.uk/download/323077184.pdf (accessed on 10th September 2025).

39. Liebermeister, W. (2013). Elasticity sampling links thermodynamics to metabolic control. http://arxiv.org/abs/1309.0267 (accessed on 10th September 2025).

40. Machado, M. F., & Saraiva, J. (2005). Modelling the kinetics of thermal inactivation of apple polyphenoloxidase. https://core.ac.uk/download/15567640.pdf (accessed on 10th September 2025).

41. Marley, M. S., & Zahnle, K. J. (2014). Methane, carbon monoxide, and ammonia in brown dwarfs and self-luminous giant planets. https://core.ac.uk/download/pdf/189597261.pdf (accessed on 10th September 2025).

42. Marley, M. S., Morley, C. V., Moses, J. I., Zahnle, K., et al. (2016). Photolytic hazes in the atmosphere of *51 Eri b*. https://core.ac.uk/download/pdf/76423622.pdf (accessed on 10th September 2025).

43. Morrison, R. E., Moser, R. D., & Oliver, T. A. (2018). Representing model inadequacy: A stochastic operator approach. http://arxiv.org/abs/1604.01651 (accessed on 10th September 2025).

44. Proppe, J., & Reiher, M. (2018). Mechanism deduction from noisy chemical reaction networks. http://arxiv.org/abs/1803.09346 (accessed on 10th September 2025).

45. Ritacco, H. A., Sanchez-Morales, J., & Sánchez, M. D. (2018). Reaction kinetics in the production of Pd nanoparticles in reverse microemulsions. Effect on particle size. https://core.ac.uk/download/552341121.pdf (accessed on 10th September 2025).

46. Whittman, C. L., Grosjean, D., Cass, G. R., Druzik, J. R., Whitmore, P. M., & Nazaroff, W. W. (1989). Protection of works of art from atmospheric ozone. https://core.ac.uk/download/71354084.pdf (accessed on 10th September 2025).

CHAPTER 2

RATE OF REACTION AND ORDER

Contents

Introduction... 41
2.1. Factors Affecting Reaction Rate 46
2.2. Measuring Reaction Rates 52
2.3. Reaction Order Explained 57
2.4. Determining Reaction Order 62
2.5. Applications of Reaction Rate and Order.. 67
2.6. Case Studies in Reaction Rates.. 71
Summary.. 75
References.. 77

INTRODUCTION

The study of reaction rates and their orders is at the forefront of chemical kinetics, a branch of physical chemistry that deals with establishing the rate of reaction and parameters that influence such rates. At the core of chemical kinetics is the kinetic analysis, which quantitatively explains the rate of chemical transformation and describes the mechanism involved. These kinetics principles are of academic interest only; they have very profound implications in numerous industrial processes and biochemical pathways. They include most reactions in the chemical manufacturing industry, where intricate reaction networks lead to multiple reactions occurring simultaneously, and the whole system's behavior is defined by all these concurrent reactions. Representation of these networks using mass action kinetics allows for a rational modeling framework with attendant implications on reaction mechanism and efficiency. It is critical in optimizing the production processes and maximizing the production yield of the target products in real-life applications (Johnston et al., 2013). The reaction order, which is the exponent to which a reactant concentration is raised in the rate law, plays a critical role in characterizing the kinetic behavior of chemical reactions.

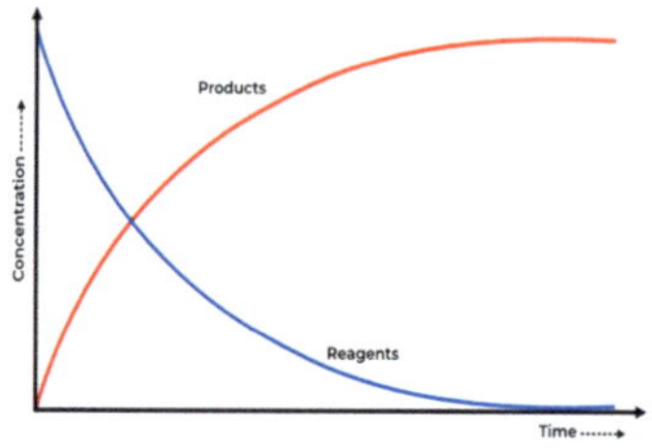

Figure 2.1. *Rate of reaction.*

Source: https://media.geeksforgeeks.org/wp-content/uploads/20221208100011/Rate-of-Reaction.jpg.

It enables chemists to predict the effect of concentration change on the rate of reaction and thereby enable accurate control of reaction conditions. Different reactions are of different orders, generally classified as zero, first, or second order, and so on. The classification into these orders is useful in applying the most suitable kinetic models that are necessary both for theoretical studies and for real-world application in laboratory and industrial operations. For example, in ethene selective oxidation, a complex kinetic model like Langmuir-Hinshelwood can describe the interaction of the adsorbed reactant with product on a catalyst and how product adsorption could be a controlling factor on the overall rate and selectivity of the reaction (Borman et al., 1996). Overall, then, the investigation of reaction rates and orders is a significant area of research that transverses abstract thought and application. By applying strict mathematical models for reaction kinetics description, scientists can successfully direct research and work on optimization. The convergence of mass action kinetics, reaction order, and other influencing factors like temperature, pressure, and concentration is key to achieving desired outcomes in natural and industrial processes. Measurement of reaction rate and orders is discussed in this chapter, along with the application of such work to real-world systems and ultimately establish the importance of reaction kinetics in developing scientific research as well as industrial processes (Borman et al., 1996; Johnston et al., 2013) (Figure 2.1).

Definition of Reaction Rate

The rate of reaction is a basic term employed in the field of chemical kinetics, which prescribes the speed at which a reaction occurs. It can be quantitatively measured as the rate of variation of a reactant or product concentration with time. The rate of reaction varies depending on different chemical reactions and is influenced by a variety of factors such as temperature, pressure, concentration, and the presence or absence of a catalyst. The subtle sensitivity of reaction rates allows a more profound understanding of the mechanisms underlying chemical processes, particularly in complex systems where more than one reaction may be occurring simultaneously. For instance, in biochemical networks, interaction of the reactants and products dynamically can have a significant role in determining the overall reaction rate, rendering such a system predictive and controllable difficult (Johnston et al., 2013). Therefore, realizing the intricate factors that govern rates of reaction is essential to theoretical research and practical uses in chemistry and all its allied disciplines. Furthermore, reaction rate calculation can be controlled by various mathematical models and kinetic orders describing concentration-reaction rate correlations. Reaction orders are classified based on the effect of reactant concentration on reaction rate, leading to zero, first, or second-order reactions, among others. Zero-order reactions, for instance, indicate a constant rate of reaction that is independent of reactant concentration, whereas first-order reactions have a linear proportionality whereby the rate increases as the concentration of a reactant increases (Feliu et al., 2013). These various orders not only shed light on the underlying mechanisms of single reactions but also enable prediction of reaction behavior in multicomponent systems that may involve complex interactions. Identification and implementation of these principles are essential in ascertaining correctly reaction rates in a variety of applications ranging from industrial manufacturing to biochemical pathways. Additionally, the study of the reaction rate is of greater importance in explaining the dynamics and stability of chemical networks, particularly those with more than one steady state. These reaction networks are amenable to being modeled as ordinary differential equation systems, with the composition of the mixtures yielding outcomes depending on the kinetic character of the involved reactions. These analyses reveal how certain of the network topologies can preclude or permit stable multi-stationarity, similar to changes in reaction order that will impact the global dynamics of the system. For example, using determinant criteria based on the Jacobian of formation rate functions for species allows researchers to determine conditions under which a reaction network will or will not possess multiple steady states (Feliu et al., 2013). This dependence of reaction rates upon network dynamics is the explanation for why a thorough understanding of reaction rates is crucial not only to theoretical kinetics but also to experimental studies in a broad variety of scientific fields (Figure 2.2).

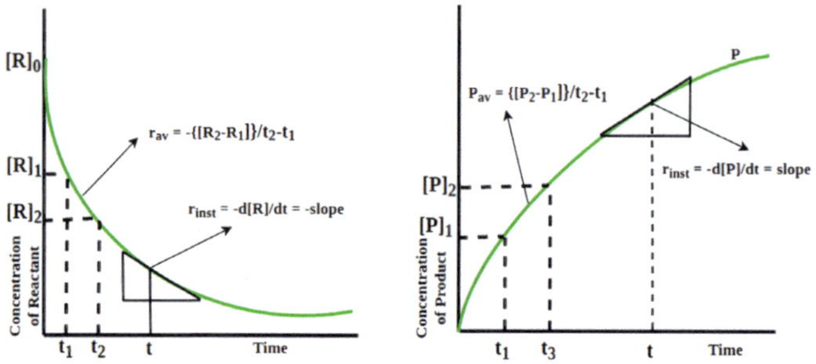

Figure 2.2. *Concept of reaction rate.*

Source: https://media.geeksforgeeks.org/wp-content/uploads/20230527124700/Rate-of-Reaction-Graph.png.

Importance of Studying Reaction Rates

Knowledge of reaction rates is crucial in understanding how chemical reactions take place and how their speeds are determined. Reaction kinetics enables chemists to quantify the rate at which reactants are converted to products, which is significant in process design and maximization of chemical processes. By studying these rates, scientists can determine the effect of variables such as concentration, temperature, and pressure on reaction efficiency. For instance, in CO_2 absorption into sterically hindered aqueous amine solutions of mass transfer operations, it is shown that rigorous numerical methods are necessary to effectively solve differential equations describing mass transfer. Such rigorous methods prevent oversimplification of complex reactions and promote more precise understanding of the involved processes as well as better predictability of reaction performance (Bosch et al., 1989). Hence, the study of reaction rates is an essential field of both theoretical and applied chemistry. Aside from their theoretical significance, the study of reaction rates has practical uses in the pharmaceutical field, environmental science, and materials technology. For instance, the pharmaceutical industry relies on kinetic research to formulate optimum drug delivery and drug formulation systems so that drugs are effective as well as safe for patients. Understanding reaction rates helps in regulating the conditions of a reaction, which is important in ensuring maximum yield and eliminating undesired byproducts. In addition, in environmental chemistry, understanding about reaction rates is important to predict the fate of contaminants and develop restoration strategies. Through studies on how contaminants degrade under different conditions, scientists are able to come up with more effective strategies in reducing harm to the environment. Hence, the relevance of reaction rates transcends academia, with immense importance in new materials and technology development and on practical applications and societal health. Moreover, ramifications for reaction rate study extend to new material and technology development. Materials science, for example, can be regulated by reaction rate upon synthesis process to produce novel material properties with implications in anything from electronics to clean energy. Reaction rates can control the microstructure, and hence material performance properties. In energy applications, an understanding of reaction kinetics can enhance catalysis processes, which are pivotal in achieving energy efficiency in fuel generation and conversion. The need to employ advanced modeling techniques to account for complex reactions further highlights the connection between kinetics and material innovation (Pinto et al., 2017). The more complex the chemical systems are, the more critical it becomes to carefully analyze reaction rates, which puts pressure on its importance not only in scientific investigation but also in technological advancement.

Overview of Reaction Order

An understanding of reaction order forms the core of the science of chemical kinetics because it describes how the rate of a chemical reaction varies as a function of the concentration of its reactants. Reaction orders are frequently classified into zero, first, second, and fractional orders, among others. Zero-order reactions are those in which the rate does not depend on the concentration of the reactants but is constant. In contrast, first-order reactions have a linear proportionality of reaction rate to the concentration of one reactant, while second-order reactions have a quadratic proportionality, often involving two reactants. The contrast between them tells us about mechanisms of chemical change and can be determined experimentally from observations of reaction rates at more than one concentration. A key aspect of reaction order is its capability to assist in the design and optimization of chemical processes. An example is when understanding the kinetic importance of species in intricate reaction networks is crucial in the modeling and prediction of product distributions in synthetic chemistry. The development of tools such as KiNetX has given a fillip to the research in kinetic networks by allowing researchers to identify and eliminate kinetically irrelevant species, thus refining their understanding of reaction pathways (Proppe et al., 2018). This functionality is particularly significant in cases involving multiple potential intermediates where traditional tools might not be able to identify dominant pathways effectively. By a meticulous study of reaction order and concentration of species, scientists can design conditions for desired products, enhancing both the efficiency and yield of chemical synthesis. Furthermore, interplay between reaction order and ambient conditions, such as temperature and pressure, further complicates the kinetic scenario. For instance, in photochemical reactions, light intensity and wavelength as variables added to the scenario introduce a variable that modulates reaction rates and orders. Current studies have established that sulfur photochemistry not only transfers H_2S to elemental sulfur but also highlights variations in reaction dynamics under varying temperatures (Marley et al., 2016). The evidence presented supports the need to adopt a comprehensive strategy to reaction order by viewing external effects on kinetic behavior. As researchers continue to study complex reaction networks and their interactions with different environmental circumstances, a deep understanding of reaction order will be essential toward the development of theoretical models and everyday applications in chemistry (Figure 2.3).

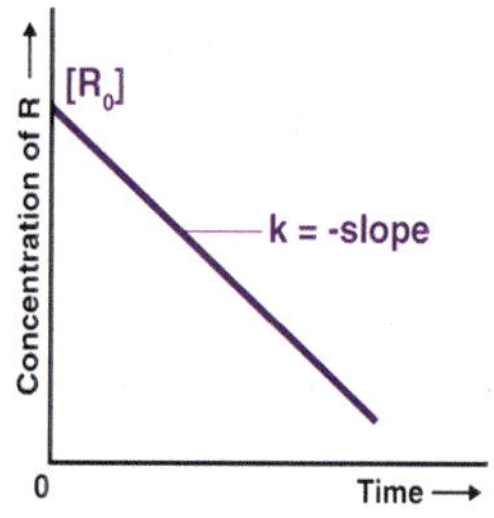

Figure 2.3. *Reaction order.*

Source: https://cdn1.byjus.com/wp-content/uploads/2020/09/Zero-Order-Reaction-1.png.

2.1. FACTORS AFFECTING REACTION RATE

Understanding what controls the rate of chemical reactions is essential to theoretical and practical applications in chemistry. Concentration is one of the most critical variables among these. As reactant concentration rises, the number of collisions between molecules will rise proportionally, thereby accelerating the rate of reaction. This is easiest to see in gas or solution reactions where the rate is directly determined by particle density. Also, temperature is crucial because temperatures increase tend to increase the kinetic energy of particles so that the molecules travel faster and will collide more intensely and often. Thus, messing around with concentration and temperature becomes significant in optimizing reaction rates, a subject that has been general discussion in high school textbooks, attesting to the use of these variables in chemistry research (Carera et al., 2012). Another significant factor that determines reaction rates is the presence of catalysts. Catalysts are substances that increase the rate of a reaction but are not permanently altered in doing so.

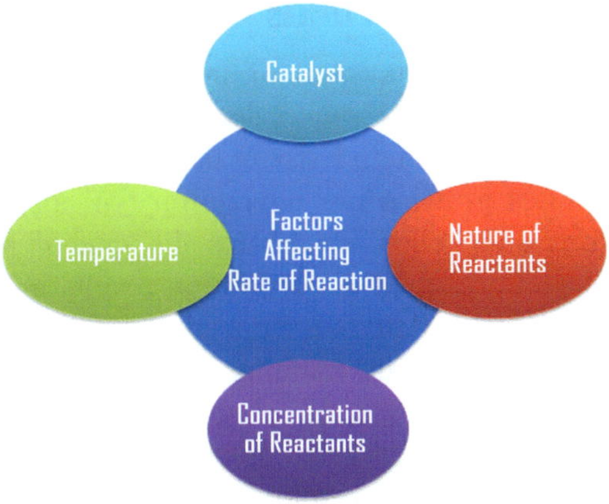

Figure 2.4. *Factors affecting rate of reaction.*

Source:https://files.askiitians.com/cdn1/images/201495-1199669-9213-capture.png.

They achieve this by providing an alternative route for the reaction with less activation energy and thus allowing more of the reactant molecules to have sufficient energy to react. The principle of catalytic effect is a highly prized concept in numerous industrial contexts, illustrated through material prepared for teaching high school students. For instance, understanding catalyst functions can demystify complex reactions and make them understandable to students, thereby enhancing their knowledge of chemical kinetics (Carera et al., 2012). Other than catalysts, surface area also plays a significant role; reactions between solid reactants accelerate when in the powdered or in the form of fragments because increased surface area allows for more collisions of the particles. Also, the influence of outside conditions, such as pressure, should not be ignored. For gas-phase reactions, changing pressure may significantly affect the reaction rate, particularly if reactants are in the gaseous state. Higher pressures cause higher concentration of gaseous reactant, which results in a higher number of collisions and therefore reaction rate. Apart from pressure, there are other factors that affect the reaction dynamics, such as the nature of the reactants and physical states (solid, liquid, gas). Knowledge of these factors was obtained through student interactions and tests, which suggest that social factors, such as teaching styles and group learning setups, also significantly contribute to influencing students understanding of these principles. Incorporation of numerous environmental factors, thus, within education systems can enhance students understanding of reaction rates and their applications in real-life scenarios (Figure 2.4).

2.1.1. Concentration of Reactants

The concentration of the reactants is one of the crucial determinants of the rate of a chemical reaction, which, in essence, decides the system's kinetics as well as the reaction mechanism. Through a rise in the concentration of the reactants, molecular collision rates grow. More interaction typically enhances the rate of effective collisions and hence accelerates the rate of product formation. Conversely, a decrease in reactant concentration results in fewer instances of such collisions, which is equivalent to a reduced rate of reaction. Such relations highlight the importance of understanding how varying concentrations of the reactants can influence the overall reaction kinetics, providing important information regarding reaction order and mechanism in analysis. Furthermore, the specific conditions in which the reactions occur—temperature and pressure, in this case—can affect how reactant concentrations impact the rates of reactions. For example, research on the chemical vapor deposition process reveals the complexity of these interactions whereby varying concentrations of gaseous reactants significantly change the dynamics of reaction. Specifically, the change in water vapor and chloride vapor concentrations has a linear correlation with the rate of velocity and efficiency of the deposition reaction, an intricate relationship between concentration and reaction competence. Empirical and theoretical results from such studies point out how concentration of the reactant finds a place in models of reaction kinetics, enabling prediction of kinetic behavior in various chemical systems (Burggraaf et al., 1992). For more than one reactant involved, competitive pathways may be determined by the concentration, as in bimolecular reactions. In reactions involving separate species, the concentration of each reactant bears a direct relationship to the percentage of reactant contact, an indicator of reaction rates. For example, when one considers a case where the reaction $A + B \rightarrow$ products is occurring, the concentrations of A and B will need to be optimized in order to allow for optimal reaction rate, as changes in one will dramatically change the dynamics of the other. The effectiveness of collisions here is crucial, particularly in reactions involving immovable reactants where access is limited, thus affecting reaction order. This such careful appreciation of how concentrations of reactants can influence bimolecular interactions ultimately aids in de-mystifying reaction mechanisms (Deem et al., 1998) (Figure 2.5).

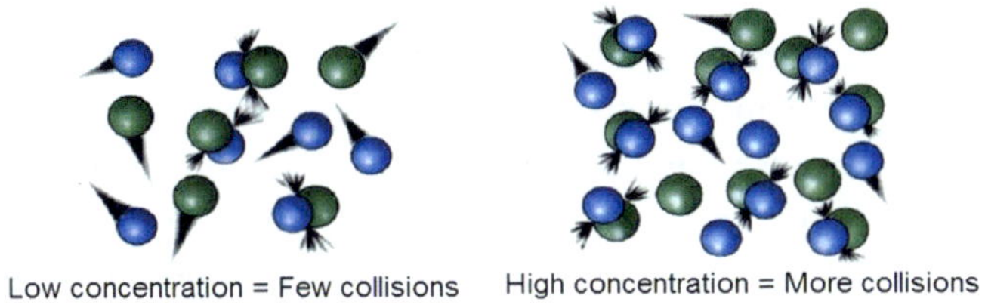

Figure 2.5. *Reaction rate tends to increase with concentration phenomenon explained by collision theory.*

Source: https://upload.wikimedia.org/wikipedia/commons/4/41/Molecular-collisions.jpg.

2.1.2. Temperature variations

Temperature differences are of utmost significance to the kinetics of chemical reactions and exert a powerful influence on the speed at which reactants are transformed into products. The relationship between reaction rate and temperature is primarily governed by the Arrhenius equation, which asserts that increasing the temperature will tend to increase kinetic energy between molecules. Increased energy enables more frequent and effective collisions between molecules, resulting in a higher reaction rate. Thus, a small increase in temperature can bring about tremendous variations in the rate constant, the determining factor of reaction velocity. This is a fundamental principle substantiated by empirical observations illustrating that many reactions are able to double their rate with just a 10°C rise. Knowledge of this temperature sensitivity is essential for industrial application and laboratory experiment alike, as control of temperature is pivotal towards optimizing reaction conditions. In addition to this, temperature variations can introduce complexity to reaction mechanisms and equilibrium positions, particularly to reactions that are reversible. One instance of how solvent and temperature influence the product of a reaction is through the aldol condensation of benzaldehyde and heptanal. Studies established that solvents can greatly influence conversion levels and product yields, and hence the need for the inclusion of activity coefficients in kinetic models. When the temperature changes, so do these coefficients, resulting in equilibrium constant changes and subsequently affecting desired product yields. Accurate modeling of temperature's influence, then, not only forecasts the response of a reaction but also optimizes the efficiency of chemical processes utilized in industries like pharmaceuticals and green chemistry, where the proper choice of solvent is instrumental in sustainable production (Heynderickx et al., 2019). Moreover, analytical methods deployed to track reactions, that is, temperature-sensitive reactions, point towards precision in kinetic studies. To give an example, through vegetable oil transesterification, a speedy analysis procedure founded on Near InfraRed spectroscopy exemplifies the complexity of tracking reactions in real-time, further amplified by temperature variations. This method allows on-line monitoring, and it is far superior to other devices such as gas chromatography since it is labor intensive and time consuming (Dubreuil et al., 2011). Reaction kinetics can vary with to temperature changes, affecting parameters such as activation energy. Similar to justifying methods against established kinetics models, researchers are able to draw more accurate conclusions on the effect of temperature on reaction rates and general process optimization. In effect, temperature variations are not only an environmental parameter; they are a constituent part that must be precisely regulated to achieve desired outcomes in chemical synthesis (Figure 2.6).

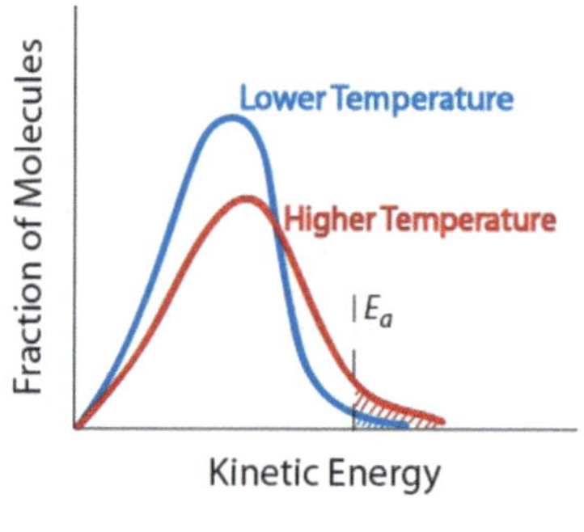

Figure 2.6. *Effect of temperature on the kinetic energy distribution of molecules in a sample.*

Source: https://opentextbc.ca/introductorychemistry/wp-content/uploads/sites/17/2021/06/Temperature-and-Kinetic-Energy-Distribution-1.jpg.

2.1.3. Presence of catalysts

The function of catalysts is very important in regulating the rate of chemical reactions, primarily through their ability to reduce the energy barrier for activation. By lowering the energy demand, the reactants can easily transform into products and thus accelerate the rate of the reaction. Traditional examples illustrate the way in which catalysts such as platinum in catalytic converters facilitate the oxidation of carbon monoxide in auto exhaust, thus minimizing toxic emissions. This catalytic activity not only lowers reaction speeds but also allows for the utilization of more favorable reaction conditions, such as lower temperatures and pressures. Interacting with reactants' molecular framework, catalysts create transient intermediates that allow for the conversion to products, and in essence, alter the dynamics of reaction pathways with the notable significance of catalysts to industrial and environmental chemistry. Also, the examination of some types of catalyst demonstrates their specific behaviors under distinct conditions, corresponding directly to the reaction rate. For instance, studies of copper catalysts in the synthesis of alcohol mixtures from syngas demonstrate the paramount role of metal composition and surface properties to catalytic activity.

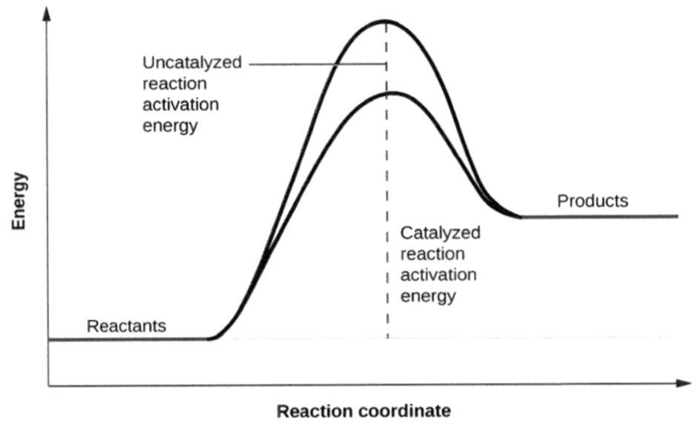

Figure 2.7. *Presence of a catalyst increasing the rate of a reaction by lowering its activation energy.*

Source: https://chem.libretexts.org/@api/deki/files/56216/CNX_Chem_12_02_Co-ordinate.jpg?revision=1&size=bestfit&width=529&height=332.

Research in leading institutions, such as the *Institut für Energieverfahrenstechnik* and Delft University, points out the comparative effectiveness of copper/cobalt catalysts over other catalysts, such as copper/zinc oxide/alumina, under specific conditions. Laboratory test experiments using pressure micropulse reactors reveal the significant differences in the efficiency of different catalysts and point out the need for high-level specificity in the choice of reaction order and throughput required in chemical processes (Drecken et al., 1994). The subtle understanding of how different catalysts influence reaction mechanisms demonstrates how they are key in optimizing both reaction rate and product yield. The role of catalysts can also be viewed in selective catalytic reduction (SCR) applications, specifically in the field of environmental technology. Research on copper oxide nanoparticles on silica focuses on their potential to remove deadly nitrogen oxides from flue gases through catalyzed reactions with ammonia. The catalytic behavior shows that such systems can maintain good selectivity for nitrogen production, temperature- and sulphation level-dependent (Edelaar et al., 1992). Through demonstration of the potential for increasing the maximum reaction temperature and improving nitrogen oxide removal efficiencies using additional sulphation, the findings exemplify the ability of site-specific catalysis to provide benefits for lowering environmental contaminants and improving industrial processes. These results not only contribute to theoretical understanding in catalytic systems but also guide practice in the development of catalysts for real applications, hence reducing the gap between application and innovation in environmental chemistry (Figure 2.7).

2.2. MEASURING REACTION RATES

Measurement of reaction rates forms the core aspect of understanding chemical kinetics and chemical reaction dynamic nature. The reaction rate may be measured in terms of the rate of change in reactant or product concentration with respect to time, which provides important information regarding the mechanism responsible for chemical processes. Some of the other measurement methods for these rates are through changes in concentration via spectroscopy, by pressure measurements in gas reactions, or by seeing the formation of precipitates or gases. All of these have their advantages, depending on reaction conditions and the desired accuracy. By applying these various methods, chemists can identify the order of a reaction, i.e., the power to which the concentration of a reactant is increased in the reaction for the rate law, basically connecting substance concentrations with reaction speed and efficiency. Technologies for measurement have recently been augmented by the inclusion of computer analysis, such as Computational fluid dynamics (CFD) modeling. These computer simulations provide insight into complex reaction environments, which facilitate a deeper analysis of reaction kinetics. With the implementation of isothermal approximations and coupled with CFD, researchers can derive analytical equations that relate kinetic parameters and flow dynamics, as demonstrated in research such as the analysis of C_{10H7Br} pyrolysis efficiency in microreactors. This process not only assists in properly predicting experimental results but also indicates definite means for the measurement of rate coefficients in various reaction conditions, thus improving significantly reaction dynamics models. Such approaches indicate the significance of both the inclusion of classical experimental techniques and utilization of sophisticated computational tools in furthering the accuracy and efficiency of reaction rate measurement. In addition to empirical estimates, theoretical models are also instrumental in the understanding of reaction rates and their applicability in a global context. One such example is the thermodynamic uncertainty relation that connects the mean current in stationary states out of equilibrium to entropy production rates. It is particularly useful when studying processes such as molecular motors and the efficiency at which they operate under an external force. By providing limits on thermodynamic efficiency in terms of experimentally measurable quantities only, it is then possible to assess reaction rates and mechanisms without being forced to investigate their intricate details. These frameworks not only offer an extremely useful analytical tool for the comparison of varying reaction paths but also add depth to our understanding of the relationship between thermodynamics and kinetics in chemical reactions and thus add richness to the field of reaction rate measurement and analysis (Figure 2.8).

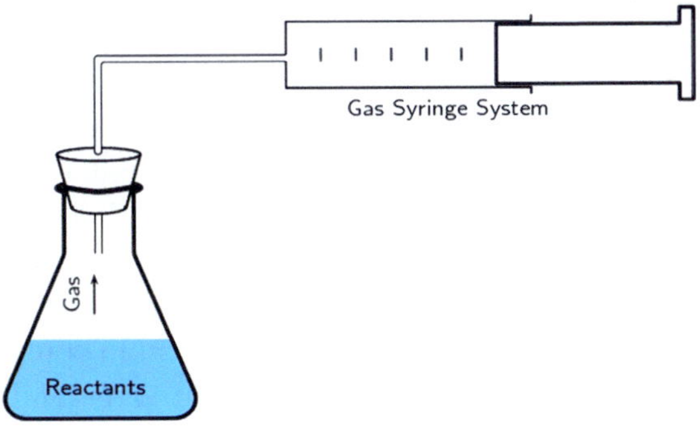

Figure 2.8. *In a reaction that produces a gas, the volume of the gas produced can be measured using the gas-syringe method.*

Source: https://assets.coursehero.com/study-guides/lumen/images/introchem/measuring-reaction-rates/cg12c3-0052.png.

2.2.1. Methods of Measuring Reaction Rates

Quantitation of chemical reaction rate is at the core of reaction kinetics, and various methods have been found to accurately quantify these rates. One of the most widely used methods is quantitation of reactant or product concentration with respect to time by spectrophotometry. Spectrophotometry is founded on quantitating light absorbance, which depends on the concentration of a given species present in a solution. By plotting concentration against time, the scientists can write the rate law, which gives both the order of reaction and the rate constant. Similarly, mass changes—either as gas production or as precipitate formation—can be tracked gravimetrically, providing additional evidence of reaction progress. These methods not only serve to elucidate underlying kinetics but also facilitate sensitivity analyses, as observed in gas-phase model works where reactions contributing to molecular abundance in the interstellar medium are highlighted (Agundez et al., 2010). Another useful method for determining reaction rates involves the use of calorimetry, which targets the heat transfer that accompanies chemical reactions. This approach is of greatest use in the case of endothermic and exothermic reactions since calorimetric information can provide clues towards reaction kinetics and changes in enthalpy. From the measurement of the changes in temperature over a period of time, the energy profile of a reaction can be calculated and its rate estimated by using heat transfer parameters. In addition, sophisticated techniques such as differential scanning calorimetry (DSC) enhance precision with the capacity to observe both thermal events and reaction rates simultaneously. The integration of calorimetric data with kinetic modeling offers scientists a better understanding of how different parameters influence reaction rates, thereby informing real-world applications in fields of materials science and thermodynamics (Adelberger et al., 2010). In addition, new technology has produced automated reaction monitoring systems that enhance precision and efficiency in reaction rate measurements. They are capable of continuously harvesting information through various sensing technologies, including mass spectrometry and chromatography, which can monitor reactant and product concentration in complicated mixtures in real time. Automated systems mitigate human error and provide higher reproducibility, which is critical in quantitative analysis. Moreover, the high throughput screening capability across different reaction conditions enables researchers to rapidly discover best parameters for specific reactions. Being combined with traditional methods like spectrophotometry or calorimetry, automated techniques not only facilitate the process of measurement but also accelerate the discovery of new chemical reactions and the development of new materials (Adelberger et al., 2010; Agundez et al., 2010).

2.2.2. Use of Spectrophotometry

Spectrophotometry application in the examination of the rate of reaction has become an indispensable use in the majority of scientific disciplines, particularly chemistry and biochemistry. Spectrophotometry helps scientists to find out the absorbance or transmittance of light by a solution and gather information concerning the concentration of the reactants or products present at any point during a reaction. When absorbance is plotted against time, the rate at which the compounds react or are generated can be determined from it, allowing kinetics of reaction to be determined. The technique has the benefit of enabling continuous measurement, and this is critical in the consideration of reactions that undergo fast change. Such techniques enable scientists to obtain detailed data necessary in the determination of reaction orders, and this ultimately results in a higher level of comprehension of the kinetics underlying various reactions. Spectrophotometry in environmental science applies to measure concentrations of pollutants to enhance monitoring of reactive processes in such compounds. An example is measuring chromium contamination in soil where various forms of chromium have diverse toxicity and reactivity. Experiments confirm that chromium(IV) is particularly toxic, hence the necessity for proper monitoring to environmental and public health ends (Palma et al., 2015). Through spectrophotometric methods, researchers can efficiently track reduction reactions, such as those facilitating the reduction of hexavalent chromium (Cr(VI)) into a less toxic form. The kinetic reaction is accessible through interpretation of spectrophotometric data, allowing order and rate constants to be established to inform remediation activities, demonstrating the immediate applicability of this method to hazardous material handling. Besides, spectrophotometry finds maximum application in everyday clinical and automotive science, where it is employed to quantify the concentration of various biological fluids or contaminants. For instance, its use in cerebrospinal fluid analysis has been quoted to enhance clinical judgment by the discrimination between various pigments that could have suggestive conditions (Keir et al., 2006). The sensitivity of the techniques makes it possible to detect trace amounts of compounds in complex matrices, a requirement for both environmental analysis and medical diagnosis. As technology progresses, the combination of spectrophotometry with automated devices may bring about quicker and more accurate results. Such a shift not only maximizes processes but also helps researchers and practitioners gain a deeper understanding of reaction kinetics, thus making optimization of methods used in various disciplines possible (Figure 2.9).

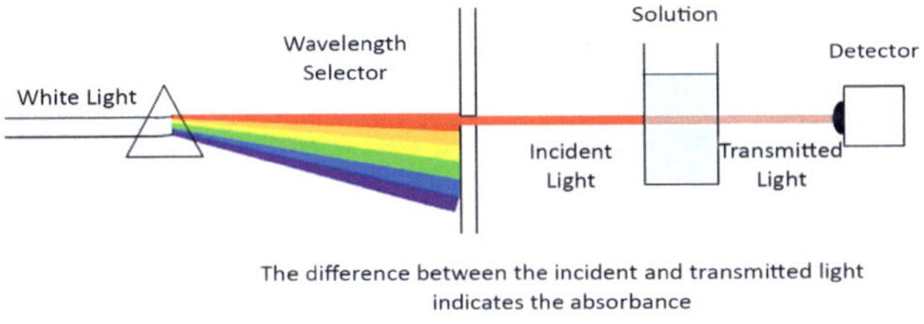

Figure 2.9. *General spectrophotometer diagram.*

Source: http://kaffee.50webs.com/Science/images/spectrophotometer.jpg.

2.2.3. Monitoring Gas Production

In monitoring gas production, sophisticated analytical techniques play a very important role in interpreting reaction kinetics and order of reaction. Effective monitoring of gas generation has the capability to expose how rates of reaction change with changing concentration and temperature. For instance, techniques such as Near InfraRed (NIR) spectroscopy have been applied in order to monitor transesterification reactions in real time, providing a convenient tool for monitoring reaction progress without the need for complex sampling regimens. This capability is significant, particularly in industrial processes where the application of fast data acquisition can be utilized to allow for adjustments to maintain optimal rates of reaction. Effective application of such methods can greatly enhance the understanding of gas production dynamics, which in turn can maximize chemical processes' efficiency (Dubreuil et al., 2011). Besides this, application of sophisticated optical sensors has revolutionized the quantification of gas production in most chemical vapor deposition (CVD) reactors. These sensors, including Fourier transform infrared spectroscopy and tunable diode laser spectroscopy, yield rich information by analyzing the *in-situ* gas phase compositions.

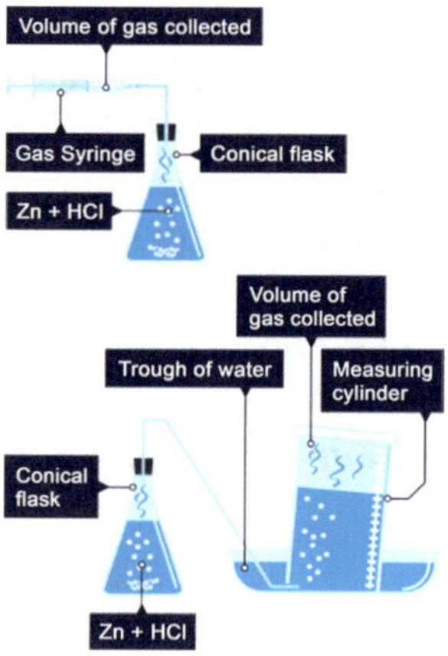

Figure 2.10. *Two different ways to measure the volume of a gas that is produced.*

Source: https://bam.files.bbci.co.uk/bam/live/content/zgvsb9q/large.

They permit real-time feedback to process control systems for real-time correction that can enhance the yield and quality of the product being synthesized. This chemistry feedback system illustrates the importance of combining real-time information so that reaction conditions can be tailored for the intended outcome. By correlating spectroscopic data with process performance, researchers have a solid platform to optimize gas production, essential to scalability and efficiency in the industrial environment (Beil et al., 2002). In addition to the increase in direct monitoring capability, integration of these technologies promotes increased understanding of reaction kinetics and their orders. The real-time data generated allows for finer differentiation of the impact of the various contributing factors to gas production rates, such as reactant concentration and catalyst activity. Moreover, activation energy calculated using such monitoring gives important insights into the energy barriers governing reaction rates. With the knowledge of such parameters, chemists can design means of optimizing reactions to increase yields and reduce waste. Ultimately, proper monitoring of gas production not only contributes to fundamental scientific knowledge but also translates into significant practical advantages in industrial applications, demonstrating the significant relationship between theory and practice within the subject of chemical kinetics (Dubreuil et al., 2011) (Figure 2.10).

2.3. REACTION ORDER EXPLAINED

Understanding reaction order is crucial in interpreting how the rate of a chemical reaction is affected by changes in concentration. Reaction order refers to the exponent within the rate law that explains the dependence of the rate of a reaction on the concentration of the reactants. Reaction order provides information about the mechanism of the reaction such that chemists can utilize it to predict how changes in conditions, either concentration or temperature, can impact reaction rates. For instance, a first-order reaction is proportional to the concentration of one reactant and leads to an evenly enhanced rate of reaction as that concentration of the reactant is raised. Thus, the determination of the reaction order is crucial in the derivation of proper rate laws and the understanding of the chemical process kinetics frequently encountered in laboratory and manufacturing environments, as testified by the large body of work that identifies unique orders under varying conditions in various chemically catalyzed operations. The complexities in the reaction order extend to other chemical regimes and systems, which betrays its dependence on the specific mechanisms involved. In certain catalytic reactions, for instance, there has been first-order dependence on the catalyst and zero-order kinetics with regard to the substrate concentration, as seen in isatin oxidation (Gowda et al., 2008). The above describes the function of the catalyst in reducing the energy barrier for the advancement of the reaction so that it can take reactants better into products irrespective of the substrate concentration beyond a critical point. Thus, an analysis of the responses demonstrates how order of reaction can be altered even under identical reaction conditions depending on catalyst concentration and presence, providing a better insight into catalytic efficiency and effectiveness in various chemical reactions. In addition, investigation of reaction order typically necessitates the utilization of elaborate methods and techniques that enhance kinetic data resolution. Techniques such as chromatographic and spectroscopic methods can offer useful information on how reaction rates respond to various conditions. For example, electron beam induced etching reveals a high understanding of kinetic processes because it illustrates that the substrate nature and the application conditions of the electron beam can control the reaction order (Martin et al., 2015). This allocation of various operational parameters reflects the intricacy in ascertaining reaction orders, as they can differ significantly based on experimental design and extraneous factors. Lastly, the knowledge of complexity in reaction order is a primary factor for researchers in planning reactions for targeted uses, paving the way towards material science and chemical engineering innovations (Figure 2.11).

Figure 2.11. *Chemical reaction orders.*

Source: https://www.thoughtco.com/thmb/p_tqJ0AsscHYtz3CCvbj6fHJRcI=/1500x0/filters:no_upscale():max_bytes(150000):strip_icc()/test-tubes-with-liquid-on-table-at-laboratory-685101663-58a0f52d3df78c4758309d27.jpg.

2.3.1. Definition of Reaction Order

In chemical kinetics, reaction order is a key parameter that explains the dependency of the rate of a chemical reaction on the concentration of reactants. Experimentally determined, reaction order may be zero, first, second, or higher. A zero-order reaction is where the rate is constant and independent of the concentration of reactants and is common in saturated catalyst reactions. On the other hand, first-order reactions show linear proportionality of rate with reactant concentration, as in the case of radioactive decay. Second-order reactions show quadratic relationships that have a very significant effect on mechanisms in chemical environments. An understanding of these distinctions provides a firm foundation for the prediction of effects of concentration changes on reaction rates and is vital in theory and in practical application in chemistry and engineering. Clarifying reaction order is not only useful to identify reaction mechanisms but also useful in building mathematical models that govern most chemical processes. Through rate laws, chemists are able to infer reaction order from experiments, often quantifying concentration against time in integrated rate laws. These correlations are crucial to interpret kinetic data in complex systems, especially where there are multiple reaction pathways. For example, the use of kinetic Monte Carlo (kMC) simulations makes it possible to project discrete reactions onto lattice models, thereby facilitating an advanced understanding of reaction dynamics across various conditions, such as temperature and pressure (Hoffmann et al., 2014). The advanced technique helps in designing reduced kinetic mechanisms that give reduced but effective representations of complex chemical behaviors, with significant impact in fields such as catalysis and combustion. Furthermore, reaction order has import beyond theoretical concepts, into the domain of real-world applications across many industries. Understanding reaction kinetics, or even better, reaction order, is directing many important processes, including drug synthesis and material preparation. An understanding of how reactant concentration can affect reaction rate is directly transferable into optimizing catalytic reaction under industrial conditions, optimizing efficiency and selectivity (Bollig et al., 1999). By careful consideration of reaction order, technologists are able to tailor conditions for maximum rates of manufacture with fewer undesirable byproducts, and more sustainable chemical manufacture. Therefore, complete familiarity with reaction order is essential, reinforcing the significance of learning reaction order in fundamental chemistry and research application that contextualizes technological and industrial innovation (Figure 2.12).

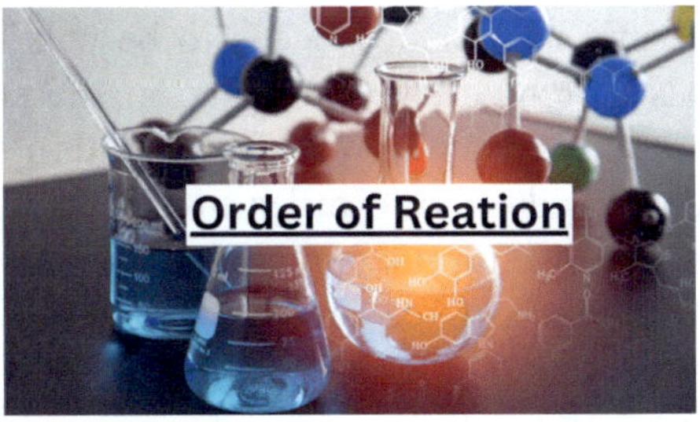

Figure 2.12. *Order of reaction.*

Source: https://www.careerpower.in/blog/wp-content/uploads/sites/2/2024/01/10124703/order-of-reaction-1.png.

2.3.2. Zero-Order Reactions

The concept of zero-order reactions is the basis for understanding chemical reaction kinetics, particularly those that do not rely on the traditional concentration dependences of the reactants. Zero-order kinetics is characterized by the reaction rate being independent of the concentration of reactants, and the reaction proceeds at a constant rate with time.

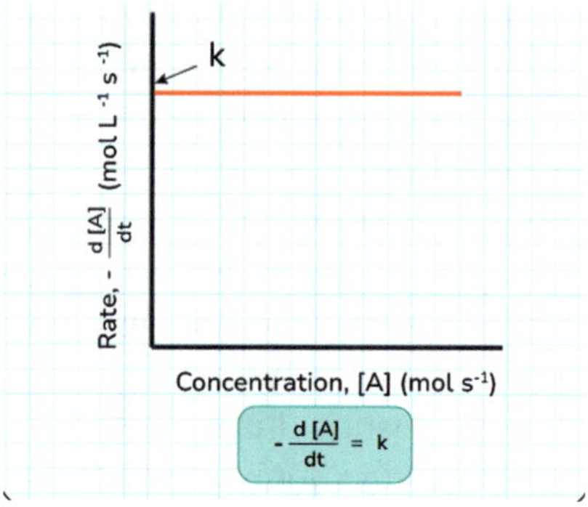

Figure 2.13. *Zero-order reactions.*

Source: https://media.geeksforgeeks.org/wp-content/uploads/20240410180939/ Zero-Order-Reaction.png.

This phenomenon is most prevalent in reactions involving catalyst saturation, which creates the scenario of increased reactant not providing a reaction rate increase. One such reaction is the disproportionation reaction of pentylamine, with hydrogen pressure zero-order kinetics, illustrating that increasing hydrogen concentration has no impact on reaction rate under a specific set of conditions. This interesting aspect illustrates the potential consequences in industrial applications, where reaction efficiency needs to be maximized. Furthermore, zero-order kinetics has a tremendous influence in the study of biochemical reactions and food preservation. Fish protein oxidation studies indicate that these reactions are described as zero-order in certain conditions at various temperatures. With increasing storage temperature, the rate constant enhances, which means a given amount of reactant protein is oxidized with time and therefore contains a greater amount of carbonyl. This information is significant to food scientists who are trying to optimize the storage life of fish foods. The zero-order oxidation rate, independent of reactant concentration, is useful in helping in planning better conditions for storage that can prevent oxidative damage. Knowledge of zero-order kinetics is necessary to devise means of maintaining food quality and rewarding food consumers and producers. Most notably, zero-order reactions are most significant in catalysis, as reactants coordinate with a catalyst at its state of saturation. For example, hydrocarbon synthesis reactions experienced using pentylamine on the catalyst display a zero-order nature based on the increased concentration of the reaction intermediates. This is to say that as long as the catalyst is saturated, hydrocarbon formation is constant, ensuring more consistent results in chemical production. The more that industries minimize reaction conditions, the better is the understanding of zero-order kinetics in terms of its capacity to guide more informed planning both reactor design and resource allocation. Through knowledge of mechanisms of zero-order reactions, chemists can better regulate reaction pathways to achieve both efficiency and greenness in chemical reactions (Figure 2.13).

2.3.3. First-Order and Second-Order Reactions

The classification of chemical reactions based on order is significant in the analysis of their kinetics, particularly to distinguish between first-order and second-order reactions. For first-order reactions, the rate is proportional to the concentration of one reactant. It is the indication that whenever the concentration of the reactant is changed, the reaction rate is changed linearly. A good example of a first-order reaction is radioactive decay, whereby the decay rate is directly proportional to the amount remaining. This straight-line dependency of rate on concentration is one of the most basic concepts in kinetics, and it enables one to have simple mathematical models with the ability to predict the behavior over time. On the other hand, second-order reactions depend on two reactants' concentrations or on one reactant squared, leading to a higher-order relationship. Understanding these differences is essential to modeling chemical behavior appropriately and optimizing reaction conditions. When discussing second-order reactions, it is worth mentioning their applications in various areas of science, such as combustion and more complex chain reactions. An example of this is the hydrolysis of *phenyl chloroacetate*, which, under certain conditions, is second order with a rate constant $k = 430\ M^{-1}\ s^{-1}$. At high concentrations of hydroxide, the reaction is pseudo-first-order, showing how the relationship between reaction order and concentration is complex. The ability to track these responses with highly advanced techniques, such as a miniaturized mixer assembly for capturing events in under a submillisecond, allows scientists to interrogate reaction mechanisms more deeply than previously possible. By demonstrating these kinetics experimentally, scientists can advance the understanding of reaction pathways and rate-determining parameters. These are critical considerations for medicinal chemistry applications through materials science, where close control over rate is often desired. Moreover, appreciation of the interplay between reaction order and reaction conditions can find broader application in chemical kinetics. For example, formulating equations that describe laminar flame reactions may entail an understanding of first-order as well as second-order kinetics, especially when there are several competing concurrent reactions. Scientists have used steady-state approximations to describe the function of free radicals in the reaction system for ozone decomposition. This helps in better comprehension of highly kinetic mechanisms, emphasizing the significance of distinguishing between different orders of reactions. With these analytical tools, scientists aim to derive explicit expressions that forecast reaction speeds in a particular scenario with accuracy. Overall, enhancing our knowledge of first-order and second-order reactions not only improves theoretical models but also enhances innovations in numerous scientific applications (Figure 2.14).

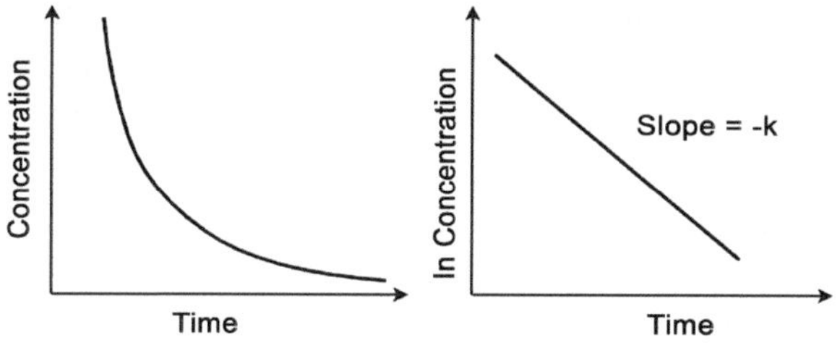

Figure 2.14. *Graphical representation of a first-order reaction.*

Source: https://media.geeksforgeeks.org/wp-content/uploads/20231130115818/First-Order-Reaction.jpg.

2.4. DETERMINING REACTION ORDER

Reaction order lies at the heart of determining the kinetics of chemical reactions, which can play an influential role in theory estimations and experimental operations in subjects such as environmental science and biochemistry. The reaction order tells us one of the concentrations of the reactants versus the rate of reaction, which may be zero, first, second, or higher. For instance, a first-order reaction will typically be characterized by a rate that varies directly with the concentration of one reactant, while a second-order reaction would be a function of two different reactants' concentrations or the square of one reactants concentration. An advanced understanding of the reaction order may allow us to construct good models for complex systems, including enzyme kinetics, for which Michaelis-Menten kinetics is often used to describe dynamic behavior adequately, thus the necessity of correct order determination in kinetic modeling (Bordbar et al., 2016). Both differential rate laws and integrated rate laws may be employed to establish the reaction order. Integrated rate laws consist of looking at concentration versus time information to determine the order through graphical techniques, i.e., plotting concentration versus time or natural logarithm plots for first-order reactions. Differential rate laws, on the other hand, look at the rate of reaction at a specific moment and give a more dynamic view of how different concentrations influence reaction velocity. These approaches not only play a key role in ordering the actual order but also in unveiling mechanistic routes.

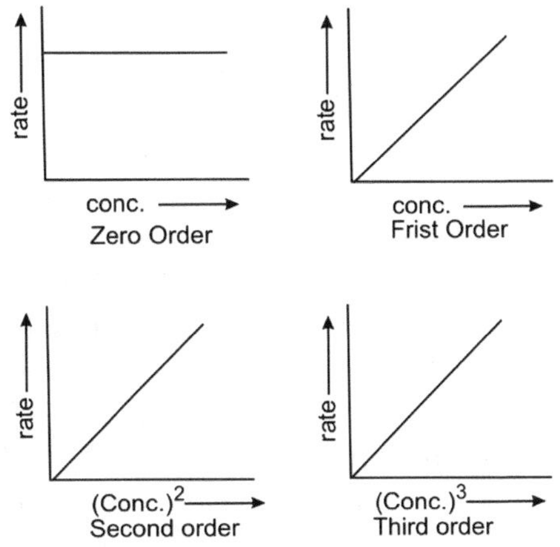

Figure 2.15. *Methods of determining reaction order.*

Source: https://cdn.entrance360.com/media/uploads/2019/10/22/image-9-1.jpg.

For instance, the application of the steady-state approximation is an appropriate means of investigating cases involving intermediates, often leading to more understandable routes for inferring complex reactions such as chain reactions in combustion reactions (Penner et al., 1954). In practical applications, the determination of the order of reaction can have crucial implications for scientific research and industrial application. Based on understanding the order of a reaction under different conditions, scientists and engineers are able to optimize procedures, for instance, in drug preparation, where reaction rates have the potential to make a substantial impact on yield as well as efficiency. In addition, reaction order significance is not limited to chemical engineering and chemical physics, but it also has importance in environmental science when establishing pollutant degradation rates that can affect regulatory policy. The focus on precise reaction order evaluation confirms models remain strong and relevant to actual use, reiterating the critical element of integrating theoretical models with experimental measurements within chemical kinetics and reaction engineering. Therefore, the determination of the reaction order is a cornerstone of theoretical as well as industrial chemistry and a portal for potential future experimental investigations and uses (Bordbar et al., 2016) (Figure 2.15).

2.4.1. Integrated Rate Laws

The theory of integrated rate laws is the core to the explanation of how the concentration of the reactants and time relate in a chemical reaction. Integrated rate laws are derived from differential rate laws, expressing how the amount of the reactants or products changes with time. Integrated rate laws are significant in explaining the kinetics of the reaction based on specific mathematical equations based on the order of the reaction—zero, first, and second. For zero-order reactions, concentration is time-independent, while for first-order reactions, concentration is proportional to time. By integrating the rate laws, chemists are able to determine the concentration of reactants at any given time during the course of the reaction, forming the theoretical basis on which more intricate simulations and experimental analyses, such as those applied to silicon nanocrystal growth, are based (Blaise et al., 2007). More detailed examination of the rate laws integrated for the first time reveals their significant role in numerous applications, particularly in industry processes where reaction control is of utmost importance. For instance, deposition rates of silicon nanocrystals onto substrate surfaces rely on a precise understanding of influences of factors upon reaction rates, as discovered in recent computational studies. These studies emphasize the mechanisms by which reactant concentration fluctuation and flow dynamics may affect transition rates and, consequently, the efficiency of deposition. The integrated rate laws allow for the measurement of these effects, thus informing practical adjustments to chemical vapor deposition techniques. These discoveries render integrated rate laws fundamental not just for scholarly purposes but also in applied practice in materials science and chemical engineering (Blaise et al., 2007). Apart from their theoretical implications, integrated rate laws also facilitate innovation in computational modeling and simulation techniques. For example, application of algorithms that calculate transition rates between subvolumes in chemical systems illustrates the integration of rate laws in simulations of reaction-diffusion processes. By applying such advanced techniques, scientists are more capable of gaining an in-depth grasp of the spatial distribution of reactants with time so that predictions of reaction outcomes are possible with more accuracy. In addition, the simulations also employ intrinsic rate laws to ensure findings validate microscopic as well as macroscopic predictions in whatever dimension is involved. The integration of rate laws and computational procedures demonstrates their generalizability and versatility in current scientific inquiry, thus solidifying their importance in the overall scheme of reaction order and kinetics (Figure 2.16).

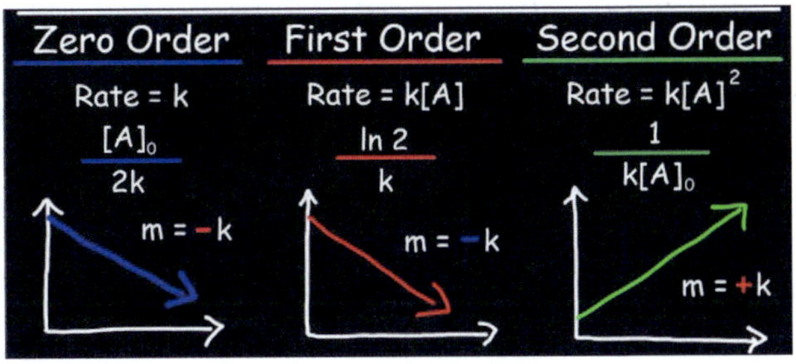

Figure 2.16. *Integrated rate laws.*

Source: https://i.ytimg.com/vi/7I0Xg92_eA4/maxresdefault.jpg.

2.4.2. Method of Initial Rates

The initial rate method is a simple method in chemical kinetics through which researchers can deduce the order of reactions against each reactant involved. By measuring the initial rate of a reaction at varying concentrations, one can have a collection of equations for the rate that tell how the rate changes with concentration. This method is particularly valuable in that it eliminates complications arising from intermediate species or reversibility and provides unequivocal information regarding the mechanism of reaction. Through ordered experimentation, the initial rate may be experimentally examined quantitatively, showing concentration dependence on reactants, and permitting mathematical relations to be set up that will predict reaction behavior in terms of conditions. Furthermore, making use of the initial rate method enables chemists to determine the kinetics of complex reactions, which is evident in various catalytic reactions. For instance, a study of biphasic hydroformylation with a rhodium catalyst indicated a first-order dependence with regard to initial rates for catalyst and reactant concentrations. However, it also indicated more complex behaviors for other reactants, e.g., hydrogen and carbon monoxide, whose impacts considerably influenced the reaction rates. This kind of complexity calls for the selection of appropriate mathematical models to describe data accurately. By dividing the initial rates into these various concentrations, chemists gain a clearer vision of the reactions dynamics and customize conditions for enhanced yield and selectivity, as has been seen in present day research techniques. Additionally, the integration of methods makes the technique of initial rates better by enabling advanced analysis otherwise not feasible with traditional means. The establishment of a simplified master equation, as in the case of lattice gas reaction research, showcases the way in which theoretical models complement experimental data. Focusing on significant reaction parameters in infinite diffusion limits allows scientists to develop efficient simulation methods which take into account adlayer equilibrium conditions and reactant mobilities. These developments in computational chemistry, coupled with initial rate methodology, allow scientists to predict the outcome of reactions and design more effective more potent catalytic systems more efficiently, as documented by Delmas et al. (2011) and Haseltine et al. (2006). Such an interdisciplinary endeavor serves to amplify the importance of the initial rate approach in advancing knowledge of chemical kinetics in both theoretical and practical contexts (Figure 2.17).

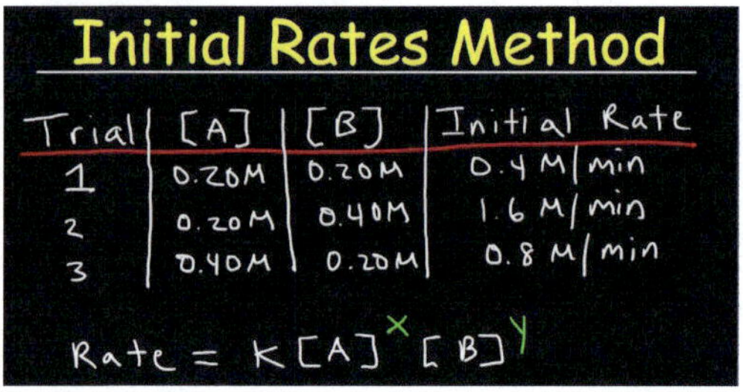

Figure 2.17. *Initial rates method.*

Source: https://i.ytimg.com/vi/oh4L2gcI5ds/maxresdefault.jpg.

2.4.3. Half-Life Calculations

It is necessary to know half-life calculations in order to study the kinetics of chemical reactions, particularly for determining the order of a reaction. Half-life of a reactant as the time taken for its concentration to reduce to half of its original value is an important parameter in first-order and second-order reactions. For first-order reactions, the half-life is independent of the initial concentration and can simplify it to make predictions about the time course of the reaction. For second-order reactions, the half-life has an inverse relationship with the initial concentration such that longer time will be required for the reactant concentration to reach half as the reaction continues. This difference emphasizes the significance of half-life in reaction kinetics since it provides a clear-cut measure to compare reaction orders and their respective dynamics. In addition, calculations of half-life are a necessary component in assessing real-world applications, e.g., origins-of-life research and combustion dynamics. In prebiotic chemistry, the concentration and stability of some compounds, such as dissolved nitrogen oxide anions (NO_x^-), have significant functions in determining pathways potentially accountable for Earth's origin of life. Research indicates that the determination of NO_x^- half-life in early Earth environments is paramount to understanding its chemical behavior during this period. These findings suggest that the kinetics of these kinds of molecules would have substantial implications on the types of prebiotic chemistries within natural environments. In doing so, a proper understanding of half-life calculations becomes an essential foundation for investigating additional questions of chemical origins and evolution early on Earth (Babbin et al., 2019). Similarly, in the field of combustion science, half-life calculations can give clues to reaction mechanism and burning process efficiency. Depending on literature reviews, strong relationships are found between combustion mechanisms and the respective half-lives of the reactive intermediates in flames. Different pressure conditions, for instance, can alter the laminar propagation of flames, and consequently the half-lives of the reacting species. As researchers investigate further the interplay of half-life behavior, reaction order, and combustion efficiency, an enhanced grasp of these effects will unfold. Eventually, the integration of half-life computations into the study of reaction rates not only enhances theoretical models but also enhances practical applicability across disciplines—ranging from chemistry to environmental science and energy generation (Penner et al., 1954).

2.5. APPLICATIONS OF REACTION RATE AND ORDER

Applications of reaction rate and order are ubiquitous across most scientific disciplines, as they provide irreplaceable information regarding the kinetics of chemical reactions. In environmental chemistry, for instance, the formation of secondary organic aerosol (SOA) is significantly influenced by both gas-phase and aerosol-phase processes. The processes of SOA formation may be studied through a general equilibrium/kinetic model based on the relative contributions of numerous factors—such as gas-phase oxidation rates and particle-phase reaction—on the SOA yield. Notably, formation rate and ultimate yield bear a strong reliance on the interaction of first- and second-order kinetics, testifying to the role of reaction rates in translating laboratory findings to atmospheric scenarios. This is particularly crucial in the development of more effective environmental policies and mitigation methods, underlining the essential role of reaction kinetics in atmospheric chemistry (Chan et al., 2007). In polymer chemistry, reaction rates dictate the chain formation process efficiency between polymer-polymer interfaces. In fact, studies have affirmed that end-functionalized chain reactions exhibit second-order kinetics, particularly under conditions of low chemical reactivity. This kinetic behavior shifts as the reactivity increases, moving to diffusion-controlled kinetics that capture certain time-dependent relations based on the entanglement or non-entanglement of chains. While second-order models hold for short-time kinetics, longer-term dynamics can be represented by first-order kinetics driven by the diffusion of more diluted species to the interface. Such reaction kinematics pinpoint how reaction order knowledge is crucial to forecast the process yield of polymer blend operations in industrial applications, which in turn indicates the range of chemical kinetics research (O'Shaughnessy et al., 1998). Moreover, the concept of reaction order is significant to advance pharmaceutical applications, where reaction rates establish the efficacy of drug formulation and drug delivery systems. Kinetic experiments make drug release profiles optimizable, which are often characterized to ensure that a therapeutic action is produced within a defined duration. The investigation of the reaction order enables researchers to engineer drug release kinetics such that it is consistent with physiological conditions for better therapeutic efficacy. Zero-order kinetics, for instance, can be employed in sustained release products and first-order kinetics in immediate release. Preferment of the rate of reaction, besides facilitating desirable drug pharmacokinetics, aids in informing drug development and formulations against regulatory standards. This indicates the enormous implications of the order and rates of reaction in industries and in medicine, as it further supports the interdisciplinary nature of chemical kinetics (O'Shaughnessy et al., 1998; Chan et al., 2007).

2.5.1. Industrial Applications

Reaction rate optimization is vital in industrial chemistry in order to boost productivity and minimize cost. A good case in point is the transesterification reaction for the manufacturing of fatty acid methyl esters (FAME) from waste cooking oil (WCO). By utilizing new process hardware like oscillatory baffled reactors and microwave reactors, industry can significantly improve the efficiency of the reaction. These advanced reactors allow for enhanced mass and heat transfer, thus increasing the rates of reaction for WCO conversion. Additionally, the application of suitable catalysts, particularly heterogeneous types, can help enhance performance in these modern reactors. As companies persistently seek out sustainable processes, scale-up of FAME production is a classic demonstration of the way reaction order and rate behavior can optimize industrial application and meet sustainability regulations, as discussed in recent summaries in chemical engineering (Aubin et al., 2013). Across all industries, there has been a need for increased efficiency, which has driven into continuous processes from traditional batch processes. For instance, a Bluestar Silicones study outlines the trend towards continuous processing technologies to improve safety and environmental profiles alongside productivity objectives. The design of a heat exchanger reactor constructed from silicon carbide plates is a prime illustration of the new trend, as it has been found to be appropriate for numerous industrial reactions. This transformation to continuous processes not only allows for better control over reaction conditions—hence influencing the order of rates of reaction—but also lowers operation risks associated with batch processes. With the growth of the industry, such technologies become necessary to ensure cleaner and more efficient processes. This study emphasizes the need for the conversion of reaction systems based on industrial needs (Elgue et al., 2009). Combining reaction order and rate considerations with industrial operations not only optimizes efficiency but also enables sustainability. For example, synergies between reaction and separation steps in operations like microstructured reactors can lead to colossal energy saving and waste reduction. Organizations can create systems that not only enhance the level of production but also comply with the environment standards and sustainability standards through innovation in design and implementation. The experience in the increase in the production of FAME and innovations in continuous processing in Bluestar Silicones brings out the point that knowledge of reaction kinetics is crucial in enhancing industrial processes. Ultimately, these advancements highlight the connection between reaction rates, order, and industrial applications as crucial in developing a cleaner and more efficient future for chemical synthesis.

2.5.2. Environmental Implications

The research into the environmental implications on reaction rates and orders does offer useful information regarding the occurrence of contaminants in aquatic systems. One example that is of special interest is hexavalent chromium formation, a toxicant that in drinking water is a considerable health risk. Chromium-bearing iron corrosion product oxidation in water systems can cause the sudden appearance of Cr(VI), hence the need to research chemical kinetics in environmental engineering. The molar composition of Fe(III), pH values, and the presence of other anions including bromide influence the rate of this reaction. Greater Fe(III) slows the reaction through inducing chlorine to decompose, and more bicarbonate levels could enable the reaction rate. Such processes introduce requirements for careful monitoring and pre-emptive controls to prevent hazardous exposures in drinking water systems (Chebeir et al., 2018). Further, nitrogen oxide anion kinetics under early earth conditions provides more framework for the interpretation of rates of reaction pertinent to environmental chemistry. Research has indicated that the stability and concentration of these dissolved species, particularly nitrates (NO_3^-) and nitrites (NO_2^-), are critical in the potential prebiotic chemical pathways. Initial models suggested these compounds were omnipresent in the prebiotic ocean;

however, recent research finds their concentration would have been much lower due to mechanisms like photolysis and reaction with reduced iron (Fe^{2+}). This shock instability is important to take into account when studying early Earth's chemical landscape, because it contributes to making certain reactions more likely to occur that might have made life possible (Babbin et al., 2019). Lastly, reactions rate research implications extend to modern environmental management methods. The capacity to comprehend order and rate of reaction allows for the development of improved pollution remediation and resource management methods. For instance, experience gained through kinetic studies allows environmental scientists to predict the fate of pollutants in various environments, thus informing treatment process design on pollutants like Cr(VI) and nitrogen oxides. Through integration with this scientific underpinning, policy makers are able to mobilize action to guarantee water quality and public health, ultimately facilitating a more sustainable human action-nature relationship. This integration of chemical kinetics into environmental policy is an illustration of the manner in which science can bring about significant change and maintain ecosystems (Babbin et al., 2019; Chebeir et al., 2018).

2.5.3. Biological Significance

The dephosphorylation and phosphorylation reactions form a central part of cellular signal transduction pathways, stimulating many biological processes that regulate metabolism, growth, and differentiation. Energy generated during the hydrolysis of ATP plays a critical role in these phosphorylation reactions, which modify the activity of proteins associated with signaling cascades. As in the kinetic model analysis of such cycles, energy availability controls sensitivity and amplification of cellular response. The coefficient of such amplification tends to be near an optimum point, as it represents the complex relationship between reaction rates and biological impact. Such dynamics function to underscore the biological significance of reaction rates and represent the way by which small deviations lead to gigantic cellular adjustments. Understanding these mechanisms not only reveals normal biological function but also implies understanding of disease signaling pathways, emphasizing the relevance of reaction order in cellular homeostasis (Bourne et al., 2002). In the realm of enzymatic reactions, dynamic disorder comes forward as an impetus influencing biological systems. Enzymes are not static structures but vary in catalytic activity that is coupled with conformational changes on various time scales. Recent advances in single-molecule enzymology have enabled one to trace such activity fluctuations and reason about their profound biological implications. An examination of a stochastic reaction system illustrates how rate fluctuations of enzymes can have a major impact on substrate concentrations in a cell. Through the formulation of equations for the relationship between the speed of such fluctuations and mean substrate concentration, researchers expose mechanisms through which the seemingly minor fluctuations in reaction rates lead to large deviations from values predicted using deterministic models. The findings highlight the biological importance of enzymatic rate fluctuations as having the potential to lie at the heart of the versatility and efficiency of processes of the cell (Gupta et al., 2017). Lastly, the interplay of reaction rates, order, and biological significance presents a multi-faceted landscape to navigate. The interdependency of enzymic reactions and their regulation highlights the necessity of quantitative approaches to unravel cellular metabolism. As biological systems are inherently stochastic, incorporating probabilistic elements into traditional kinetics models can potentially be the solution to unraveling cellular behavior, especially in perturbed conditions. Studying these dynamics not only gives greater insight into fundamental biological processes but can also yield therapeutic insights for the treatment of disease manifesting via dysregulated signaling pathways. Examination of reaction rate and order is thus not purely theoretical; it is critical in uncovering the relevance of biology to life-sustaining processes on both the cellular and organismal level (Bourne et al., 2002; Gupta et al., 2017).

2.6. CASE STUDIES IN REACTION RATES

The study of reaction rates through case studies is highly informative in terms of the complexity of chemical kinetics. The rates of reaction are defined by a myriad of variables, including concentration, temperature, and catalysts, all of which have the capacity to impact upon the manner in which reactants are being converted into products. A critical examination of two alternate kinetic schemes illustrates such subtleties, illustrating how subsequent processes and simultaneous reactions are capable of producing apparently identical kinetic profiles. Specifically, the subtle actions of aforementioned mechanisms, when extended to Reactions under conditions of experimentation, illustrate how they may appear to follow pseudo-first-order reactions. This observation implies that traditional kinetic studies might overlook intricate details inherent in reaction mechanisms, thus highlighting the significance of comprehensive studies in the areas of chemical kinetics and equilibrium measurements, as reflected in the research of Balogh et al. (2015). In addition, the evolution of mixing models significantly contributes to the understanding of reaction rates under turbulent conditions. The new continuous model of mixing resolves the issue of time discontinuities for chemical processes. It enables total coupling between mixing of fluids and chemical reactions to create a probability density function (pdf) that closely approximates a Gaussian distribution, unlike conventional models. The development is significant in that it impacts reaction rate calculation and prediction, particularly in systems experiencing homogeneous turbulence decay. Knowledge of these interactions is essential to successful kinetic modeling, particularly in industrial and environmental contexts such as in chemical engineering and environmental science, where flow and concentration fluctuations could lead to varying behavior in reactions, as emphasized by Hsu et al. (2025). Reaction mechanism study applications also extend beyond theory models as they are industrially and environmentally applicable. Other chemical reactions—anything from pollutant breakdown to drug synthesis—are inherently associated with the rates at which they go on. By examining case studies detailing the impact of alterations in conditions such as agitation and temperature on reaction rates, researchers can plan optimized protocols that optimize efficiency and yield. Furthermore, as shown in kinetic analysis of reactions catalyzed by simultaneous processes, an accurate understanding of reaction order and its properties is crucial to the prediction of the influence of changes in one component on reaction dynamics. This is knowledge which is vital to chemists and engineers in systems design for controlling reaction rates efficiently for successful outcomes in laboratory experiments as well as practical applications.

2.6.1. Example of a Zero-Order Reaction

The characteristics of zero-order reactions yield worthwhile insight into chemical kinetics and reaction mechanisms. The rate of a zero-order reaction is constant with time and independent of the reactant concentration. Such behavior is liable to occur in instances where a reactant is in excess, thereby saturating a surface of a catalyst or active site. A common example is accompanied with enzyme-catalyzed reactions, particularly in situations where the provision of substrate exceeds the capacity of the enzyme to metabolize it. Here, the rate is determined by the catalyst activity rather than by substrate concentration, as would be postulated by the quasi-steady-state approximation (QSSA) in various reaction mechanisms (Haseltine et al., 2007). Understanding such responses is vital in fields like pharmaceuticals, where drug release rates are controlled under zero-order kinetics. The characteristics of zero-order reactions find meaning that goes beyond theoretical interest into industrial and biological applications. A case in point is the importance of zero-order kinetics in controlled drug delivery systems, where therapeutic drugs are delivered at a constant rate independent of concentration. This maximizes drug effectiveness with few side effects since continuous exposure gives optimal therapeutic levels in the long term. In biochemical reactions, zero-order kinetics is an intrinsic principle describing substrate interactions and limitations by enzymes. By applying the QSSA principles, scientists are able to characterize complex biological processes with a reasonable degree of accuracy, for example, in classical texts on enzyme kinetics where Michaelis-Menten principles are applied under conditions that simulate zero-order conditions (Haseltine et al., 2007). Additionally, zero-order reactions also indicate the fine balance between concentration dynamics and external parameters such as temperature and pressure. As in diffusion-limited reactions, zero-order kinetics can be indicative of more complex underlying processes. For instance, the Lotka-Volterra model provides a subtle description of these reactions, illustrating how variations in reactant quantities may have a profound impact on overall reaction kinetics. This effect is frequently caused by discrete stochastic behavior that differs from classical expectations, highlighting the relevance of integration of both deterministic and stochastic thinking into reaction studies. With ongoing advancements in science, the relationship between zero-order reactions and emerging kinetic models will undoubtedly revolutionize our knowledge of reaction rates, facilitating a greater understanding of the complexity of chemical systems.

2.6.2. Example of a First-Order Reaction

Understanding first-order reaction kinetics is important in chemical kinetics, as it forms a major part of many industrial and laboratory processes. First-order reactions, with the rate equal to the concentration of one reactant, allow for simple modeling and reaction behavior prediction. A great example of a first-order reaction is hydrogen peroxide (H_2O_2) decomposition into water and oxygen gas. The rate is explainable by the rate equation: Rate = $k[H_2O_2]$, where k is the rate constant. The simplicity highlights the reaction's sensitivity to the hydrogen peroxide concentration, and hence it serves as an evident point of reference in reaction mechanism studies and their corresponding order. In addition to hydrogen peroxide decomposition, first-order reactions occur frequently in various applications, such as biological. Enzyme-catalyzed reactions usually exhibit first-order kinetics under substrate concentrations that are low, and under these conditions, the enzymes' saturation does not occur. The association emphasizes the importance of reaction order in biochemistry and gives an account of how living organisms metabolize substances. For instance, the initial breakdown of a substrate by an enzyme can be modeled as a first-order reaction until the point of saturation, thus providing information on metabolic rates as well as enzyme efficiency. Understanding this correlation between reaction order and biological processes, as evidenced in enzyme kinetics, improves the overall understanding of the dynamics of reaction and its use in various fields. The influence of reaction order on mass transport phenomena sheds light on the intricacies of correspondence between the kinetics of reaction and physical transport phenomena. First-order reactions are elementary building blocks in the description of system behavior in complex multicomponent systems, e.g., those under study by sophisticated modeling methods. Corresponding relevance is similarly applicable in reactions such as esterification, where kinetic and equilibrium factors control the reaction in liquid phases. Multiresolution schemes included help solve challenging reaction-diffusion systems, particularly when steep gradients or phase boundaries exist. Such elaborate theoretical frameworks, according to Baier et al. (2008), demonstrate how the integration of reaction kinetics with diffusion models advances the predictive accuracy of simulations. As such, the exploration of first-order reactions not only increases fundamental chemical understanding but also drives the operational performance of most industrial processes in chemical engineering, cited by Ahmadi et al. (2010).

2.6.3. Example of a Second-Order Reaction

The concept of second-order reactions is key in the understanding of various chemical processes, particularly where the rate of the reaction depends on the concentrations of two reactants. A perfect illustration of a second-order reaction is when hydrazine undergoes decomposition. In this kind of reaction, the rate is directly proportional to the product of the concentration of the hydrazine and the catalyst used. This second-order reaction shows a fundamental characteristic of second-order kinetics, whereby the concentration of both the reactants needs to be considered in order to properly predict the rate of reaction. Notably, second-order processes' straight-line relationship between reaction rate and concentration form the basis for the construction of systems of higher order, as in instances involving chain reactions, such as combustion phenomena (e.g., the burning of fuel), where detailed kinetic work reveals rich information on reaction mechanisms (Penner et al., 1954). Another well-known example is ozone decomposition kinetics, where the reaction relies on the use of the steady-state approximation in the free concentrations of the radicals. In this reaction order, it relies on the concentration of the radicals and the ozone formed in the process of the reaction. The need to employ such approximations helps in explaining the reaction dynamics in a less complicated way yet in accordance with second-order kinetics rules. The method not only simplifies the reaction mechanisms but also conforms to standard models used in chemical kinetics, thereby enabling better predictions of the flame speed in combustion reactions involving ozone (Penner et al., 1954). Further, the investigation of these mechanisms compels us to better understand chain reactions in complicate systems yet further notes the importance of second-order reactions in a wide range of chemical applications. Recent studies show the intricacies involved in modeling reactive transport in porous media and detail the importance of second-order bimolecular reactions in such a setting. Within studies that take a Lagrangian approach, particles representing solute fractions travel through porous domains as they react simultaneously according to second-order kinetics. This modeling approach emphasizes how reaction rates may become sophisticated as influenced by various concentrations and activities of reactants. As established in the literature, incorporating techniques that consider sophisticated kinetic behaviors expands the applicability of second-order models for reactions to diverse chemical systems. By applying Kernel Density Estimators, researchers were able to revise reaction probabilities from particle distributions and hence refine simulations of reactive transport and check the validity of second-order reactions in advanced chemical kinetics research.

SUMMARY

Lastly, an understanding of order and rate of reaction provides significant knowledge about the chemistry of chemical reactions as it allows us to understand the several factors that influence reaction kinetics. One of them is the effect of the structure of the catalyst, especially in systems using acidic zeolites, which have been shown to play a huge role in catalytic activity. The delicacy of the catalyst structure versus the inherent rate coefficients emphasizes the necessity for measurement and modeling accuracy. By studying the cracking of the alkanes under controlled conditions, researchers have elucidated that the kinetic parameters are extremely sensitive to catalyst structure and reaction conditions, they exhibit a complex relationship that must be unraveled (Bell et al., 2018). Besides, new advances in computational methods, as demonstrated by KiNetX, have revolutionized kinetic analysis of complex reaction networks. These tools are designed to ease the study of reaction paths by automatically identifying and eliminating extraneous species systematically, thereby enhancing transparency of kinetic mechanisms. The ability to construct correct approximations of sensitivity of species concentrations to changes in rate constants is a significant move toward enhanced predictive model development in chemistry. Not only does this function help to reduce complex networks but also ensures crucial kinetic information is not left behind, making it even more useful in reaction rate and order research (Proppe et al., 2018). Lastly, the subject of reaction rate and order is not merely a theoretical undertaking; it has significant practical implications, particularly in the fields of catalysis and materials chemistry. The subtleties of reaction rates and molecular order can help in the development of more efficient catalysts and reaction conditions, developing green chemistry and sustainable energy technology. Because ongoing research becomes better with the inclusion of new computer aids and test techniques, there are substantial opportunities for optimizing chemical reactions. Thus, ongoing research in reaction kinetics is crucial to ongoing scientific and industrial progress. Inlearning reaction rate and order, one needs to possess an understanding of the fundamental principles that govern the character of chemical reactions. Reaction rate has to do with how quickly reactants are converted into products and varies based on a number of variables including concentration, temperature, and pressure. Increasing the concentration of the reactant typically leads to a faster reaction rate due to the increased chances of collision among the reactant molecules. Additionally, temperatures play a significant role by providing reactant molecules with energy needed to overcome an activation barrier of a reaction. Understanding these dynamics is crucial, and it was properly demonstrated by research targeting the intricate entanglement of reaction conditions and the coefficients for rates that describe such processes (Bjork et al., 2017). It is knowledge that paves the way for the prediction and regulation of chemical behavior in numerous applications. The application between rate laws and the order of reaction is another major point to observe in describing the rates of reaction. The reaction order provides data on how the rate is affected by reactant concentration. A first-order reaction, for example, is directly proportional to the concentration of one reactant, while a second-order reaction is proportional to the concentration of two reactants or to the square of the concentration of one reactant. This classification helps chemists in reaction kinetics modeling and is describable by deterministic ordinary differential equations, a macroscopic limit of more refined probabilistic models (Desmond et al., 2007). In giving a systematic framework within which to analyze various chemical reactions, understanding reaction order enables researchers to predict the behavior of complex systems under various conditions. Besides mere observations, reaction kinetics is also an area that extends beyond mere observations to encompass advanced mechanistic details obtained by modern spectroscopic techniques. For instance, time-resolved frequency comb spectroscopy has been instrumental in understanding complex kinetics of significant reactions, such as the interaction of carbon monoxide and hydroxyl radicals. This technology allows scientists to see the transient species involved in these reactions, providing a complete perspective on reaction dynamics like branching and isomerization kinetics. The potential to measure reaction intermediates and products in real-time adds sophistication to reaction rate research, demonstrating that even seemingly straightforward reactions are complex (Bjork et al., 2017). These advances provide evidence of the ongoing progress in reaction kinetics and underscore the need for merging experimental data with the theoretical models for exhaustive investigation.

REFERENCES

1. Adelberger, E. G., Balantekin, A. B., Bemmerer, D., Bertulani, C. A., et al. (2010). Solar fusion cross sections II: The pp chain and CNO cycles. https://core.ac.uk/download/397660.pdf (accessed on 10th September 2025).
2. Agúndez, M., Cuppen, H. M., Geppert, W., Herbst, E., et al. (2010). Reaction networks for interstellar chemical modelling: Improvements and challenges. http://arxiv.org/abs/1011.1184 (accessed on 10th September 2025).
3. Ahmadi, H., Agreda, V., Barbosa, D., Chakravaty, S., Rouzineau, D., Danckwerts, P., et al. (2010). Rigorous multicomponent reactive separations modelling: Complete consideration of reaction-diffusion phenomena. https://core.ac.uk/download/12043245.pdf (accessed on 10th September 2025).
4. Ahmed, M., Azyazov, V. N., Kaiser, R. I., Mebel, A. M., et al. (2018). Functional relationships between kinetic, flow, and geometrical parameters in a high-temperature chemical microreactor. https://core.ac.uk/download/294625778.pdf (accessed on 10th September 2025).
5. Aubin, J., Mazubert, A., & Poux, M. (2013). Intensified processes for FAME production from waste cooking oil: A technological review. https://core.ac.uk/download/18540047.pdf (accessed on 10th September 2025).
6. Babbin, A. R., Ranjan, S., Rimmer, P. B., Sasselov, D., et al. (2019). Nitrogen oxide concentrations in natural waters on early Earth. http://arxiv.org/abs/1902.07740 (accessed on 10th September 2025).
7. Baier, R. R., Bendahmane, M., Bürger, R., Schneider, K., et al. (2008). Adaptive multiresolution schemes with local time stepping for two-dimensional degenerate reaction-diffusion systems. http://arxiv.org/abs/0807.1359 (accessed on 10th September 2025).
8. Balogh, Á., Fábián, I., Kalmár, J., Lente, G., et al. (2015). Reaction schemes that are easily confused with a reversible first order reaction. https://core.ac.uk/download/42942858.pdf (accessed on 10th September 2025).
9. Barato, A. C., Pietzonka, P., & Seifert, U. (2016). Universal bound on the efficiency of molecular motors. http://arxiv.org/abs/1609.08046 (accessed on 10th September 2025).
10. Beil, S., Brennfleck, C., Spee, C. I. M. A., Sheel, D. W., Gordon, M., Holdsworth, P., Hopfe, N., et al. (2002). *In-situ* monitoring for CVD processes. https://core.ac.uk/download/104382.pdf (accessed on 10th September 2025).

11. Bell, A. T., Head-Gordon, M., Janda, A., Lin, S., et al. (2018). Understanding Brønsted-acid catalyzed monomolecular reactions of alkanes in zeolite pores by combining insights from experiment and theory. https://core.ac.uk/download/158345582.pdf (accessed on 10th September 2025).

12. Bjork, B. J., Bui, T. Q., Changala, P. B., Nguyen, T. L., et al. (2017). Direct measurements of DOCO isomers in the kinetics of OD+CO. https://core.ac.uk/download/211372083.pdf (accessed on 10th September 2025).

13. Blaise, P., Caussat, B., Djafari Rouhani, M., Estève, A., et al. (2007). Towards multiscale modeling of Si nanocrystals LPCVD deposition on SiO_2: From ab initio calculations to reactor scale simulations. https://core.ac.uk/download/12039837.pdf (accessed on 10th September 2025).

14. Bökenkamp, D., Desai, A., Marzluff, E. M., Mayo, S. L., et al. (1998). Microfabricated silicon mixers for sub millisecond quench-flow analysis. https://core.ac.uk/download/4889488.pdf (accessed on 10th September 2025).

15. Bollig, M., Liñán Martínez, A., Lázaro Gómez, B., Sánchez Pérez, R., et al. (1999). Reduced kinetic mechanisms for modelling LPP combustion in gas turbines. http://oa.upm.es/1483/ (accessed on 10th September 2025).

16. Bordbar, A., Dräger, A., Flamholz, A., Joshi, A., Kinoshita, A., Canelas, A. B., Dräger, A., et al. (2016). Evaluation of rate law approximations in bottom-up kinetic models of metabolism. https://core.ac.uk/download/323065299.pdf (accessed on 10th September 2025).

17. Borman, P. C., Schouten, E. P. S., et al. (1996). Influence of reaction products on the selective oxidation of ethene. https://core.ac.uk/download/pdf/11452680.pdf (accessed on 10th September 2025).

18. Bosch, H., van Swaaij, W. P. M., & Versteeg, G. F. (1989). Gas-liquid mass transfer with parallel reversible reactions—I. Absorption of CO_2 into solutions of sterically hindered amines. https://core.ac.uk/download/pdf/11473934.pdf (accessed on 10th September 2025).

19. Bourne, H., Ferrell, J. E., Goldbeter, A., Gresser, M. J., Hartwell, L. H., Hill, C., et al. (2002). Thermodynamic and kinetic analysis of sensitivity amplification in biological signal transduction. https://core.ac.uk/download/25352781.pdf (accessed on 10th September 2025).

20. Branco Pinto, D., Constales, D., Marin, G., Yablonsky, G. S., et al. (2017). Kinetic intersections as a source of information on rate coefficients. https://core.ac.uk/download/91288153.pdf (accessed on 10th September 2025).

21. Burggraaf, A. J., & Lin, Y. S. (1992). CVD of solid oxides in porous substrates for ceramic membrane modification. https://core.ac.uk/download/pdf/11474501.pdf (accessed on 10th September 2025).

22. Cardy, J., Doi, M., Bettelheim, E., Lee, D., Lotka, A. J., Shnerb, N. M., Agam, O., et al. (2000). "Quantum phase transitions" in classical nonequilibrium processes. http://arxiv.org/abs/cond-mat/9908450 (accessed on 10th September 2025).

23. Carera, I. P., & Dasna, I. W. (2012). *Pengembangan Bahan Ajar Kimia Rintisan SMA Bertaraf Internasional Kelas XI Materi Laju Reaksi*. https://core.ac.uk/download/236389447.pdf (accessed on 10th September 2025).

24. Chan, A. W. H., Kroll, J. H., Ng, N. L., Seinfeld, J. H., et al. (2007). Kinetic modeling of secondary organic aerosol formation: Effects of particle- and gas-phase reactions of semivolatile products. https://core.ac.uk/download/4876769.pdf (accessed on 10th September 2025).

25. Chebeir, M., & Liu, H. (2018). Oxidation of Cr(III)-Fe(III) mixed-phase hydroxides by chlorine: Implications on the control of hexavalent chromium in drinking water. https://core.ac.uk/download/323067167.pdf (accessed on 10th September 2025).

26. Deem, M. W., & Park, J. M. (1998). Field theoretic approach to long range reactions. http://arxiv.org/abs/cond-mat/9811102 (accessed on 10th September 2025).

27. Delmas, H., Deshpande, R. M., Julcour-Lebigue, C., Kelkar, M., et al. (2011). Kinetics of hydroformylation of 1-octene in ionic liquid–organic biphasic media using rhodium sulfoxantphos catalyst. https://core.ac.uk/download/12043525.pdf (accessed on 10th September 2025).

28. Di Palma, L., Gueye, M. T., & Petrucci, E. (2015). Chemical reduction of hexavalent chromium (VI) in soil slurry by nano zero valent iron. https://core.ac.uk/download/74319583.pdf (accessed on 10th September 2025).

29. Drecken, C. B. von der, Hoffmann, U., Höhlein, B., Kunz, W., et al. (1994). Development and screening of selective catalysts for the synthesis of clean liquid fuels. https://core.ac.uk/download/pdf/11474581.pdf (accessed on 10th September 2025).

30. Dubreuil, B., Li, Y., Prat, L. E., Richard, M., et al. (2011). On-line monitoring of the transesterification reaction between triglycerides and ethanol using near infrared spectroscopy combined with gas chromatography. https://core.ac.uk/download/12042713.pdf (accessed on 10th September 2025).

31. Edelaar, A. C. S., Kiel, J. H. A., & Prins, W., et al. (1992). Performance of silica-supported copper oxide sorbents for SOx/NOx-removal from flue gas II. Selective catalytic reduction of nitric oxide by ammonia. https://core.ac.uk/download/pdf/11453359.pdf (accessed on 10th September 2025).

32. Elgue, S., Ferrato, M., Gourdon, C., Jolivet, P., et al. (2009). Silicon carbide equipments for process intensification of silicon reactions. https://core.ac.uk/download/12041266.pdf (accessed on 10th September 2025).

33. Feliu, E., & Wiuf, C. (2013). Power-law kinetics and determinant criteria for the preclusion of multistationarity in networks of interacting species. http://arxiv.org/abs/1202.3621 (accessed on 10th September 2025).

34. Fernández García, D., Rodríguez Escales, P. F., Solé Marí, G., Sánchez Vila, X., et al. (2017). A KDE-based random walk method for modeling reactive transport with complex kinetics in porous media. https://core.ac.uk/download/148622164.pdf (accessed on 10th September 2025).

35. Gowda, N. M. N., Jagadeesh, R. V., Puttaswamy, Vaz, N., et al. (2008). Ruthenium catalyzed oxidative conversion of isatins to anthranilic acids: Mechanistic study. https://core.ac.uk/download/72806412.pdf (accessed on 10th September 2025).

36. Gupta, A., Khammash, M., & Milias-Argeitis, A. (2017). Dynamic disorder in simple enzymatic reactions induces stochastic amplification of substrate. https://core.ac.uk/download/518754491.pdf (accessed on 10th September 2025).

37. Hadi, L., Marthafera, P., & Melati, H. A. (2018). *Deskripsi Pemahaman Konsep Siswa Pada Materi Laju Reaksi*. https://core.ac.uk/download/289710804.pdf (accessed on 10th September 2025).

38. Harmayani, E., Hidayat, C., Husain, R., Suparmo, D., et al. (2017). *Kinetika Oksidasi Protein Ikan Kakap (Lutjanus sp) Selama Penyimpanan*. https://core.ac.uk/download/290124194.pdf (accessed on 10th September 2025).

39. Haseltine, E. L., Mastny, E. A., & Rawlings, J. B. (2006). Stochastic simulation of catalytic surface reactions in the fast diffusion limit. https://core.ac.uk/download/4874657.pdf (accessed on 10th September 2025).

40. Haseltine, E. L., Mastny, E. A., & Rawlings, J. B. (2007). Two classes of quasi-steady-state model reductions for stochastic kinetics. https://core.ac.uk/download/4876324.pdf (accessed on 10th September 2025).

41. Heynderickx, P. (2019). Activity coefficients for liquid organic reactions: Towards a better understanding of true kinetics with the synthesis of jasmin aldehyde as showcase. https://core.ac.uk/download/237011261.pdf (accessed on 10th September 2025).

42. Higham, D. J., & Martínez-Urreaga, J. (2007). Modeling and simulating chemical reactions. https://core.ac.uk/download/9023225.pdf (accessed on 10th September 2025).

43. Hoffmann, M. J., Matera, S., & Reuter, K. (2014). *kmos*: A lattice kinetic Monte Carlo framework. https://core.ac.uk/download/267951745.pdf (accessed on 10th September 2025).

44. Hsu, A. T. (2025). *PDF Turbulence Modeling and DNS*. https://core.ac.uk/download/pdf/42809917.pdf (accessed on 10th September 2025).

45. Johnston, M. D. (2013). Translated chemical reaction networks. http://arxiv.org/abs/1305.5845 (accessed on 10th September 2025).

46. Keir, G., Petzold, A., & Sharpe, L. T. (2006). Spectrophotometry for cerebrospinal fluid pigment analysis. https://core.ac.uk/download/1687791.pdf (accessed on 10th September 2025).

47. Makrakis, D., & Noel, A. (2018). Algorithm for mesoscopic advection-diffusion. https://core.ac.uk/download/161511935.pdf (accessed on 10th September 2025).

48. Marley, M. S., Morley, C. V., Moses, J. I., Zahnle, K., et al. (2016). Photolytic hazes in the atmosphere of *51 Eri b*. https://core.ac.uk/download/pdf/76423622.pdf (accessed on 10th September 2025).

49. Mars, P., & Sonnemans, J. (1974). The mechanism of pyridine hydrogenolysis on molybdenum-containing catalysts: III. Cracking, hydrocracking, dehydrogenation and disproportionation of pentylamine. https://core.ac.uk/download/pdf/11471725.pdf (accessed on 10th September 2025).

50. Martin, A. A. (2015). Gas-mediated electron beams induced etching. https://opus.lib.uts.edu.au/bitstream/10453/36971/1/01front.pdf (accessed on 10th September 2025).

51. O'Shaughnessy, B., Vavylonis, D., Doi, M., Friedman, B., Kotomin, E., Scott, C., et al. (1998). Reactions at polymer interfaces: Transitions from chemical to diffusion-control and mixed order kinetics. http://arxiv.org/abs/cond-mat/9809027 (accessed on 10th September 2025).

52. Penner, S. S. (1954). Spectroscopic studies of premixed laminar flames. https://core.ac.uk/download/4883608.pdf (accessed on 10th September 2025).

53. Penner, S. S., & von Kármán, T. (1954). Fundamental approach to laminar flame propagation. https://core.ac.uk/download/4883607.pdf (accessed on 10th September 2025).

54. Proppe, J., & Reiher, M. (2018). Mechanism deduction from noisy chemical reaction networks. http://arxiv.org/abs/1803.09346 (accessed on 10th September 2025).

CHAPTER 3

TEMPERATURE DEPENDENCE OF REACTION RATES

Contents

Introduction.. 83
3.1. Theoretical Background.. 87
3.2. Kinetic Molecular Theory... 93
3.3. Experimental Evidence.. 98
3.4. Arrhenius Equation.. 102
3.5. Catalysts and Temperature... 107
3.6. Real-World Applications... 111
Summary.. 115
References... 118

INTRODUCTION

The study of temperature sensitivity to rates of reaction lies at the heart of chemical kinetics and thermodynamics. Temperature plays a key role in determining the rate at which reactions occur, affecting both the frequency of collision among the reactants and the energy with which these collisions are begun. As temperature rises, the kinetic energy of the molecules is greater, and thus there is a higher tendency to overcome the activation energy barrier that can impede reaction progress. This phenomenon is described by the Arrhenius equation, which quantitatively defines the correlation between a reaction rate constant and temperature. By dissecting this correlation, researchers can infer reaction mechanisms and inherent properties of materials, opening the door to advances in fields such as materials science, biochemistry, and environmental chemistry. Further reading of the literature sharpens focus on the complexity involved with temperature-dependent reaction rates. Account must be taken not only of the idealized assumptions of the Arrhenius scheme but also of real-world factors that can influence reaction rates. For example, in nucleosynthesis processes, astrophysical uncertainties in reaction rates are brought about by temperature fluctuations affecting interactions between nuclei within stellar surroundings.

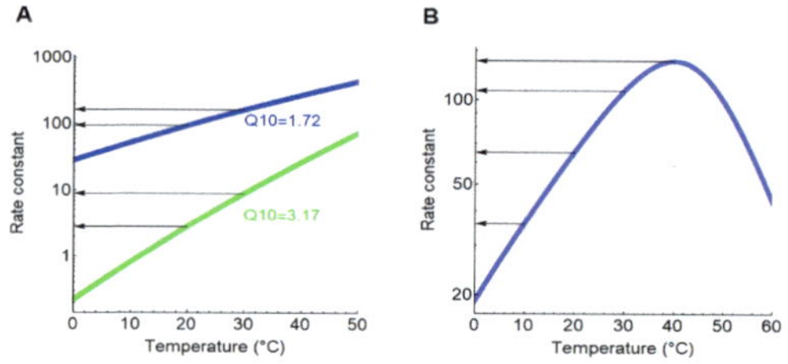

Figure 3.1. *Temperature dependence of reaction rates. (A) Reaction rates calculated according to the Arrhenius formula. The blue curve represents an enzymatic reaction (E_a = 40 kJ/mol [9.5 kcal/mol]), whereas the green curve denotes the uncatalyzed counterpart with higher activation energy (E_a = 85 kJ/mol [20.3 kcal/mol]). The corresponding Q_{10} values are shown for a temperature shift between 20 and 30°C. (B) Temperature dependence of an imaginary enzymatic reaction rate based on Peterson et al. that takes enzyme denaturation into account, with a melting temperature of 45°C, in equilibrium condition. The Q_{10} values indicate a gradual decrease with temperature: Q_{10} (10/20°C) = 1.81, Q_{10} (20/30°C) = 1.65, and Q_{10} (30/40°C) = 1.27.*

Source: https://www.researchgate.net/publication/362564763/figure/fig2/AS:1143 1281082882827@1662222073663/Temperature-dependence-of-reaction-rates-A-Reaction-rates-calculated-according-to.png.

Examining these temperature-dependent errors can have significant implications for understanding elemental abundances in the universe. The interaction between thermal excitation of nuclei and the impact of simultaneous reaction pathways requires a multifaceted model to explain variances outside the scope of experimental variability. As has been discussed in relevant research, for example in the case of the s-process in massive stars, thermal dynamics is found to impact outcomes in ways not originally anticipated (Cescutti et al., 2017). In the meantime, with the progress of research work, the use of computational modeling has increasingly become significant in simulating dynamic responses in reaction systems under various thermal conditions. More refined techniques, as are being practiced at the Center for Simulation of Dynamic Response of Materials, have indicated the capability to predict material behavior at extreme temperatures above typical thermal limits. The ability to digitally recreate a broad variety of reactions allows researchers to test hypotheses that can reveal yet unseen trends in the rate of reaction as dictated by changes in temperature. This multi-disciplinary approach not only offers predictive accuracy in theoretical models but also improves verification of test results, bridging the gap between theory and practice for chemical kinetics. In general, temperature dependence is crucial in comprehending the intricacies of chemical reactions and optimizing applications in various fields of science (Aivazis et al., 1999) (Figure 3.1).

Definition of Reaction Rates

In the area of chemical kinetics, the definition of reaction rates can include the rate at which the reactants are being converted to products during a chemical reaction. This may be measured using the change in concentration of products or reactants per unit time. At the center of understanding reaction rates lies the distinction between instantaneous and average rates; the latter is used to describe the rate at a specific point in time, and the former gives a measurement of the overall change during the entire course of the reaction. The temperature at which reactions occur strongly influences these rates because it controls the kinetic energy of molecules. In general, with increasing temperature, the average molecular velocity also rises, thereby boosting the rate of effective collisions among reaction molecules. This is the foundation for the Arrhenius equation, which defines how temperature relates to the rate of reaction through its effect on activation energy, ultimately to the intricacies of temperature dependence (Marley et al., 2014). Additionally, temperature's relationship with reaction rates extends beyond straightforward molecular activity. The effect usually lies in the description given by the activation energy term, which plays a key role in specifying to what degree a reaction takes place. When temperatures are high, they usually lower the activation energy barrier, allowing reactants easily to convert into products. This not only applies to terrestrial chemistry but also to astrochemical conditions, where temperature variations can establish the prevalence of given molecular species. For instance, in planetary atmosphere research, temperature was found to influence the equilibrium between gases such as methane and carbon monoxide. Depending on temperature variations, different reaction paths become favorable, hence the equilibrium of these compounds is changed in accordance (Haile et al., 2010). This illustrates temperature as a variable that influences the control over reaction dynamics in different environments. In addition to the direct control over molecular kinetics, reaction rates can exhibit complicated behaviors controlled by other factors, such as concentration, pressure, and the availability of catalysts. Temperature dependence is irrevocably linked to all these factors, making them couple differently with reaction mechanisms. For example, rate laws for the progression of chemical reactions often include temperature as a key variable to allow chemists to make predictions about reaction behavior under a variety of conditions. Moreover, theoretical models such as the Transition State Theory offer a conceptual basis by which temperature facilitates the transition from reactants to product via an activated complex. Hence, an overall comprehension of the definition of reaction rates encompasses not only kinetic observation but also the intricate function played by temperature in defining reaction pathways and outcomes under controlled as well as natural conditions (Marley et al., 2014) (Figure 3.2).

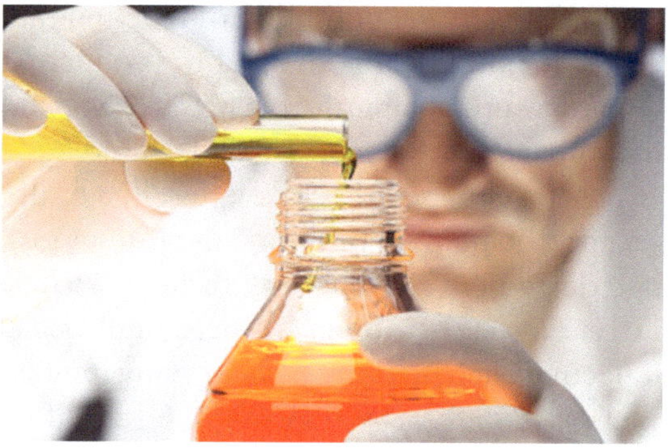

Figure 3.2. *Definition of reaction rates.*

Source: https://www.chemistryexplained.com/photos/reaction-rates-3402.jpg.

Importance of Studying Temperature Effects

Understanding temperature dependence of reaction rates is significant in a number of science fields, including biology, environmental science, and chemistry. Temperature decides the speed at which the reactants can form products, in some cases determining the path of a reaction. Increasing temperature usually leads to more kinetic energy between molecules, thereby enhancing the likelihood of successful collisions. This simple concept can have impact from ordinary chemical reactions to industrial applications, where reaction rate control is paramount to efficiency and safety. The capability to predict and control such rates has the power to allow researchers and practitioners to optimize conditions to desired effects, ultimately resulting in research and technological breakthroughs. It is thus important to proceed systematically through the manner in which temperature influences reaction because it will increase our grasp of subtle systems and direct applications into useful problems in a lot of science fields. The applicability of studying the influence of temperature extends further to knowledge of various phenomena encountered in energy production and combustion science. For instance, research on ignition of primary reference fuel (PRF) mixtures demonstrates how various conditions of temperature can influence behavior in reactions. Experiments using rapid compression machines (RCMs) indicated the complexity of ignition characteristics under pressures and temperatures. The first-stage energy release and total ignition delay time measured provide hints toward reaction mechanisms, which might be overlooked when temperature parameters are not considered. Such an interaction necessitates simulating reactions for a broad set of temperatures to yield an all-around description of their kinetics. By studying such processes, we are able to more effectively design energy systems optimally with fewer harmful emissions, illustrating the practical applications of learning about reaction rate control by temperature (Allen et al., 2017). Furthermore, temperature effects are of central significance in applications to astrophysics when nucleosynthesis processes beyond iron are taken into account. Different reactions, i.e., the *s*-, *r*-, γ-, and *vp*-processes, face different challenges related to temperature variations and stellar setting uncertainty. The temperature variations influence not only the reaction rates but also decide nucleosynthesis processes, necessitating improved understanding of their thermal characteristics. Condition uncertainties like thermal occupation of excited states highlight the requirement for precise temperatures to model and theory better. These insights feed back into experimental methodologies that close the gap between laboratory results and stellar observations, ultimately deepening our knowledge of cosmic processes. Therefore, exploring temperature effects provides key inputs to both theoretical models and pragmatic experimental designs, identifying its fundamental contribution to the development of our scientific understanding (Fowler et al., 2014).

3.1. THEORETICAL BACKGROUND

The understanding of temperature dependence of reaction rates calls for a firm theoretical foundation that accounts for the interconnection between molecular motion and energy interaction. One of the underpinnings of such a foundation is the Arrhenius equation, which postulates that reaction rates increase exponentially with increasing temperature. This is premised on the notion that rising temperature provides reactant molecules with additional kinetic energy, thus elevating their possibility of surmounting the activation energy barrier to enable reactions. Activation energy is one of the variables of significance employed in this formula and is determined by temperature in accordance with the intricate dynamics of molecular collisions and energy levels needed for optimal interaction.

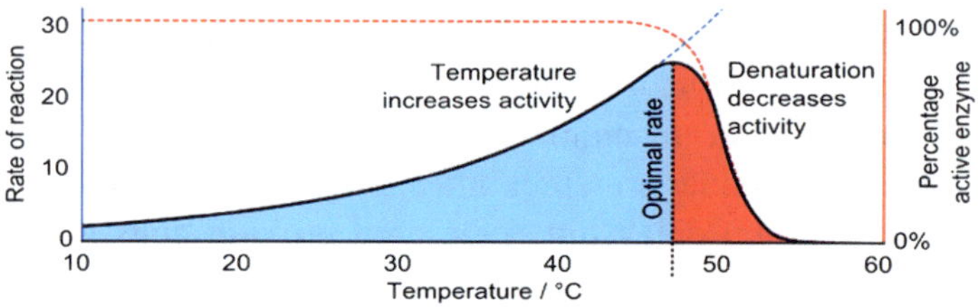

Figure 3.3. *Temperature influence on rate of reaction.*

Source: https://d20khd7ddkh5ls.cloudfront.net/temp.png.

This principle underlies understanding of numerous processes in various areas of science, from biochemical reactions to industrial processes where temperature management can have significant effects on product yield and reaction rate. For astrophysics, temperature dependence analysis is even more complex. As shown in the study of nuclear reactions, for instance, in the synthesis of elements via stellar processes, temperature is the critical parameter that controls reaction pathways and reaction rates. This can be illustrated in studies on the astrophysical gamma process that explains the impact of temperature variations on cross-section measurements and subsequently on reaction rates for some isotopes such as 130Ba and 132Ba. The use of statistical model calculations to predict these rates highlights the intrinsic link between temperature and energy states in systems that are undergoing reaction kinetics. Through investigations into such astrophysically relevant energy windows, scientists can obtain key information regarding how temperature changes affect element development within the universe, highlighting the overall uses of temperature dependence across a wide range of scientific disciplines. Theoretical advances rely on empirical checks to a significant degree, particularly in the highly complicated fields of particle physics and cosmology. As science evolves in exploring the time variation of the fundamental constants in early ages of the universe, temperature and reaction rates' correlation still goes into new frontiers. For instance, variation of basic constants such as the fine structure constant and the Higgs vacuum expectation value can tell us about how the elementary particles interact at various temperatures. These kinds of relationships again emphasize the necessity for an overarching theoretical framework that encompasses both classical thermodynamic principles and sophisticated quantum mechanics. These integrative approaches not only explain the temperature dependence of reaction rates but also enhance our knowledge of the universes evolution and interactions governing it, thereby enhancing the scientific community discourse (Figure 3.3).

3.1.1. Collision Theory and Reaction Rates

The fundamental postulates of collision theory are that for a chemical reaction to occur, reacting particles must collide with sufficient energy and proper orientation. Collision theory provides us with a context for understanding how temperature influences the rate of reaction. As temperature rises, there is greater kinetic energy of molecules, with resultant increased collisions between the particles of the reactants. The higher energy not only increases the frequency of collision but also the fraction of collisions with the energy necessary for effective reactions, commonly known as the activation energy. For this reason, reaction rate rises with temperature, demonstrating how conditions of temperature might govern the kinetic parameters of essential chemical reactions. Collision theory also describes the molecular orientation factor in effective collisions. It maintains that not all collisions will lead to a reaction, but only those with the correct spatial atom orientation can shatter and bond. As temperature increases, increased vibrational motion among molecules results in more ability to shift into collision orientations favorable to reaction. This effect is particularly crucial in complex reactions involving multiple reactants where the orientation of reactants plays a major role in determining the final product. Therefore, an in-depth explanation of collision theory provides significant inputs on how changes in temperature dramatically affect mechanisms and rates of reactions. In experimental investigations of combustion kinetics, collision theory provides a major basis to explore how combinations of reactants interact under varying temperatures. For instance, research with combustion spectroscopy have analyzed the combustion process in different regimes of pressure and have shown how temperature affects ignition processes indirectly through alteration of particle collision character in a flame (Penner et al., 1954). Ionization reaction research in shock-heated gases also shows how concentrations of impurities regulate initial reaction rates using collision dynamics (Sturtevant et al., 1966). These findings highlight the intricate relationship between collision theory, temperature, and reaction rates, the wide impact of thermal fluctuations on chemical reactions across different environments (Figure 3.4).

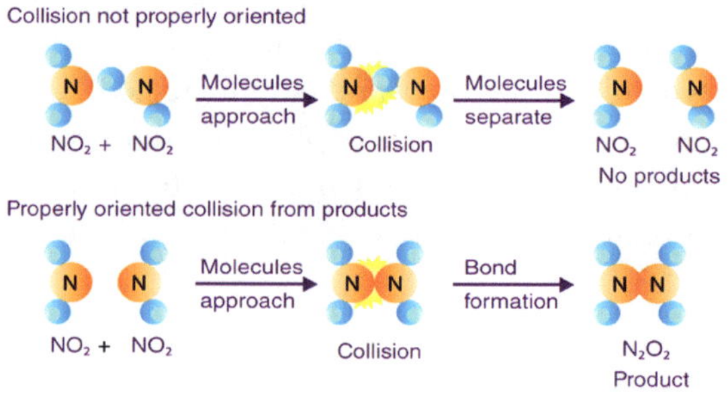

Figure 3.4. *Collision theory of chemical reactions.*

Source: https://cdn1.byjus.com/wp-content/uploads/2015/12/The-Collision-Theory-Of-Chemical-Reactions-.png.

3.1.2. Activation Energy and Its Role

Activation energy is the key to understanding reaction rate dependency on temperature because it is the energy barrier that reactants need to cross in order for a chemical reaction to occur. Activation energy, in effect, controls the rate of a reaction by defining the threshold energy for the formation of transition states—intermediate structures along the course of the reaction. As temperature increases, so does the average kinetic energy of molecules, so that more molecules find it easier to overcome this energy barrier. As a result, increasing temperature is typically followed by increasing reaction rate, as predicted by the Arrhenius equation, a mathematical expression of this relationship. This equation illustrates that reaction rates are exponentially dependent on temperature due to the exponential term related to activation energy, thus explaining the intrinsic importance of activation energy in chemical kinetics. In addition, the intricate relationship between activation energy and reaction conditions is revealed by experiments demonstrating how reaction conditions like pH, ionic strength, and the presence of catalysts affect activation energy barriers. For instance, recent studies on catalytic mechanisms indicate that specific conditions can significantly change the activation barriers with regards to electron transfer reactions on semiconductor surfaces. Aschaffenburg et al. (2019) highlights the occurrence of typical activation pathways based on reaction-dependent variables and shows how variations in temperature and other conditions affect both the activation energy height and the reaction kinetics. This versatility reflects the dynamic nature of chemical reactions and illustrates the central role activation energy plays in regulating not only whether a reaction is feasible but also its rate under various conditions. In practical situations, control over activation energy has far-reaching implications for industry, particularly in the design of efficient catalysts that optimize reaction rates. For example, in the development of water oxidation catalysts, the activation barriers have to be interpreted in order to increase the oxygen evolution reactions, as in recent research on *n*-$SrTiO_3$ surfaces. According to Aschaffenburg et al. (2019), the way the thoughtful selection of reaction conditions helps chemists to choose among the various routes for O_2 evolution, ultimately improving the efficiency of the catalysts. Furthermore, experimental techniques that determine thermooxidative stability, such as for PMMA/silica composites, illustrate the significance of understanding activation energy in determining the performance of materials subjected to heat stress, as elaborated in. Through comprehensive examination of activation energy across various applications, researchers are able to define new strategies for optimizing reaction rates and material stability (Figure 3.5).

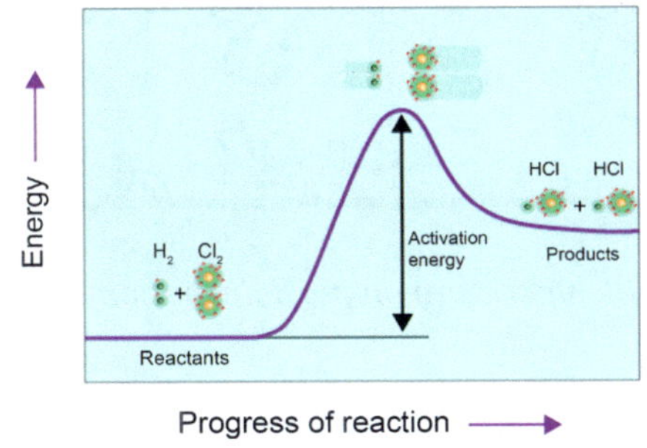

Figure 3.5. *Activation energy.*

Source: https://cdn1.byjus.com/wp-content/uploads/2015/12/The-Collision-Theory-Of-Chemical-Reactions-01.png.

3.1.3. Temperature's Influence on Molecular Motion

Temperature plays a crucial role in assessing molecular movement that influences reaction rates. At the molecular level, temperature is a measure of average kinetic energy, driving translational, rotational, and vibrational particle movement. When temperature rises, the kinetic energy of the molecules is greater, resulting in rapid motion and a high frequency of collision between the reactants. This increased activity typically aligns with increased reaction rates, as molecules are inclined to act and form products. The underlying connection between temperature and molecular motion is central to thermodynamics and reaction kinetics, illustrating how thermal energy is the cause of accelerated chemical reactions. It is significant to comprehend how temperature affects molecular motion in practical applications across various fields, including chemistry and materials science. For instance, carrying out reactions under elevated temperatures can optimize product yields in addition to shortening reaction times.

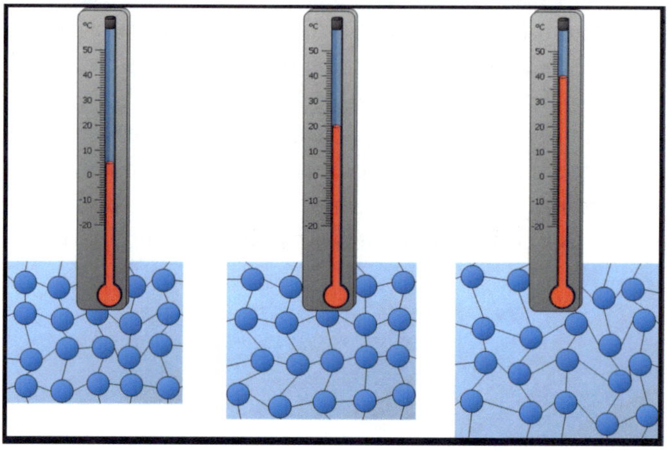

Figure 3.6. *Influence of temperature on particle motion and thermal expansion.*

Source: https://www.tec-science.com/wp-content/uploads/2021/05/en-particle-model-of-matter-temperature-thermal-expansion.jpg.

Experimental evidence has confirmed that when the temperature is raised, not only do the number of collisional interactions increase, but so does the collision energy, which facilitates breaking and forming bonds with ease. The discussion goes up to the level of complex systems, such as biological processes, where the temperature influences the enzymes and the stability of metabolites. Temperature control is therefore very important in achieving the desired outcome in synthesis and analysis. In addition, the influence of temperature on molecular movement can be explained better by quantum mechanics and advanced simulations. Mechanical motions in molecular ensemble studies have been discovered, under certain conditions, to significantly enhance excitation transfer efficiency with the benefit of cooperative molecular interactions (Asadian et al., 2010). These quantum signatures refer to the distinctive nature of molecule behavior at various temperatures, which emphasizes the precise differences between classical and quantum energy transfer processes. In addition, the structure dynamics in ultracold gas experiments reveal information about the binding energies' influence and collisional dynamics based on temperature fluctuations. This information provides new directions for application in precision spectroscopy and quantum chemistry, illustrating the remarkable influence of temperature on molecular systems (Gerritsma et al., 2020) (Figure 3.6).

3.2. KINETIC MOLECULAR THEORY

Kinetic Molecular Theory (KMT) describes the behavior of gases, particularly the dependence of reaction rates on temperature. According to KMT, gas molecules are always in motion, and temperature is a measure of their kinetic energy. As temperature increases, the average kinetic energy of gas molecules will rise, resulting in increased frequency and intensity of collisions between them. This increased rate and energy of collisions is crucial for overcoming activation energy barriers in chemical reactions. Therefore, understanding of KMT explains why reaction rates generally increase with raised temperatures, a relationship which is still fundamental to theoretical as well as practical application in chemistry. The implications of KMT extend beyond ideal gas behavior, influencing other areas of research including ultracold atom and ion dynamics. Recent cold ion-atom hybrid experiments have showed the effect of low-temperature conditions on molecular interactions. Specifically, collisions at very low temperatures can lead to new phenomena such as elastic and inelastic collisions that violate conventional predictions (Willitsch et al., 2014). The study shows how the effect of temperature not only affects accelerational factors but also enables new reaction mechanisms and pathways. In practice, the study of temperature dependence based on KMT allows scientists to investigate new mechanisms for controlling and manipulating chemical reactions in quantum and classical systems. To better illustrate the concept, advanced theoretical studies have investigated the dynamics of trapped ions that undergo interactions in ultracold atomic gases. These experiments demonstrate the rates of collisions to be very close to Langevin collision dynamics under some binding energies, displaying crossover behavior where molecular ion formation is significant (Gerritsma et al., 2020). These temperature-dependent dynamics exhibit the interactions between molecular interactions, kinetic energy, and the resulting reaction rates. By extrapolating KMT to ultracold temperatures, researchers are in a position to engineer conditions that offer good reaction rates with accuracy, thus broadening the scope of useful applications to quantum chemistry and spectroscopy. Ultimately, KMT not only describes current phenomena but also guides future research towards the optimization of reaction conditions based on temperature variations (Figure 3.7).

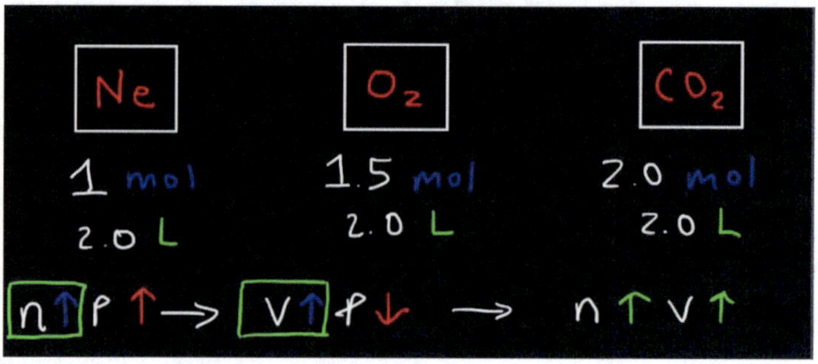

Figure 3.7. *Kinetic molecular theory.*

Source: https://i.ytimg.com/vi/iAsP-9m2aH0/maxresdefault.jpg.

3.2.1. Explanation of Kinetic Molecular Theory

Kinetic molecular theory (KMT) provides a fundamental framework of concepts for the behavior of gases and their interaction at a molecular level. Gas molecules, according to KMT, are always moving around in a random manner, and their kinetic activity is greater with rising temperature. With the rise in temperature, the kinetic energy of the molecules grows higher, leading to more energetic and more frequent collisions between them. This simple principle accounts for the effect of temperature on reaction rate directly; because kinetic energy enhances the motion of reactant molecules, the likelihood of effective collisions, which are responsible for chemical reactions, also increases. Second, KMT dictates that gas molecules occupy an infinitesimally small volume compared to the volume between them, emphasizing that molecule-molecule interactions during a reaction derive mostly from their kinetic activity rather than volume. In examining kinetic molecular theory's application, note how variations in temperature can be an important factor in reaction rates of various chemical processes. For instance, reactions in heterogeneous catalysis are typical in demonstrating an Arrhenius-like temperature dependence, an aspect which can be elucidated through the context of KMT. According to theory, an increase in temperature means the molecules receive increased amounts of energy and consequently traverse the activation barriers with ease. It is here that the role of compensation phenomena steps in; shifts in surface coverage of the reactants can alter the effective activation energy, thereby influencing the rate of chemical reactions. The temperature-reactant rate relationship as described by KMT assists in making predictions in theoretical as well as practical chemistry applications, for example, catalytic conversions that are based on the surface-active sites of various catalysts (Farra et al., 2012). Second, the significance of kinetic molecular theory is not limited to gases and transcends the other states of matter, such as liquids and solids, where motion of molecules is also applicable. In solid-state physics, for instance, KMT assists in describing processes such as diffusion, wherein atomic jumps are accountable for reaction or alteration of solid materials. This diffusion can also be described in terms of activation energies, or the minimum amount of energy with which a molecule must change from one state to the next. Recent advances indicate a high correlation between temperature-dependent activities and diffusivities in solid media, and implications for theoretical models and experimental confirmations (Stoneham et al., 1989). As a whole, KMT is an essential theory that unifies temperature influences, molecular movement, and reaction kinetics across a number of areas of chemistry with a well-defined foundation for additional scientific study and technological advancement (Figure 3.8).

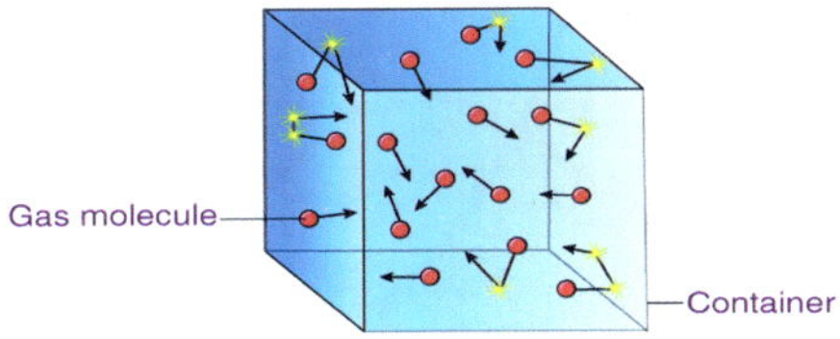

Figure 3.8. *Kinetic theory of gases.*

Source: https://cdn1.byjus.com/wp-content/uploads/2020/09/Kinetic-Theory-Of-Gases-Assumptions.png.

3.2.2. Relationship Between Temperature and Kinetic Energy

The relationship between kinetic energy and temperature is fundamental to understanding the rates of reaction. With increasing temperature, the kinetic energy of molecules also increases, directly corresponding with the frequency and force of molecular collisions. High-temperature molecules have greater velocities, which means a greater likelihood of overcoming activation energies necessary for reactions. This phenomenon accounts for the fact that many chemical reactions accelerate as temperatures rise, effectively raising reaction rates and promoting product formation. Further, the influence of temperature will also alter the equilibrium of the majority of the reactions, tilting toward reactant or product formation according to Le Chatelier's principle. Lastly, the interaction of kinetic energy with temperature provides the basis for reaction dynamics and kinetics, leading to more complex considerations regarding reaction mechanisms and pathways. In addition to common kinetic considerations, the relationship between reaction dynamics and temperature can be investigated using complex mechanisms such as heterogeneous catalysis. In this case, temperature influences not only the kinetic energy but also the dynamics of charge-trapping intermediates at the surface of catalysts, as has been revealed in recent studies. For instance, certain conditions—i.e., pH and temperature variations—can influence activation energy barriers for these intermediates, which in turn can alter trajectories to produce products. These are significant findings since they reveal more information about the advanced ballet of molecular interactions that trigger reactions.

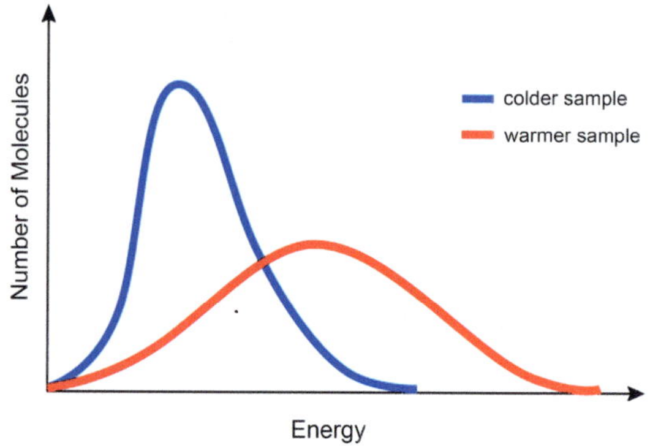

Figure 3.9. *A distribution of molecular kinetic energies as a function of temperature. The blue curve is for a low temperature, whereas the red curve is for a high temperature.*

Source: https://flexbooks.ck12.org/flx/show/THUMB_POSTCARD/image/user%3AY2sxMnNjaWVuY2VAY2sxMi5vcmc./98045-1364448537-05-30-C-IntCh-04-01-05-Kinetic-Energy-Distribution.png.

The ability to manipulate temperature to modulate such catalytic processes suggests that reaction rates can be tuned not only by varying kinetic conditions but also by manipulating the environmental conditions that control the energy landscape of chemical systems, as exemplified in experiments on charge-trapping intermediates and reactivity (Aschaffenburg et al., 2019). Also, temperature plays a special role in the ignition characteristics of various fuels, illustrating its influence on the kinetics as well as the energetics of combustion reactions. In a controlled experimental setup, scientists have found that upper temperature limits correspond to characteristic ignition behavior, especially through two-stage ignition in primary reference fuel mixtures. The sensitivity coefficient analysis based on several different parameters highlights the impact of variations in temperature on energy release and ignition delay times throughout combustion. Specifically, it was observed that high peak heat release rates in ignition may lead to variations of reaction efficiency when the measurement is taken for different ranges of temperatures. This accentuates not only the significance of understanding the rudimentary correlation between kinetic energy and temperature but also how this can be applied to practical applications, rendering energy processes more efficient and effectual (Allen et al., 2017) (Figure 3.9).

3.2.3. Implications for Reaction Rates

The reaction temperature plays a significant role in reaction rates, varying the rate and the reaction mechanism. Higher temperatures mean molecules possess more kinetic energy, increased frequency of collisions, and more energetic collisions. The increased energy enables molecules to overcome activation energy barriers, leading to reactions that happen faster. The implications of these temperature-induced changes are profound across various fields, including material science and biochemistry. For instance, in the context of neutron star astrophysics, as illustrated in the kaon condensation mechanism, reaction rates directly affect stellar evolution dynamics and neutron star stability as temperatures change with neutron star formation (Akmal et al., 1999). In addition, temperature-reaction interaction may also lead to significant changes in product molecule content. In atmospheric chemistry, for example, variations in temperature may alter the equilibrium gas concentrations of gases like carbon monoxide and methane. As discussed in studies of brown dwarfs and bright planets, kinetic parameters governing these reactions are sensitive to atmospheric conditions like temperature gradients and vertical mixing processes. The resulting effects on rates of reaction translate into our understanding of atmospheres on planets and the potential for life (Marley et al., 2014). The sensitivity to conditions of the environment points to the pivotal position of temperature in defining the mechanistic mechanisms of chemical reactions. Ultimately, the effects of reaction rates as they are defined by temperature have consequences reaching from bench chemist to applications in technology and environmental science. Quicker responses at elevated temperatures will be more catalytic, but at the expense of restraining undesirable side reactions that are caused by greater energy levels. Hence, prediction and control of these rates are essential in the design of efficient chemical processes in industry. Also, recognizing how temperature changes will influence reaction pathways not only advances technological development but also enables us to appreciate our overall comprehension of the environmental effects and stability of various chemical systems that occur in nature (Akmal et al., 1999; Marley et al., 2014).

3.3. EXPERIMENTAL EVIDENCE

Experimental evidence for temperature dependence of reaction rates is crucial in the study of chemical kinetics. Temperature is a dominant force in determining the rate at which reactions occur, and its influence is partly due to its impact on molecular collisions and energy. Numerous experiments have established that temperature increase causes an increase in reaction rates, according to the Arrhenius equation, which qualitatively relates temperature to rate constants. For example, careful measurements of nuclear reaction rates in the context of stellar environments show that high temperatures strongly control fusion processes, suggesting subtle interaction between thermodynamic principles and reaction kinetics. Comparison of measured solar neutrino fluxes has proven interesting, with dynamics following reaction rates as a function of temperature, though there are some discrepancies with standard model expectations. These experiments indicate the relevance of temperature in characterizing the behavior of reactions under varying conditions as a foundation for further research in this field (Adelberger et al., 1998).

In addition, rate theories have been generalized to model effects that are unaccounted for by conventional models, particularly for non-equilibrium conditions. In radiation-induced precipitation reactions in solids, it was found that the *quodon* lattice soliton density plays a significant role in reaction rates. *Quodon* density approaches levels comparable to phonon gases with increasing irradiation flux, and they are as responsive to temperature. These findings are contrary to classical rate theory, which could not correctly calculate parameters of precipitation in complex systems like FeCu binary alloys. Experimental data indicate that the modified rate theory can predict concentration change and precipitate growth, suggesting that temperature and other factors like irradiation can alter reaction behavior in unexpected ways (Turkin et al., 2014).

This indicates the requirement for experimental data to be included in theoretical models for an integrated understanding of reaction kinetics. In addition, the experimental research emphasizes the universal significance of temperature effects on reaction rates in contexts beyond customary chemical reactions. The impact of temperature on reaction kinetics cascades to domains such as atmospheric chemistry and environmental science, where changes in temperature can significantly affect reaction mechanisms. For instance, temperature variations may influence the atmospheric degradation rates of pollutants and be instrumental in dictating urban air quality and ensuing public health effects. Quantitative experimental measurements of such temperature-sensitized degradation rates enable improved predictive modeling of the fate of various chemical species under varying environmental regimes. These results indicate that the extensive investigation of reaction rates, though based on basic thermodynamic postulates, is constantly being revised as new experimental information updates and refines prevailing theories. By meticulous observation and examination, scientists are constantly revealing the subtle interplays between temperature and the kinetics of reactions (Adelberger et al., 1998; Turkin et al., 2014).

3.3.1. Overview of Key Experiments

In order to properly comprehend the temperature dependence of reaction rates, it is important to consider first-hand experimental studies that illuminate the connection. Experimentally, laser fusion yields are a specific example of how reaction rates depend on varying temperatures, with the yield proportional to lasers power input. These experiments demonstrate that temperature influences not only the reaction rate but also plasma density and plasma particles' projected range in the target medium. Thus, the derived scaling laws have a robust predictive capability for fusion yields in yet unexplored nuclear reactions. These observations underscore the temperature as the most important factor in the catalytic efficiency of reactions catalyzed by extreme conditions (Bonasera et al., 2012) and hence provide a full description of the involved complex reaction rate dynamics. Besides fusion yields, mechanical unfolding experiments on biomolecules also provide evidence of the influence of temperature on reaction kinetics. Employing laser optical tweezers and atomic force microscopy (AFM), researchers work on one molecule to quantify the effect of temperature fluctuations on their unfolding curves. Such experimental devices allow scientists to measure the roughness of the energy landscape intrinsic in biomolecular structures regardless of assumed reaction coordinates. These methods provide valuable information regarding the thermodynamic parameters involved, with temperature being a factor that has a direct impact on reaction rates and transition state positions. The findings advance the concept of the energy landscape by providing a rich interplay among temperature, energy barriers, and molecular stability (Hyeon et al., 2006). Lastly, the concept of transition states becomes pivotal in the interpretation of the data from these temperature-dependent experiments. Of particular interest are the issues of challenges to identifying the transition state in mechanical unfolding, which define the nuances in analyzing experiments. The position of the transition state is both a function of temperature and force, precluding derivation-based determination of reliable reaction kinetics. Identification of the difference in behavior of the transition state for soft and hard biomolecules is essential to accurate modeling of reaction rates. These variations underscore the importance of employing comprehensive analytical models involving temperature and other factors to fully benefit from the experimental data at hand. As a whole, principle experiments in the field not only lend credence to our understanding of reaction kinetics but also illuminate the rudimentary role of temperature in shaping such processes.

3.3.2. Data Analysis and Interpretation

The analysis and interpretation of data for reaction rate dependence on temperature are necessary to enhance our understanding of chemical kinetics. Experimental measurements underway, several factors must be tackled, one of which is the influence of temperature on reaction mechanism and rate coefficients. For example, increased temperatures will lead to enhanced molecular motion that may lead to a rise in collision rates between the reactants. The temperature dependence of the reaction rates is often plotted on Arrhenius plots with reaction rate logarithm on one axis and the reciprocal of the temperature on the other. In this manner, the activation energy can be read off from the graph and can provide some hints on the energy barrier that needs to be broken down before the reaction can be performed effectively. Data-driven explanations are essential to the development of predictive models in chemical engineering and other manufacturing processes. When studying the specifics of reaction rates when modifying the conditions of temperature, data analysis tends to borrow advanced techniques to enhance accuracy and reliability. Computational fluid dynamics and isothermal assumptions have been effective in good explanations of experimental results. For instance, the use of CFD simulations allows for the pyrolysis performance of chemicals like $C_{10}H_7Br$ to be tested in high temperature microreactors. Experimental data can be used to arrive at valid correlations between kinetic parameters, gas flow rates, and reactor geometrical configurations using such advanced modelling methods. These relationships form the basis for the establishment of an in-depth understanding of reaction dynamics and allow for the formulation of a plan of strategy for measurement of bimolecular and unimolecular rate coefficients in complex surroundings (Ahmed et al., 2018). The influence of temperature on the reaction rate can also extend beyond straightforward chemical reactions, giving insight into astrophysical phenomena. In astrophysical conditions such as stellar cores, the coupling between density and temperature parameters is observed to be revealing nonideal plasma behavior and the potential for statistical particle behavior in a nonextensive system. The nonideal interaction speaks of a departure from classical models of thermodynamic equilibrium and hints at a metastable regime with power law behavior in ions. Highly sophisticated data interpretation methods are required in such an analysis since deviations from the Maxwell-Boltzmann distribution speak of complicated interactions at play. These results not only enrich our knowledge of reaction kinetics under extreme conditions but also indicate the scope of method needed to provide accurate data analysis and interpretation in science (Lavagno et al., 2005).

3.3.3. Case Studies Demonstrating Temperature Effects

The temperature-reactant rate relationship can be described using various case studies, and that is the reason combustion chemistry makes a perfect example. One of these case studies is the case of butanol isomers, which have been extensively studied since they are under consideration as second-generation biofuels. The study demonstrated how laminar flame speed and autoignition delay times of these isomers are highly dependent on temperature fluctuations. The skeletal mechanisms derived for these isomers of butanol showed that with temperature increase, fuel mass fraction consumption rates also increased, producing accurate information about their combustion characteristics. The well-validated chemical kinetic models used here confirm that the understanding of temperature varying reaction kinetics and influencing the efficiency of different types of fuels in real-world applications such as homogeneous compression-ignition engine simulation is essential, with important consequences both for optimization and fuel design (Brady et al., 2017). Another informative case study is dedicated to the catalytic activity of acidic zeolites in alkane cracking reactions and takes a subtle look at temperature dependence. This research highlights the significance of considering zeolite structure in assessing catalytic performance, particularly in the face of variable temperatures. By studying the kinetics of dehydrogenation and cracking of alkane at low levels of conversion, the researchers developed the way temperature affects the intrinsic reaction rate coefficient as well as the adsorption equilibrium constant. The study proved that temperatures rise when reaction rates increase, thereby enhancing the overall catalytic process. Furthermore, the ability to accurately compute adsorption constants by employing advanced simulations like configurational-bias Monte Carlo serves to deepen understanding of the role of temperatures since it becomes apparent that reaction rates exhibit a deep relationship with the structural features of catalysts (Bell et al., 2018). The cumulative evidence realized in these case studies directs toward the central tenet of chemical kinetics: temperature directly influences reaction rates in a range of systems. In the research of butanol isomers, temperature-induced variations in combustion characteristics are illustrated to have a profound impact on biofuel production. Similarly, research on zeolites under acidic conditions in the cracking of alkanes presents the significance of interactions between catalyst architecture and temperature, demonstrating capability to revise catalytic procedures in industry through adapting these factors. Since temperature continues to be a determinant of reaction mechanisms and response, ongoing research in this field has the possibility of further pushing both theoretical and applied solutions for chemical reactions and material science. Through ongoing research, the full extent of temperature influence over reaction rates can be further understood, leading to innovation in the field of energy production and catalytic technologies.

3.4. ARRHENIUS EQUATION

Arrhenius Equation is a basis of temperature dependence upon the rates of reaction, measuring the influence of temperature on reaction. Arrhenius developed the equation in 1889 that relates the rate constant of a reaction with the absolute temperature and activation energy. The equation shows that with an increase in temperature; the rate of reaction will be more. The equation can also be represented as $k = Ae^{-E_a/(RT)}$, where A is the pre-exponential factor to be used in compensating for molecular collision frequency and direction. The Arrhenius Equation is hence vital in determining reaction kinetics, thereby allowing chemists to control conditions in order to obtain maximum yield and efficacy in most chemical reactions. Moving deeper into the Arrhenius equation implications, it is clear that the activation energy E_a is essential in defining the rate of a reaction. The reactants have to cross this energy barrier for them to produce the products, and its value varies greatly depending on the different chemical reactions. A lower activation energy will usually indicate a reaction will be able to occur at lower temperatures, while a higher activation energy will signify the requirement of higher thermal energy so large reaction rates can be achieved. For example, the range of temperatures in carbon monoxide to methane conversion under varying atmospheric conditions undergoes drastic variations because of E_a values, as can be seen in studies of solar composition brown dwarfs and planetary atmospheres (Marley et al., 2014). These observations reflect the versatility of the Arrhenius Equation in diverse scientific studies. In addition to data obtained from the Arrhenius Equation, various experimental methods can assess reaction kinetics with effectiveness. Thermogravimetry, for the sake of illustration using one example provided in the paper on thermo-oxidative stability of polymer materials, employs heating rates to determine induction period, a measure of composite material stability under heat stress (Čechová et al., 2012). Applying isoconversional methods, researchers can derive parameters that describe material behavior under different temperatures, exemplifying the applicability of kinetic parameters derived from the Arrhenius Equation in actual applications. Such versatility in arriving at chemical behavior attests to the broader applicability of temperature dependence investigation in fields like materials science, environmental chemistry, and astrochemistry, thus addressing the design and optimization of reaction routes in a variety of applications (Figure 3.10).

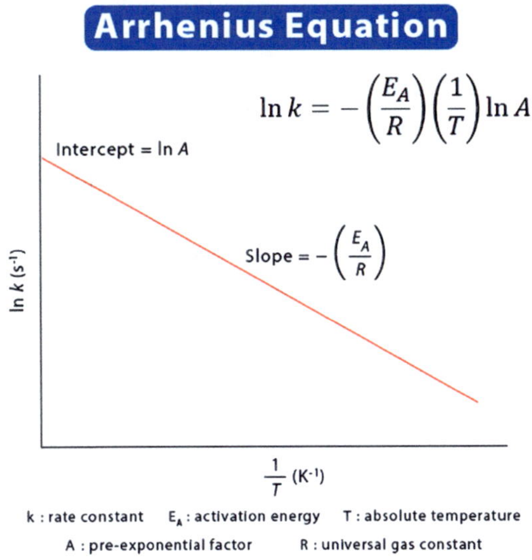

Figure 3.10. *Understanding Arrhenius equation.*

Source:https://www.chemistrylearner.com/wp-content/uploads/2022/04/Arrhenius-equation.jpg.

3.4.1. Derivation of the Arrhenius Equation

The Arrhenius equation is a pillar in the description of temperature dependence of reaction rates for chemical kinetics. Its derivation relies on the expectation that the rate of a chemical reaction exponentially depends on the temperature and activation energy, the minimum energy barrier that the reactants must overcome to transform to products. Establishing this relationship involves the addition of particle collision theory, whcrcin rcactions occur because an adequate number of molecular collisions possess sufficient energy and proper orientation. This theoretical postulate translates into the fact that increased temperature enhances the motion of molecules, hence the frequency of collisions. Thus, the Arrhenius equation can be expressed in the form $k = Ae^{-E_a/(RT)}$, where k is the rate constant of reaction, A is the pre-exponential term, E_a is the activation energy, R is the universal gas constant, and T is the absolute temperature, thus briefly encapsulating the quintessence of the influence of temperature on the kinetics of reaction. The significance of the Arrhenius equation extends beyond its theoretical development; it strongly agrees with empirical data in heterogeneous catalysis. Experiments have revealed a consistent temperature dependence of reaction rates exhibiting Arrhenius-like behavior, which demonstrates the robustness of the equation in applications. Compensation phenomenon investigations for heterogeneous reactions also add depth to the understanding of interconnection between activation energy and the rate coefficients as a function of temperature. For instance, while investigating reaction rates on catalyzed surfaces, researchers noted that changes in surface coverage significantly impact both the apparent activation energy and the pre-exponential factor, as noted by Farra et al. (2012). This observation speaks to the complexity of catalyzed processes while affirming the original hypothesis that temperature is a focal element in controlling reaction rates. Besides, the Arrhenius equation relates to broader thermodynamic principles as discussed in other studies looking at its underlying assumptions. The correlation between the rate coefficients, equilibrium constants, and thermodynamic values forms an integral scheme that is governed by the rules of thermodynamics. The Arrhenius equation development may be taken as a representation of the overarching paradigm, where understanding Gibbs free energy and such equilibrium issues illuminates our comprehension of chemical kinetics. The research findings presented by Heng et al. (2015) indicate to us how all these notions come together to provide us with a comprehensive insight into the dynamics of reaction, encompassing analytical modeling in specified chemical systems. Underlying the Arrhenius equation as something greater than a mere formula but an innate reality of chemical kinetics and equilibrium, one is able to explore further the temperature dependence of reaction rates and its use in different branches of science (Figure 3.11).

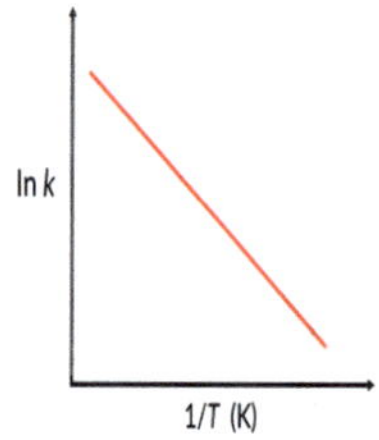

$$\ln k = -\frac{E_a}{R}\left(\frac{1}{T}\right) + \ln A$$

Figure 3.11. *Graph of Arrhenius equation.*

Source: https://www.jove.com/files/media/science-education/science-education-thumbs/11380.jpg.

3.4.2. Role of Temperature in the Equation

The effect of temperature on reaction rates is a basic principle in chemical kinetics that is central to explaining reaction behavior under different thermal conditions. Temperature is a critical factor that not only influences the rate of collisions among reactant molecules but also the force with which reactant molecules collide. This energy has to be high enough, or greater than the activation energy, in order for a reaction to occur. In essence, with rising temperature, the molecules' kinetic energy is greater, so there is a greater chance of surmounting this energy barrier of activation. As a result, the rate of reaction usually rises with temperature, illustrating the exponential function given by the Arrhenius equation.

Therefore, understanding the influence of temperature on reaction rates is necessary in terms of predicting and planning chemical processes across various fields of application, from industrial production to environmental chemistry. In determining the dependency of reaction rates on temperature, several methods can be employed to quantify this relationship. Examples include differential thermal analysis and thermogravimetry, which are methods that allow researchers to study how substances respond to increasing temperature. Through the observation of the degradation mechanisms of various compounds, i.e., the thermooxidative stability of poly(methyl methacrylate) composites filled with silica particles, scientists can determine the impact of temperature on stability over time (Čechová et al., 2012). Through the results, it is apparent that high silica particles can destabilize the polymer matrix, while lower nanoparticles with low content have minimal influence. This refined appreciation is compounded by the imperative for accurate temperature profiling and subtle control in maintaining material integrity in chemical operations, which ultimately governs design considerations across various applications. Temperature-dependent reaction complexities further ensue when pressure and molecular interactions are considered. A case in point is the reaction kinetics of the hydroxyl radical (OH) with carbon monoxide (CO), a key atmospheric and combustion chemistry model. The temperature behavior of the rate coefficient is closely related to the dynamics of the transient intermediates that are stabilized and broken up under varying conditions. Studies using advanced techniques like time-resolved frequency comb spectroscopy have illuminated branching and isomerization kinetics of such reactions, thus enabling quantification of products and reactants as a function of pressure and temperature (Bjork et al., 2017). Inserting the above into a larger picture of temperature's role gives the challenges faced by chemists in being able to predict reaction behavior within the conditions of real-world environments.

3.4.3. Applications of the Arrhenius Equation in Predicting Rates

The Arrhenius equation is a cornerstone in the understanding of the influence of temperature on reaction rates during different chemical reactions. Measuring the effect of temperature on the rate of reaction, the equation yields valuable information concerning kinetic behavior. The nucleus of the equation is rooted in the concept of activation energy, which is the least energy a reaction requires to proceed. With an increase in temperature, the ratio of molecules with this minimum energy also rises, resulting in faster reaction rates.

This concept is not only theoretical but can be applied practically, e.g., predicting the behavior of chemical reactions in industrial processes. For instance, firms apply the Arrhenius equation to optimize conditions for reactions, thereby improving efficiency as well as reducing costs of energy used in chemical production. Besides, applicability of the Arrhenius equations could also be found in particular disciplines such as materials science and nanotechnology. A good example is in metallic nanoparticles preparation in microemulsions.

Researchers have demonstrated that control of particle size relies mostly on the reaction rate, something that one can manipulate through a range of conditions, including temperature and surfactant properties. The kinetics concerned show that as reaction rates rise, so does nanoparticle size, showing a linear relationship between the two. This observation fits within the Arrhenius paradigm, in which temperature is an important influence on the kinetics of the associated processes. The ability of the Arrhenius equation to predict such outcomes validates its relevance to the advancement of technologies that depend on the precise control of chemical reactions (Ritacco et al., 2018).

Moreover, new applications of the Arrhenius equation are not limited to straightforward kinetics; it also encompasses sophisticated modeling techniques, such as transfer function representations. With closed-form solutions of the Arrhenius equation, researchers can quantify the bandwidth of reaction rates and relate these to thermophysical properties of materials. This technique has proven particularly beneficial in explaining complex reactions, such as crystallization in phase-change material. The ability to predict crystallization rates and other major kinetic behaviors based on the viewpoint of the Arrhenius equation allows researchers to engineer materials for specific purposes, designing enhanced functionality. In the end, the versatility of the Arrhenius equation across various scientific fields demonstrates its initial importance in describing temperature dependence for reaction rates as well as the development of improvements in theoretical as well as applied chemistry (Aziz et al., 2013).

3.5. CATALYSTS AND TEMPERATURE

The interaction between temperature and catalysts is of utmost importance when it comes to dealing with the nuances of the rate of reaction. Catalysts are promoters that catalyze reaction speed without undergoing any alteration themselves. However, their effectiveness is likely dependent on temperature. For instance, as the temperature increases, the kinetic energy of reactant molecules rises, leading to more often active site hits of the catalyst with the reactants. This effect allows reactions to move faster because greater numbers of molecules gain sufficient energy to surmount the activation barrier. The delicate interplay between target catalytic activity and the temperature fluctuations therefore becomes critical in fine-tuning reaction conditions for commercial application. Moreover, the role of temperature becomes inextricably linked with catalytic processes. Other catalysts, like in the instance of the *NNN*-pincer ligand systems, react to varying reaction dynamics when subjected to varying temperatures. Rise in temperature might accelerate or decelerate catalytic processes based on the catalyst and the respective reaction environment involved. The mechanism by which increased solvent basicity accelerates catalysis, as in the IPI-Ru(II) complex reaction, is one aspect of how temperature determines the speed of the catalyzed reactions. By precise control over temperature, researchers are able to change the catalytic selectivity and efficiency of reactions and thereby optimize the yield and specificity of chemical synthesis (Goddard et al., 2010). Moreover, thermal stability of catalysts under various thermal conditions is critical in maintaining constant reaction performance. Alkali metal-promoted Co-based mixed oxides display promising results in direct NO decomposition reactions. However, their activity and stability are otherwise controlled significantly by calcination temperature during their synthesis. Higher calcination temperature can increase potassium stabilization, which has a direct impact on the basicity and reducibility of the catalyst. These are essential characteristics for maximum catalytic activity toward decomposition of nitrogen oxide, citing the nuanced interplay among temperature, catalyst synthesis, and reaction efficiency (Balabánová et al., 2019). These interactions being realized and controlled allow chemists not only to build efficient catalysts but also optimized ones to function in precise temperature ranges, optimizing efficiency in any chemical reaction.

3.5.1. Definition and Function of Catalysts

The concept of catalysts is the key to understanding the temperature trend for reaction rates in chemical reactions. Catalysts can be thought of as substances that increase the rate of a chemical reaction without permanently altering themselves. They function by providing an alternate route for the reaction that has a lower activation energy compared to the non-catalyzed reaction. This ability to speed up reactions more efficiently is especially vital in industrial processes, where rates of reaction can have significant impacts on productivity and cost-sufficiency. Chemists are able to regulate conditions of the reaction, such as temperature, with knowledge of how catalysts function in order to achieve desired outcomes, which is extremely relevant as reaction rates are generally temperature sensitive. One of the most influential factors that dictate the effectiveness of a catalyst is its physical phase, and this is usually categorized into two forms: homogeneous and heterogeneous. Homogeneous catalysts are of the same phase as the reactants, which usually leads to more interaction and selectivity towards desired reactions. In contrast, heterogeneous catalysts are of a different phase, commonly solid, and reactants are either in the liquid or gaseous phase.

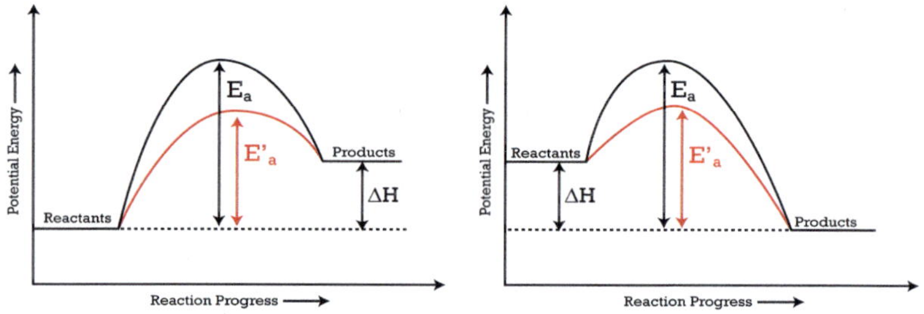

Figure 3.12. *The addition of a catalyst to a reaction lowers the activation energy, increasing the rate of the reaction. The activation energy of the uncatalyzed reaction is shown by* E_a*, while the catalyzed reaction is shown by* E'_a*. The heat of reaction (ΔH) is unchanged by the presence of the catalyst.*

Source:https://dr282zn36sxxg.cloudfront.net/datastreams/f-d%3A0ed555a979b3c01aa733cba151aae23ef1592d3c90179c1f8cc0d79c%2BIMAGE_TINY%2BIMAGE_TINY.1.

The individual interactions present on the surface of heterogeneous catalysts might be complex, necessitating the need to understand involved reaction pathways and intermediates. Developing insights into such interactions is essential if catalytic processes are to be optimized, whose impact in turn governs how temperature dependencies establish themselves in diverse reaction environments, according to Voecks et al. (2025). Catalyst operation and design continue to evolve, particularly as researchers seek to improve efficiency. Advanced technologies, such as quantum chemistry and computer modeling, allow researchers to learn more about catalyst interaction at the molecular level. These methods allow researchers to simulate reaction conditions, predict catalyst behavior under changes in temperature, and identify what aspects to modify to achieve optimal performance. For instance, vanadium oxide over titanium dioxide supports has been shown to exhibit a complex interrelationship of surface properties and catalytic activity, and the effectiveness of the catalyst depends greatly on the specificity of the molecular structure and resulting layers formed in synthesis, as outlined by Bond et al. (1986). This information search not only assists in the optimization of chemical reactions but is also strongly applicable to industrial sustainability and resource management (Figure 3.12).

3.5.2. How Temperature Affects Catalyst Efficiency

The effectiveness of catalysts in the enhancement of chemical reactions is greatly dependent upon temperature since it directly affects reaction kinetics and molecular interactions. With the increase in temperature, the average kinetic energy of the reactant molecules also increases, resulting in increased frequency of molecular collisions. It is this increase in collision frequency that becomes decisive, as it brings about an increase in the possibility of crossing over the activation energy barrier for a reaction to occur. Thus, catalysts can seriously speed up the rates of reactions at high temperatures. Nevertheless, note that efficiency in catalysts and temperature is not directly proportional; beyond a certain optimum temperature, catalysts can begin to lose efficiency due to thermal deactivation or structural deterioration. Structural stability in a catalyst can be undermined at high temperatures, adversely affecting efficiency. Catalysts, particularly heterogeneous catalysts, are also temperature-sensitive in that they respond to changes in temperature that can alter their surface structure and active sites. For instance, sintering may occur at higher temperatures, where catalytic particles agglomerate, resulting in a loss of the surface area and the consequent catalytic activity. This is a reflection of the subtle balance between optimizing reaction rate by increasing temperature and maintaining the catalyst structurally intact for its functional purposes. In line with research on plasmon-enhanced electrochemical conversion, it has been reported that even slight differences in temperature cause significant effects on the performance of the catalyst, a scenario that requires tight temperature control in practical application (Corson et al., 2020). Besides, temperature sensitivity for catalyst efficiency is not uniform for all catalytic systems and reactions. Some catalysts demonstrate considerable increase in efficiency with temperature, but some may experience little or even a decline following the pass of some specific threshold. Such variability highlights necessity to study the thermodynamic and kinetic laws regulating each separate catalytic system. In photoelectrochemical reaction process, for example, influencing of temperature by light conditions makes the dynamics of catalytic efficiency even more complicated. As explained by the distinction between normal photoelectrochemistry and plasmon-enhanced electrochemical conversion, temperature fluctuations can dictate how well these systems respond to outside conditions (Pinto et al., 2017). Hence, temperature optimization is necessary to provide customization of catalyst performance in various chemical processes.

3.5.3. Examples of Temperature-Sensitive Reactions with Catalysts

Catalyst temperature-sensitive reaction behavior is important to the study of their effectiveness and use in chemical reactions. A very much observable illustration can be seen in heterogeneous catalysis, where variations in temperature are very determinative for the relevant reaction rates. Research has revealed that compensation phenomena usually accompany an Arrhenius-type of temperature dependence, which reflects a linear relationship between apparent activation energy and the logarithm of apparent pre-exponential factor. This type of relationship highlights the way surface coverage alteration can differentially influence reaction kinetics. This effect is particularly evident when hydrochloric acid (HCl) is oxidized to chlorine gas (Cl_2) with the assistance of a RuO_2 catalyst, illustrating the profound interplay between surface occupation and temperature on reaction rates. Understanding these factors opens up doors for the optimization of catalytic processes under various thermal conditions, ultimately contributing to enhancing catalytic performance needed for industrial application. In addition to heterogeneous catalysts, catalytic design advancements have generated temperature-sensitive reactions with specific sensitivities to temperature, particularly in electrochemical applications. Research of novel dual plasma coating processes has shown the production of platinum-free catalysts for oxygen reduction reaction in fuel cells. Mechanisms of catalytic activity in these systems respond significantly to temperature variations since they impact the electrochemical character of the catalysts. By adjustment of plasma process conditions through modulation, researchers have demonstrated cataclysmic changes in catalytic activity and the subtleties that go with optimizing the reactions. The ability for control of the creation and activity of the catalysts with temperature is similar to the underlying reaction kinetics and illustrates the strong effect of temperature on reaction rate and the merit of continued research in engineering these non-catalytic systems into useful operation, as shown in existing research with unconventional dual plasma strategies (Brüser et al., 2012). Furthermore, the relationship between catalytic activity and temperature is demonstrated by a number of transition metal reactions whose activities depend strongly on temperature changes. A most applicable case in point would be the catalytic activity of metal oxides, where reactive species coverage at the surface varies considerably with temperature. This aspect is quite important, as demonstrated in experiments on coverage of oxygen species within specific reactions. The ensuing compensation effects reveal that the changes in the apparent activation energy can be counteracted by the entropic consequences arising from the configurational alteration of surface species. Such an insight is led to a new paradigm for the dynamics of temperature-sensitive reactions in heterogeneous catalysis. With ongoing research, this finding comes across as emphasizing the need for a comprehensive knowledge of catalyst thermal behavior to maximize their application efficiency across various industries like petrochemicals and environmental control, laying a foundation for future innovations in catalyst science and engineering (Farra et al., 2012).

3.6. REAL-WORLD APPLICATIONS

Temperature dependence of reaction rates has very significant implications in real-world applications, particularly in biology and biochemistry. Kinetics of chemical reactions have very significant roles in those life-saving processes such as enzyme function and metabolic pathways. As an example, enzymes increase their activity with temperature to a certain point at which denaturation occurs. This principle is of great significance in biotechnology, where the best temperature conditions can enhance enzyme-catalyzed reactions in industrial processes like fermentation and biocatalysis. It is by understanding such temperature-reactivity correlations that scientists and engineers can create more efficient systems that maximize product yield with lower energy consumption, eventually resulting in sustainable processes in various industries. One of the key applications of temperature dependence of reaction rates is in the field of materials science. The properties of materials, particularly polymers and composites, can vary substantially with changes in temperature. As temperature rises, the rate of chemical reactions like degradation and cross-linking could rise, and this affects the material's properties and lifespan. Manufacturers have to consider these factors while manufacturing to make the materials strong enough to endure severe conditions, particularly in aerospace, automotive, and electronics industries. By applying knowledge of reaction kinetics and the effects of temperature, researchers can create more efficient and even more powerful materials. Also, controlled temperature during production based on thermodynamics can lead to the creation of lighter and yet stronger material as needed for specific purposes. Aside from this, the temperature dependence laws are also at the core of energy production, and they have implications for catalytic converter optimization. Catalytic converters in car applications perform best when the operating temperature is high, in which case the reaction rates for converting poisonous pollutants into cleaner substances considerably improve. System design thus involves the optimal choice of the working temperature range to achieve efficiency as well as environmental compatibility. As research into alternative fuel and energy sources continues, study results such as the role of ribonucleotides in hydrothermal vents and biomimicry for PCR technology illustrate some of the avenues that can follow from the investigation of temperature-dependent reaction kinetics. By bridging the study of thermodynamics with catalytic systems, engineers can further advance the development of solutions to reduce emissions as well as optimize energy efficiency in modern transportation systems (Bracco et al., 2010; Palasek et al., 2013).

3.6.1. Industrial Processes and Temperature Control

In industrial processes, temperature control is crucial in determining the efficacy and efficiency of chemical reactions. By varying temperature, yields in the production process are maximized by manufacturers enhancing or suppressing reaction rates. This applies particularly to processes such as catalysis, where kinetics of reactions are greatly controlled by the range of temperature variation. For instance, increasing temperature tends to increase reaction rates by providing reactant molecules with sufficient energy to overcome activation barriers. However, industries need to maintain a delicate balance; extremely high temperatures might lead to unwanted side reactions or destruction of sensitive materials. Thus, the inclusion of proper mechanisms for temperature control, like thermostats or temperature measurement systems, is important in order to produce desired performance in many industrial processes, eventually facilitating safe and sustainable processing of materials. The interaction between temperature and reaction rates is also exemplified in thermal degradation processes, where temperature is a direct parameter affecting product stability and performance. In materials science, for example, studies have shown that the thermooxidative stability of materials like poly(methyl methacrylate) (PMMA) composites may be significantly altered through temperature variations during industrial processing. Literature also indicates that thermal conditions influence the stability of PMMA composites filled with silica; high silica particles destabilize at elevated temperatures, while nanosilica particles would enhance stability at low levels (Čechová et al., 2012). These observations underscore the importance of precise temperature control during production to enable engineers to more accurately predict material response and lives, thereby facilitating the design of safer and more efficient products within an industrial setting. The ability to accurately predict reaction rates from temperature changes is also vital in achieving effective quality control in industrial processes. The manufacturing environment demands a strong appreciation for the interaction between temperature fluctuation and reaction kinetics to avoid loss and enhance product homogeneity. As established in assessing atmospheric conditions, recognizing how temperature affects the spatial distribution and reaction characteristics of different materials constitutes one aspect of the sophisticated knowledge of industrial processes (Geller et al., 2025). By employing advanced thermal analysis techniques and temperature-dependent operations, industries are able to optimize their operating processes, reduce costs of production, and increase overall output quality. Therefore, the application of temperature control to industrial processes extends beyond simple efficiency; it dictates the very nature of efficient and sustainable production processes across a wide variety of industries.

3.6.2. Biological Reactions and Temperature Dependence

Understandings of temperature dependence of biological reactions are of crucial significance to account for elementary biochemical phenomena. Temperature is responsible for controlling the reaction rate through the regulation of molecular motion and kinetic energy of reactants. Upon an increase in temperature, molecules vibrate more vigorously, leading to an increased number of collisions and increased chances for reaction events to happen. This is especially significant in enzyme-catalyzed reactions, where enzyme formation and activity are significantly dependent on temperature. Enzymes typically possess an optimal operating temperature range within which they operate most efficiently, and departures therefrom may produce lower catalytic performance or denaturation. The study of temperature dependence of biological reactions teaches us about organismal adaptation mechanisms, especially under fluctuating environmental conditions, thereby revealing the manner in which life sustains itself through dynamic biochemical reactions. The mechanisms of electron transfer (ET) reactions, which form part of the majority of biological reactions, best illustrate the influence of temperature on biochemical kinetics. Respiration is a perfect illustration of a series of ET processes involved in crucial ATP synthesis, which is immensely influenced by temperature changes. The efficiency of electron tunneling between redox cofactors may dramatically vary with temperature and thus affect cellular energy conversion rate. For instance, in proteins like photosystem II and ribonucleotide reductase, precise temperature control can accelerate or decelerate electron transfer, thus modulating overall metabolic yield. Such sensitivity of ET reactions to temperature underscores the delicate balance that biological systems attain for delivering optimal function in energy conversion, which points towards the crucial role of temperature as a parameter in velocities of biological reactions (Dempsey et al., 2010). Furthermore, hydrolysis thermodynamics of ATP unmasks another dimension of the influence of temperature on biological mechanisms, namely in signal transduction cascades fundamental to cellular communication. Because the free energy change of ATP hydrolysis is temperature-dependent, it has the capability to regulate the dynamics of phosphorylation and dephosphorylation cycles that control a variety of cellular processes. This interaction dictates the sensitivity and amplification of these responses, with a raised phosphorylation potential that maximizes responsiveness in cellular signaling systems. Temperature, energy usage, and biochemical processes interaction demonstrates the role of thermal conditions in modulating kinetic profiles of reactions like protein biosynthesis and other important signaling pathways in cells (Bourne et al., 2002). Thus, understanding these temperature dependences is essential to the interpretation of the complexity of biological reactions and their respective rates.

3.6.3. Environmental Implications of Temperature Effects on Reactions

Temperature plays a determining role in the rates of chemical reactions, significantly influencing processes within the environment. As temperature increases, molecular kinetic energy also increases, typically causing collisions to occur more frequently and more efficiently. This principle significantly involves reactions at atmospheric conditions, such as the ozonolysis of organic compounds. α-Hydroxyalkyl-hydroperoxides (α-HHs) are reaction products and provide examples of the influence of temperature on pathways of reaction and its corresponding environmental impact. These α-HHs, resulting from ozonolysis of monoterpenes, are extremely heterogeneous with respect to stability and decomposition rate against temperature and solvent conditions. Consequently, α-HHs possess e-folding times that greatly depend on molecular structure and environmental conditions, meaning that even small changes in temperature may greatly alter their environmental stability and kinetics, which, in return, affect secondary organic aerosol formation, as explained by Colussi et al. (2020). Furthermore, temperature also affects the availability and stability of dissolved chemical species of relevance to biological activity. In prebiotic chemistry, for instance, the stability of nitrogen oxide anions (NO_x^-), like nitrate and nitrite, is practically temperature-sensitive. It has been observed that under higher temperatures, reactions with these anions are bound to be unfavorable due to competing processes like UV photolysis and reduction with reducing agents like Fe^{2+}. This instability suggests that historical presuppositions of prebiotic ocean NO_x^- concentrations need to be reevaluated. In particular, the low measured concentrations of these species may limit conditions for key prebiotic chemical reactions, and more work needs to be done exploring reaction kinetics and environmental factors for varying thermal regimes, as discussed by Babbin et al. (2019). The environmental importance of temperature's influence on reaction rates extends to climate change arguments, particularly to the enhanced greenhouse effect. An increase in global temperatures allows a range of reactions that produce the very greenhouse gases that cause climate change. For example, higher temperatures promote the oxidation reactions that produce nitrous oxide (N_2O), a potent greenhouse gas, from nitrogenous compounds in soils. Secondly, warming temperatures influence the assembly of organic matter in ecosystems, which releases carbon dioxide (CO_2) and methane (CH_4). The dynamics ultimately find larger expressions in atmospheric chemistry, marine environments, and terrestrial biospheres. Therefore, temperature dependence of reaction rates is crucial for future projections of our planets climate as well as planning effective mitigation and adaptation strategies in response to warming temperatures.

SUMMARY

With the careful consideration of temperature dependence of reaction rates, it is evident that temperature plays an important role in reaction kinetics in many chemical reactions. With a rise in temperature, molecules gain kinetic energy, hence increasing the frequency and intensity of collisions between reactants. This accounts for the Arrhenius equation, which is a basic principle underlying the temperature dependence of reaction rates. The intricate interaction of temperature and molecular dynamics forms the basis of theoretical models as well as directs technical applications in catalysis and industrial chemistry. By examining temperature sensitivity of reaction mechanisms, researchers can make chemical processes efficient, and eventually, innovations can come about in fields ranging from pharmaceuticals to materials science. Probing deeper into particular reaction mechanisms reveals insight into the catalytic activity depending upon changes in temperature. For instance, zeolite-catalyzed reaction studies illuminate the role of structural features in determining reaction rates. Our research on zeolite matrix cracking shows that temperature alone does not affect equilibrium constants and intrinsic rate coefficients but that the crystalline character of the catalyst plays a very significant role in influencing these parameters. Essentially, the temperature dependence of reaction rates is intertwined deep within catalytic activity and is expressed via the relationship between the apparent rate coefficient and adsorbate interactions. Disentangling such complex interactions allows one to acquire profound insight into catalytic efficiency and industrial process optimization, which forms the foundation of the necessity for sufficient exploration in this area (Bell et al., 2018). Furthermore, temperature dependence affects some nuclear processes, illustrating the universal applicability of this rule across from simple chemical reactions. The study of reactions involving the ^{7}Be nucleus is a very vivid illustration of the effect of temperature on reaction rate within a nuclear context and its influence on the kinetic of solar fusion reactions. For example, nonradiative ^{8}B formation in solar plasma is very sluggish as opposed to binary reactions, showing the influence of temperature and reaction dynamics on nuclear interactions (Ajzenberg-Selove et al., 1997). More generally, the conclusion drawn from this study highlights the importance of understanding temperatures role in influencing various reaction mechanisms, ultimately leading to progress in theoretical knowledge as well as practical applications in a range of scientific disciplines.

Temperature is a critical factor that influences the rate of chemical reactions, usually because of its effect on molecular kinetic energy. As temperature rises, molecules gain more kinetic energy, which means more frequent and intense collisions. The Arrhenius equation measures this through the explanation of how an increase in temperature increases reaction rates. Other factors, such as concentration and catalysts, also have effects; however, temperature is a significant factor. Thus, it is significant to know about temperature dependence of the rates of reaction to foresee how chemical systems will behave under varying conditions of temperature, highlighting its significance in theoretical and practical applications in environmental science and industrial chemistry. The influence of temperature on reaction rates is established both empirically and theoretically. For instance, experiments with the Porter-Thomas distribution demonstrate its application to estimate thermonuclear reaction rates, whose temperature fluctuations affect the resonance of some reactions by a great extent. From the study of the mean reduced widths for compound nuclei within a specified mass range, scientists can make accurate predictions regarding reaction rates vital to understanding stellar processes, for instance, those that lead to the nucleosynthesis of heavy elements during supernovae explosions. These findings emphasize the intricate interaction of temperature with the underlying process governing reaction kinetics, and they reaffirm the imperative of strict temperature control in experimental application to optimize reaction efficiency and yield. Furthermore, scaling laws derived from laser-plasma interactions provide another dimension to the understanding of temperature dependence of reaction rates. The resulting reaction products of laser-driven reactions exhibit linear dependence on laser power, strongly correlating with plasma temperature increase. Observed scaling can be explained based on the influence of temperature on reaction rates, plasma density, and the extent of particles propagated in the target medium. This forecasting capability allows extrapolation of fusion yields in hitherto unexplored nuclear reactions, showing how temperature not only affects traditional chemical kinetics but also extends the range of the latter to advanced energy production techniques, e.g., inertial confinement fusion. As such, elucidation of temperature sensitivity of reaction rates is central to comprehensive understanding of both elementary chemistry and new technologies.

REFERENCES

1. Adelberger, E. G., Akhmedov, E. K., Alexander, T., Bahcall, J. N., Monakhov, D. E., Degl'Innocenti, S., Foldy, L. L., et al. (1998). The solar neutrino problem–An update. https://core.ac.uk/download/25243883.pdf (accessed on 10th September 2025).
2. Ahmed, M., Azyazov, V. N., Kaiser, R. I., Mebel, A. M., et al. (2018). Functional relationships between kinetic, flow, and geometrical parameters in a high-temperature chemical microreactor. https://core.ac.uk/download/294625778.pdf (accessed on 10th September 2025).
3. Aivazis, M., Goddard, B., Meiron, D., Ortiz, M., et al. (1999). *The 1999 Center for Simulation of Dynamic Response in Materials Annual Technical Report*. https://core.ac.uk/download/4890872.pdf (accessed on 10th September 2025).
4. Ajzenberg-Selove, F., Bahcall, J. N., Brauner, D. E., Monakhov, D. E., Degl'Innocenti, S., Foldy, L. L., et al. (1997). Triple collisions (e+p+Be7) in solar plasma. http://arxiv.org/abs/nucl-th/9712020 (accessed on 10th September 2025).
5. Akmal, A., Nelson, A. E., Friman, B. L., Lee, C. H., Batty, C. J., Kaplan, D. B., et al. (1999). Nonequilibrium weak processes in kaon condensation II–Kinetics of condensation. https://core.ac.uk/download/25269695.pdf (accessed on 10th September 2025).
6. Allen, C., & Wilson, D. (2017). A comparison of sensitivity metrics for two-stage ignition behavior in rapid compression machines. https://core.ac.uk/download/213086672.pdf (accessed on 10th September 2025).
7. Asadian, A., Popescu, S., Briegel, H. J., Caruso, F., Chin, A. W., Datta, A., Huelga, S. F., Plenio, M. B., et al. (2010). Motional effects on the efficiency of excitation transfer. http://arxiv.org/abs/1002.0346 (accessed on 10th September 2025).
8. Aschaffenburg, D. J., Chen, X., & Cuk, T. (2019). Selecting between two transition states by which water oxidation intermediates on an oxide surface decay. http://arxiv.org/abs/1903.01675 (accessed on 10th September 2025).
9. Aziz, M. M., & Wright, C. D. (2013). A transfer function approach to reaction rate analysis with applications to phase-change materials and devices. https://core.ac.uk/download/16463625.pdf (accessed on 10th September 2025).
10. Babbin, A. R., Ranjan, S., Rimmer, P. B., Sasselov, D., et al. (2019). Nitrogen oxide concentrations in natural waters on early Earth. http://arxiv.org/abs/1902.07740 (accessed on 10th September 2025).

11. Balabánová, J., Bílková, T., Fridrichová, D., Jirátová, K., et al. (2019). Co–Mn–Al mixed oxides promoted by K for direct NO decomposition: Effect of preparation parameters. https://core.ac.uk/download/237018678.pdf (accessed on 10th September 2025).
12. Bell, A. T., Head-Gordon, M., Janda, A., Lin, S., et al. (2018). Understanding Brønsted-acid catalyzed monomolecular reactions of alkanes in zeolite pores by combining insights from experiment and theory. https://core.ac.uk/download/158345582.pdf (accessed on 10th September 2025).
13. Bjork, B. J., Bui, T. Q., Changala, P. B., Nguyen, T. L., et al. (2017). Direct measurements of DOCO isomers in the kinetics of OD+CO. https://core.ac.uk/download/211372083.pdf (accessed on 10th September 2025).
14. Bonasera, A., & Kimura, S. (2012). Hydrodynamic scaling analysis of nuclear fusion driven by ultra-intense laser–plasma interactions. http://arxiv.org/abs/1109.5056 (accessed on 10th September 2025).
15. Bond, G. C., Bosch, H., Flamerz, S., Gellings, P., et al. (1986). Structure and reactivity of titania-supported oxides. Part 1: Vanadium oxide on titania in the sub- and super-monolayer regions. https://core.ac.uk/download/pdf/11473217.pdf (accessed on 10th September 2025).
16. Bourne, H., Ferrell, J. E., Goldbeter, A., Gresser, M. J., Hartwell, L. H., Hill, C., et al. (2002). Thermodynamic and kinetic analysis of sensitivity amplification in biological signal transduction. https://core.ac.uk/download/25352781.pdf (accessed on 10th September 2025).
17. Bracco, A., Chomaz, P., Gaardhøje, J. J., Heenen, P. H., et al. (2010). *Perspectives of Nuclear Physics in Europe: NuPECC Long Range Plan 2010.* https://core.ac.uk/download/19502483.pdf (accessed on 10th September 2025).
18. Brady, K. B., Hui, X., Niemeyer, K. E., Sung, C. J., et al. (2017). Reduced chemistry for butanol isomers at engine-relevant conditions. http://arxiv.org/abs/1706.02043 (accessed on 10th September 2025).
19. Branco Pinto, D., Constales, D., Marin, G., Yablonsky, G. S., et al. (2017). Kinetic intersections as a source of information on rate coefficients. https://core.ac.uk/download/91288153.pdf (accessed on 10th September 2025).
20. Brüser, V., Kummer, K., Quade, A., Vyalikh, D. V., et al. (2012). Using a dual plasma process to produce cobalt–polypyrrole catalysts for the oxygen reduction reaction in fuel cells–Part I: Characterization of the catalytic activity and surface structure. http://arxiv.org/abs/1208.2574 (accessed on 10th September 2025).

21. Čechová, E. (2012). Thermooxidative stability of PMMA composites. https://core.ac.uk/download/30296004.pdf (accessed on 10th September 2025).
22. Cescutti, G., Hirschi, R., Murphy, A. St. J., Nishimura, N., et al. (2017). Abundance uncertainties obtained with the PizBuin framework for Monte Carlo reaction rate variations. https://core.ac.uk/download/586804140.pdf (accessed on 10th September 2025).
23. Colussi, A. J., Enami, S., Liang, Z., Qiu, X., et al. (2020). Stability of monoterpene-derived α-hydroxyalkyl-hydroperoxides in aqueous organic media: Relevance to the fate of hydroperoxides in aerosol particle phases. https://core.ac.uk/download/287633876.pdf (accessed on 10th September 2025).
24. Corson, E. R., Creel, E. B., Kostecki, R., McCloskey, B. D., et al. (2020). Important considerations in plasmon-enhanced electrochemical conversion at voltage-biased electrodes. https://core.ac.uk/download/287621950.pdf (accessed on 10th September 2025).
25. Dempsey, J. L., Gray, H. B., & Winkler, J. R. (2010). Proton-coupled electron flow in protein redox machines. https://core.ac.uk/download/4887683.pdf (accessed on 10th September 2025).
26. Farra, R., Hevia, M., Knop-Gericke, A., López, N., et al. (2012). *In situ* surface coverage analysis of RuO_2-catalyzed HCl oxidation reveals the entropic origin of compensation in heterogeneous catalysis. https://core.ac.uk/download/210619064.pdf (accessed on 10th September 2025).
27. Fowler, W. A., Rayet, M., & Somorjai, E. (2014). Challenges in nucleosynthesis of trans-iron elements. https://core.ac.uk/download/29840285.pdf (accessed on 10th September 2025).
28. Geller, M. A., Hampson, R. F., Stolarski, R. S., Watson, R. T., et al. (2025). Present state of knowledge of the upper atmosphere: An assessment report; processes that control ozone and other climatically important trace gases. https://core.ac.uk/download/pdf/42841035.pdf (accessed on 10th September 2025).
29. Gerritsma, R., Groenenboom, G. C., Hirzler, H., Lous, M., et al. (2020). Controlling the nature of a charged impurity in a bath of Feshbach dimers. https://core.ac.uk/download/333546183.pdf (accessed on 10th September 2025).
30. Goddard, W. A., III, Hashiguchi, B. G., & Periana, R. A., et al. (2010). Acceleration of nucleophilic CH activation by strongly basic solvents. https://core.ac.uk/download/4886415.pdf (accessed on 10th September 2025).

31. Haile, S. M., Hightower, A., & Louie, M. W. (2010). Nanoscale electrodes by conducting atomic force microscopy: Oxygen reduction kinetics at the Pt|$CsHSO_4$ interface. https://core.ac.uk/download/4884969.pdf (accessed on 10th September 2025).

32. Heng, K., Lyons, J. R., & Tsai, S. M. (2015). Atmospheric chemistry for astrophysicists: A self-consistent formalism and analytical solutions for arbitrary C/O. https://core.ac.uk/download/212343442.pdf (accessed on 10th September 2025).

33. Howard, C., Iliadis, C., Longland, R. L., Mitchell, L., et al. (2013). Mean proton and alpha-particle reduced widths of the Porter–Thomas distribution and astrophysical applications. https://core.ac.uk/download/41772182.pdf (accessed on 10th September 2025).

34. Hyeon, C., Thirumalai, D., Derenyi, I., Evans, E., Fisher, T. E., Hummer, G., Lide, D. R., et al. (2006). Measuring the energy landscape roughness and the transition state location of biomolecules using single molecule mechanical unfolding experiments. http://arxiv.org/abs/cond-mat/0612433 (accessed on 10th September 2025).

35. Lavagno, A., & Quarati, P. (2005). Nonextensive statistical effects on nuclear astrophysics and many-body problems. http://arxiv.org/abs/nucl-th/0504080 (accessed on 10th September 2025).

36. Marley, M. S., & Zahnle, K. J. (2014). Methane, carbon monoxide, and ammonia in brown dwarfs and self-luminous giant planets. https://core.ac.uk/download/pdf/189597261.pdf (accessed on 10th September 2025).

37. Palasek, S. (2013). Primordial RNA replication and applications in PCR technology. http://arxiv.org/abs/1305.5581 (accessed on 10th September 2025).

38. Penner, S. S. (1954). Spectroscopic studies of premixed laminar flames. https://core.ac.uk/download/4883608.pdf (accessed on 10th September 2025).

39. Raftery, A. E., Martínez Fiorenzano, A. F., Sanchez, A. G., Boesgaard, A. M., Scóccola, C. G., Vucetich, H., Mosquera, M. E., et al. (2008). Early universe constraints on time variation of fundamental constants. https://core.ac.uk/download/487177087.pdf (accessed on 10th September 2025).

40. Ritacco, H. A., Sanchez-Morales, J., & Sánchez, M. D. (2018). Reaction kinetics in the production of Pd nanoparticles in reverse microemulsions: Effect on particle size. https://core.ac.uk/download/552341121.pdf (accessed on 10th September 2025).

41. Somorjai, E., Gyürky, G., Farkas, J., Rayet, M., Rauscher, T., et al. (2012). Investigation of alpha-induced reactions on *130Ba* and *132Ba* and their importance for the synthesis of heavy *p* nuclei. https://core.ac.uk/download/9840533.pdf (accessed on 10th September 2025).
42. Stoneham, A. M. (1989). *Defect Migration in Solids–Microscopic Calculation of Diffusion Rates*. https://core.ac.uk/download/1728264.pdf (accessed on 10th September 2025).
43. Sturtevant, B. (1966). Application of a magnetic mass spectrometer to ionization studies in impure shock-heated argon. https://core.ac.uk/download/4879846.pdf (accessed on 10th September 2025).
44. Turkin, A. A., Barashev, A. V., Meslin, E., Christien, F., Auret, F. D., Russell, F. M., Abrasonis, G., et al. (2014). Theory of a quodon gas. With application to precipitation kinetics in solids under irradiation. http://arxiv.org/abs/1406.4314 (accessed on 10th September 2025).
45. Voecks, G. E. (2025). *ECUT: Energy Conversion and Utilization Technologies Program. Heterogeneous Catalysis Modeling Program Concept*. https://core.ac.uk/download/pdf/42849346.pdf (accessed on 10th September 2025).
46. Willitsch, S. (2014). Ion-atom hybrid systems. https://core.ac.uk/download/33303287.pdf (accessed on 10th September 2025).

CHAPTER 4

COMPLEX REACTIONS

Contents

Introduction 123

4.1. Types of Complex Reactions 128

4.2. Mechanisms of Complex Reactions 132

4.3. Factors Influencing Complex Reactions 136

4.4. Applications of Complex Reactions 140

4.5. Case Studies of Complex Reactions 146

4.6. Challenges in Studying Complex Reactions 151

Summary 155

References 158

INTRODUCTION

The chemistry discipline is, by nature, driven by complex reactions that govern the interactions and transformations of matter. These reactions are not mere one-off events; rather, they are existent as networks of complexity at the heart of the diverse expression of chemical and biological processes. It is a system that requires an understanding against a context that appreciates the systematic organization of reactions, including their evolutionary significance.

As metabolic pathways provide modularity and hierarchy in biochemical systems, they illustrate how complex organisms are capable of adapting to their surroundings. Metabolic structure of living organisms can be explained by the perspective of complexity in which interacting reactions reveal underlying orders of organization and function, as discussed by Alberts et al. (2012). This research opens the door for more research on the mechanism of chemical reactions and how they shape ecosystems and life itself. The secret to complex reactions is the modularity hypothesis, which acts to facilitate easy comprehension of sophisticated biochemical pathways. Modularity refers to recurring sets of chemical reactions and functions with conserved architecture in various biological contexts.

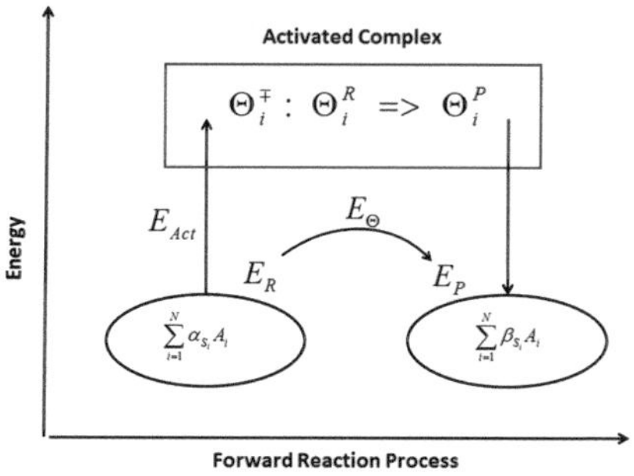

Figure 4.1. *A complex reaction mechanism involving energy barriers and transition states.*

Source: https://cdnintech.com/media/chapter/56807/1512345123/media/F1.png.

This phenomenon allows the decomposition of the intricacy of metabolic pathways such that researchers can identify significant interactions that are conserved among different networks. Conservation of enzymes and cofactors is significant in maintaining the stability and efficiency of such modular networks. As described by Alberts et al. (2012), the boundaries demarcating such modules tend to overlap with the most complex mechanisms of reactions. By analyzing these modules, scientists can follow the evolutionary record of metabolic processes and determine patterns that contribute to the origin of life's diverse forms, thus enhancing our appreciation of chemical reaction dynamics. Lastly, where chemical theory meets practical applications, new avenues of research into intricate reactions have been illuminated through going beyond classical biochemical considerations. For instance, chemically inspired interaction modeling has been proved beneficial for consensus protocol in wireless sensor networks. The novel approach utilizes chemistry principles to determine the rules of interaction that drive network node behavior and therefore shows the applicability of chemistry's fundamental principles towards the enhancement of technology. According to, the application of this chemical model facilitates the construction of robust systems that are capable of solving real-world problems based on decentralized information flow. Ultimately, such interdisciplinarity not only fosters theoretical discussion of complex reactions but also gives real solutions, highlighting the utility of chemical interactions in both biological and technological realms (Figure 4.1).

Definition of Complex Reactions

Complex reactions are fundamentally characterized by their multistep mechanisms and profuse interdependencies among reactants, products, and intermediates. Unlike simple reactions, which are usually a single-step conversion of reactants to products, complex reactions are frequently an intertwined collection of processes and may consist of many different steps, including initiation, propagation, and termination. Complex reactions can be non-linear and sensitive to a wide range of external inputs, such as concentration, temperature, and pressure. The study of complex reactions not only helps in the understanding of fundamental chemical principles but also has extensive applications in fields such as environmental science, material synthesis, and pharmaceutical production, where the outcome of chemical reactions can significantly affect product efficacy and safety. For example, a non-science major chemistry course places special emphasis on the relevance of chemical knowledge to society, focusing on how mathematical principles direct one's understanding of reaction dynamics and equilibria (Shillady et al., 2002). In atmospheric chemistry, intricate reactions are especially important since they determine the behavior of molecules in different planetary atmospheres. The interactions among small molecules such as carbon monoxide and methane are perfect demonstrations of the role of complex reaction networks in different physical conditions. For instance, the study of atmospheric models shows that in equilibrium conditions, CH_4 is favored over CO, but under most natural conditions, there is disequilibrium with a preference for CO with descriptors like vertical mixing and temperature gradients (Marley et al., 2014). This duality underlines the subtlety of complicated reactions, emphasizing the necessity for dynamic modeling that accommodates multiple kinetic paths and environmental conditions. These dynamics must be understood to predict the chemical profiles of celestial bodies and judge their habitability or ability to harbor life. Moreover, the profound implications of complex reactions influence down-to-earth applications across a broad spectrum of disciplines. Synthesis of active pharmaceutical ingredients in the pharma industry is most often a delicate reaction process that demands expert optimization to generate the desired outcomes. Control of reaction conditions allows chemists to explore numerous routes—a process that involves showing the nuances present in any gas-phase and solution-phase reactions. Additionally, as technology advanced to make it simpler to study these reactions, more studies reveal the connection of reaction kinetics with thermodynamics, leading to further catalysis and material science advancements. As scientists accept the complexities of this intricate reaction, not only do they improve chemistry's knowledge, but they also enhance their capacity to design processes that meet the demands of modern society (Shillady et al., 2002) (Figure 4.2).

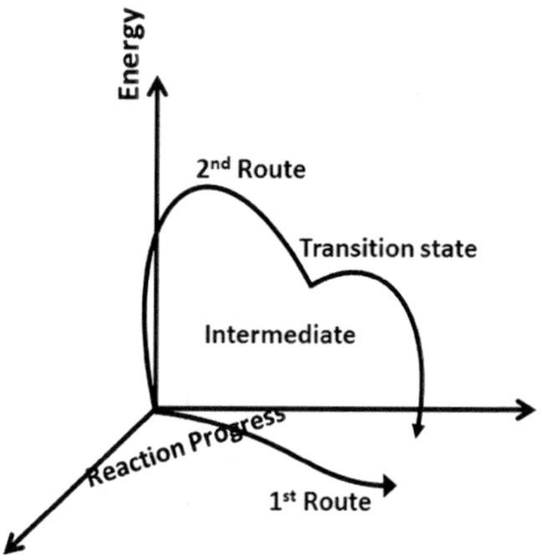

Figure 4.2. *A complex chemical reaction passes through different transition states and adopts different completion routes before yielding the products.*

Source: https://cdnintech.com/media/chapter/56807/1512345123/media/F2.png.

Importance in Chemical Processes

The contribution of chemical reactions in the domain of multifaceted reactions cannot be overemphasized, especially in energy storage devices such as lithium-ion batteries. Chemicals such as nickel-rich layered oxide cathode materials, such as $LiNi_{0\cdot8}Mn_{0\cdot1}Co_{0\cdot1}O_2$ (NMC-811), symbolize the intricate relationship between performance and chemical stability, which technological progress necessitates. Such materials' potential to provide increased battery energy density is countered by problems such as capacity retention and thermal stability. When temperatures rise, structural degradation of NMC-811 can trigger undesirable side reactions that are usually exothermic in nature. Studies on the thermal performance of such cathodes provide insight into the high-level processes, including lattice phase transitions and cation diffusion, of critical importance in the context of ensuring battery safety and efficiency (Alvarado et al., 2020). Beyond storage of energy, the reduction of greenhouse gases is also crucial in managing environmental issues, highlighting the importance of chemical processes in mitigating climate change. Acid gas treatment, particularly CO_2 removal from flue gas, has increased with the pressure of stringent environmental controls.

Figure 4.3. *Importance of chemical processes.*

Source:https://www.thoughtco.com/thmb/twNTURXdou1OiMtp9ibSlxJT0gk=/1500x0/filters:no_upscale():max_bytes(150000):strip_icc()/GettyImages-548134701-589cece95f9b58819c706790.jpg.

The wide differences in absorption efficiency with atmospheric pressures and the dilution from combustion air necessitate advanced methodologies for optimization of gas-liquid contactors. The interaction between types of packing and pressure drop becomes a design consideration of prime importance in optimizing absorption performance. Encouragingly, such conditions under which interfacial areas in these types of systems can be reliably quantified have been found through research to be necessary for the optimization of chemical absorption processes and, by extension, their efficiency at reducing dangerous emissions (Raynal et al., 2007). Moreover, the complexity of reaction involved in both energy storage and environmental clean-up clearly illustrates the interdisciplinary nature of modern chemical processes. With greater understanding, the interdependence of reaction kinetics, structural stability, and thermal stability becomes more apparent, serving as the foundation of technological progress. For instance, the development of an extended absorption model can facilitate the calculation of interfacial areas under varying conditions while generating accurate measurements without intensive initial knowledge of parameters such as liquid-side mass transfer coefficients. This approach not only paves the way for superior designs but also emphasizes maximizing chemical processes in terms of performance and sustainability in the environment. Therefore, the relevance of these complex reactions far exceeds theoretical models, extending to real applications directly affecting energy solutions and environmental conservation (Figure 4.3).

4.1. TYPES OF COMPLEX REACTIONS

Complex reactions involve many different types that play significant roles in biological and chemical processes. Chain reactions, in which a cascade of reactions has the product of one step as the starting material for the next step, is a significant branch. These types of reactions play extremely significant roles in polymerization and burning, in which a first triggered event can cascade into a significant release of energy or formation of large molecular units. The chain reaction kinetics are important in emphasizing the role of reaction conditions, such as temperature and concentration, which play a crucial role in influencing the efficiency and result of the process. Understanding these parameters is key to applications ranging from industrial synthesis to environmental chemistry, where the reactions must be controlled for specific results. The concerted reaction is an additional key type of complex reaction, where there is concurrent bond breaking and forming. This is one form of reaction that goes through one transition state and no intermediates, highlighting a characteristic peculiarity in chemical kinetics. Concerted mechanisms are usually found in pericyclic reactions, like electrocyclic reactions and cycloadditions, which are central in organic synthesis. The detailed analysis of concerted reactions can elucidate the role played by stereochemistry and orbital interactions, which are crucial in reaction outcome prediction. Moreover, the computational modeling of such reactions illuminates their mechanisms and aids in the design of novel reactions that produce high selectivity and efficiency, illustrating the interaction between theory and application in chemical research. Last but not least, the kinetics of complex reactions may be influenced by environmental forces, with significant implications for modeling and characterizing chemical behavior. For pertinent example, responses in the interstellar medium demonstrate how sensitivity analyses have the ability to point out the importance of specific reaction rates in predicting molecular abundances in space. The models must include numerous processes including ion-neutral and neutral-neutral reactions that also play vital roles in molecule synthesizing in space outside Earth. By studying the intricacies present in these varied types of reaction, researchers are able to better understand fundamental chemistry as well as apply it to practical cases within real life, especially cosmochemistry and astrobiology (Agundez et al., 2010; Regev et al., 2009).

4.1.1. Reversible Reactions

Reversible reactions are a fundamental characteristic of chemical dynamics and the driving force behind the functionality of complex reactions in numerous biological and industrial processes. Because these reactions are reversible, they move in both the forward and backward directions, establishing dynamic balance between reactants and products. Dynamic equilibrium facilitates adjustment of systems to changes in conditions such as temperature and concentration, consequently influencing reaction pathways. For instance, the reversible nature of the reactions involved in enzyme catalysis can be employed to illustrate the manner in which metabolic pathways are finely regulated by biological systems in order to maintain homeostasis. This ability to reverse reactions increases the functional efficiency of chemical systems, highlighting the subtle interactions that guide reaction behavior and product formation within complex reaction networks. Examination of reversible reactions is likely to reveal underlying mechanisms that are crucial to designing and improving synthetic pathways. Experiments indicate that systems, like the one from Stoltz and Grubbs, use triethyl silane and potassium tert-butoxide in order to gain reversible silylation and bond cleavage transformations. Their mechanistic investigations depict a world where there are many intermediates and reaction pathways, often determined by the radical species and single electron transfer dynamics. Understanding these reversible reactions not only makes it easier to design more efficient chemical reactions but also establishes that intermediates and reaction conditions play crucial roles in determining the final yields and product selectivity in complicated reaction schemes, as emphasized in relevant literature. With improvements in computational methods, our understanding of reversible reactions has further been enriched along with their position in chemical reaction networks (CRNs). The development of algorithms for computing dynamically equivalent weakly reversible realizations is especially important since it enables chemists to explore reaction systems in greater detail. Using concepts in graph theory and linear programming, researchers can compute optimal paths and check for the feasibility of specific reaction networks. In addition to enabling more accurate prediction of intricate reactions, this makes it possible to investigate systematically structural details influencing reaction dynamics. Therefore, such computational techniques in conjunction with experimental studies give rise to an arena of new areas of research that are helpful in understanding reversible reactions and their inherent role in the canvas of intricate chemical interactions, as supported by existing research in the field (Banerjee et al., 2017; Hangos et al., 2011) (Figure 4.4).

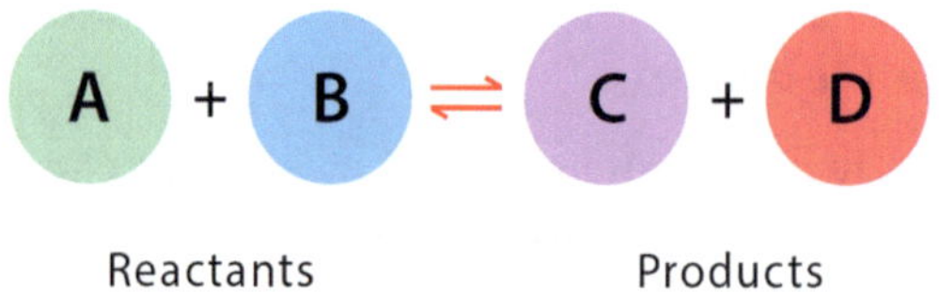

Figure 4.4. *Reversible reactions.*

Source: https://www.chemistrylearner.com/wp-content/uploads/2021/05/Reversible-Reaction.jpg.

4.1.2. Catalytic Reactions

Catalytic reactions are the core of complex reactions, hastening the processes required in chemical synthesis and industry. By facilitating the transformation of reactants to products with higher efficiency, catalysts significantly decrease the activation energy required for a reaction to occur. Here, arene-ruthenium complexes demonstrate the manner in which specific catalytic agents can form new pathways and products. Investigation of further nuclear arene-ruthenium-sulfur clusters has yielded not only new structures but also the intricate mechanisms of their formation, which are vital to be realized in order to understand how the catalysts function. Valence electron effects and solid-state interactions, in particular, through techniques such as X-ray crystallography, need to be understood in order to decipher the complexity of such catalytic reactions and thereby link basic research to material science and catalysis applications (Guo et al., 1999). In addition to traditional catalytic reactions, newer developments have enabled new catalysts based on polymers that optimize efficiency in complicated reaction environments. For instance, vinylic addition polynorbornenes (VA-PNB) have been utilized as multifunctional scaffolds to immobilize catalysts, widening the scope in catalytic chemistry even further. The natural stability of VA-PNBs makes it possible to introduce numerous functional groups which can be readily altered post-polymerization, with the facility to tailor their reactivity for specific catalytic applications. This versatility has been investigated in various applications, ranging from organocatalysis to metal-supported catalysts, enhancing the activity and recyclability of such systems. Therefore, the study of VA-PNBs not only solves problems related to the recovery of traditional catalysts but also increases the sustainability of catalytic processes, which is of utmost concern in modern chemical manufacture (Albéniz et al., 2019). Lastly, the study and application of catalytic reactions embody the essence of innovation in intricate reactions. With scientists continuing to discover the structural and mechanistic features of catalysts like arene-ruthenium complexes and VA-PNBs, they find new ways challenging conventional synthesis approaches. This confluence of fundamental science and practical application not only advances understanding of chemical change but also spurs innovation that has significant implications for the chemical industry. By pushing the boundaries of what is known about catalytic processes, scientists are not only optimizing chemical reactions but also addressing the broader challenge of sustainable practice in chemistry. Hence, ongoing exploration of catalytic reactions remains a common thread in the pursuit of innovative solutions to tough problems within the field (Albéniz et al., 2019; Guo et al., 1999).

4.1.3. Chain Reactions

Chain reactions in chemistry are an interesting form of elaborate reaction which is self-perpetuating in nature. Compared to the linear nature of simple reactions, chain reactions are a series of dependent steps where the reactive species carry out a chain of reaction to give a chain process. The effect is particularly observed in combustion and polymerization reactions. For instance, combustion of hydrocarbons normally initiates by means of the generation of radicals that enable continuous breakdown of fuel with release of energy. This intricate interplay is accountable for a lot in terms of natural and industrial phenomena and highlights the significance of comprehending chain reactions in everything from energy production to materials science. Study of chain reactions is the main point of their mechanism, especially on the base of elementary reactions as components governing the entire process. The most typical example of chain reaction usage is present research in atmospheric chemistry, such as the formation of complex hydrocarbons from simpler precursors. For instance, the detection of benzene in the atmosphere of Titan has provided impetus for theories proposing the role of polycyclic aromatic hydrocarbons (PAHs) as haze-nucleating agents. The PAH processes underlying their formation are complex, however. Experimental evidence proposes that model PAHs like anthracene and phenanthrene form through barrierless mechanisms of naphthyl radicals and vinylacetylene. Such understanding illuminates the process of how seemingly simple precursor interactions become built up to form intricate chain reactions that contribute to planetary atmospheric chemistry, thus emphasizing the connectedness of simple reactions within systems. Secondly, the implications of chain reactions extend beyond theoretical studies, with significant implications for practical applications across industries. For example, polymer chemistry borrows heavily from chain reaction mechanisms to be able to synthesize all sorts of polymers and impact plastics, drugs, and many other things. An understanding of the radical mechanisms involved greatly enhances the performance of the reactions to be applied to control the outcome of the synthesis. The results concerning PAH synthesis in the atmosphere of Titan highlight the imperative re-examination of conditions needed for PAH molecules to be formed. By accepting that PAH molecules may be produced by low temperature barrierless reactions, scientists can now refine their approach to gaining and controlling chemical processes. As studies advance, the explanation of chain reactions will continue to be central, holding far-reaching implications for disciplines as diverse as environmental science and catalysis to name just a few.

4.2. MECHANISMS OF COMPLEX REACTIONS

Understanding the machinery of complex reactions is essential for understanding the principles underlying chemical processes. Complex reactions will involve multiple steps, substrates, and intermediates that interact in complex networks. Such complexity is best exemplified by the modularity and hierarchy built into metabolic pathways, which are segregated to specific functional units that can be investigated independently. These modules have saved internal structure that enables repeated utilization in divergent pathways to encourage biochemical flexibility and effectiveness. Therefore, examining these mechanisms is useful in identifying how evolution has built metabolic systems, with the focus being on their resilience and flexibility in responding to environmental change. This modular view of organization not only provides a window into biochemical function but also underlies the comprehension of the evolutionary integration of complex reactions in the biosphere, emphasizing the delicate interplay of chemical diversity and functional conservation (Alberts et al., 2012). Another feature of advanced reaction mechanisms is the role of cofactors, which play a crucial role in the process of bringing substrates and enzymes together. Cofactors may be regarded as keys unlocking the potential of some organic reactions, greatly influencing the control and economy of metabolic processes. The advanced interactions normally occur at the junctions of metabolic modules, with sophisticated cofactors being involved in controlling more sophisticated reaction mechanisms and simpler cofactors employed in simple biochemical conversions. This tiered approach highlights the significance of cofactors in the reconciliation of metabolic processes and indicating their evolutionary advantage to facilitate plasticity and innovation. The interaction between substrate complexity and cofactor involvement further elucidates the underlying architecture of biochemical systems, reconciling micro-scale interactions to macro-scale consequences in the systems of sophisticated reactions (Alberts et al., 2012). Recent literature further illustrates the application of detailed chemical kinetic models to interpret complicated reactions, such as in the case of biofuels such as butanol. Production of skeletal mechanisms from validated models illustrates a systematic reduction in the complexity of highly detailed reaction pathways without reduction in predictability. For instance, comparison of skeletal mechanisms derived for various butanol isomers with complex models was found to be in good agreement for predicting global combustion phenomena such as autoignition delay times and laminar flame speeds. The simplification makes the case for tackling complex reactions with simpler models that retain the significant features and behavior. Besides, through the observation of butanol isomer performance against traditional gasoline surrogates, researchers have uncovered relative efficiencies of different fuels in real use and, therefore, in the long term to a more informed understanding of reaction mechanisms and effects in the new energy resource trends.

4.2.1. Reaction Pathways

Reaction pathways are also very important in explaining complex chemical reactions, particularly in the areas of biochemistry and molecular biology. Reaction pathways are collections of elementary reactions that co-operate to produce the conversion of reactants to products. These pathways usually get explained with specific kinetic models, such as the Michaelis-Menten scheme, representing enzyme kinetics in a simple manner. These models provide researchers with a numeric representation of various aspects of these reactions, illuminating the intricate dialogue between the reactants in question. The mathematical representation of these routes is significant not only for theoretical calculations but also for practical applications, like biotechnological innovations. As highlighted by recent studies, the prevalence of cooperative interactions leaves a significant effect on the dynamics of such pathways, leading to diverse phenomenological behaviors and necessitating an advanced understanding of reaction mechanisms (Agliari et al., 2018). The realization of the complexity of reaction pathways also includes the understanding of how molecular programming is tasked with controlling chemical events. This approach integrates molecular and chemical systems in order to replicate autonomous behavior, hence allowing control mechanisms to be put into the reactions themselves. Through the use of strand displacement reactions, researchers can create cascades of reactions that exhibit particular dynamics akin to coupled chemical reactions. These systems can replicate properties of a variety of behaviors, ranging from oscillatory to chaotic dynamics, hence effectively simulating the equilibrium between stability and reactivity exhibited typically in natural processes. This capacity for mimicry gives valuable insights into biological systems, revealing the process by which complexity is established from the simple molecular interactions (Andrianantoandro et al., 2009). As scientists are refining ever more complex molecular systems, the value of knowing reaction pathways becomes increasingly clear. While studying reaction pathways, it is observed that they are not isolated phenomena but affect and interact with numerous external factors and internal dynamics to regulate the direction and yield of chemical reactions. For instance, changing substrate concentration may transform a system from positive to negative cooperative behaviors, rendering reaction kinetics challenging to comprehend. This interface demands a general approach, calling upon the precepts of both physics and chemistry, to analyze and simulate these complex systems effectively. The structural analogies between reaction kinetics and classical mechanics offer powerful tools for the purpose, which shed light on the paradigms that govern each discipline. Recognition of these analogies can lead to enhanced predictive ability regarding the behavior of chemical systems, especially for cases that exhibit untypical dynamical nature (Agliari et al., 2018). Therefore, studies on pathways in reactions do not only enhance our knowledge about complex reactions but also offer windows to advances in chemistry and biotechnology.

4.2.2. Transition States

Transition states are at the center of the study of complex reactions, as the most important intermediates that dictate the directions and rates of chemical conversion. They are the high-energy structures that reactants need to evolve into products, and they embody a compromise between product and reactant stability. Their identification and characterization are the central issues of kinetic study since they decide not just the activation energy that is required for the reactions but also how likely different routes of reaction are. With this information, one can make models that will predict the behavior and rates of reaction, an issue of importance in fields such as catalysis and biochemistry where the efficiency of a reaction can be crucial to many applications. Transition states have consequences that transcend classical chemistry as they may be analyzed from the perspective of advanced theoretical schemes. For instance, placing transition state theory in conjunction with relativistic quantum mechanics rules has shed light contradicting reaction rates according to classical perception. The examination of kinetic energy fluctuation for traveling observers places emphasis on the nuances of chemical kinetics within the relativistic context, ultimately leading to the derivation of laws that will accommodate various reference systems. These advances place emphasis on transition states as critical for accurate prediction of reaction outcomes, particularly in regimes in which relativistic effects cannot be overlooked, emphasizing their significance in both theory and practice (Baig et al., 2017). Moreover, recent computational approaches have enabled deeper understanding and exploration of transition states in complex reaction manifolds. Tools like KiNetX are characteristic of automated kinetic behavior analysis techniques, effectively overcoming the complexities inherent in multivariate chemical landscapes. By focusing on the kinetic importance of species and systematic identification of those that play a role in reaction dynamics, KiNetX enables more thorough studies into reaction mechanisms. It is this integrative approach that allows chemists to effectively survey large networks of chemicals, with resultant insight into dominant reaction routes and product yields. The capacity to transfer uncertainty in free energy calculations similarly emphasizes the potential of modern computational tools to further constrain predictions of transition states and, by proxy, the nature of complex reactions (Proppe et al., 2018) (Figure 4.5).

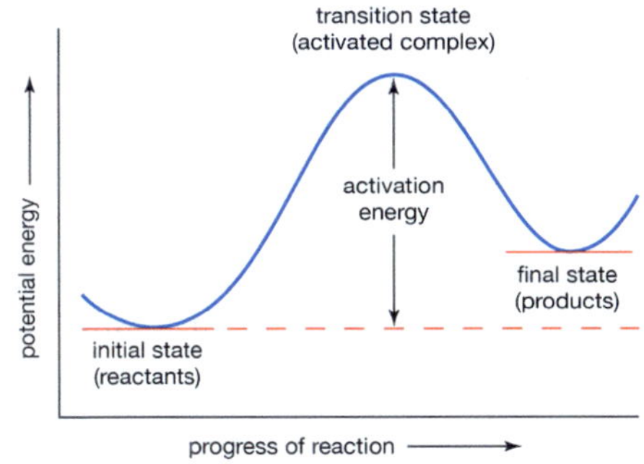

Figure 4.5. *Potential-energy curve. The activation energy represents the minimum amount of energy required to transform reactants into products in a chemical reaction. The value of the activation energy is equivalent to the difference in potential energy between particles in an intermediate configuration (known as the transition state or activated complex) and particles of reactants in their initial state. The activation energy thus can be visualized as a barrier that must be overcome by reactants before products can be formed.*

Source: https://cdn.britannica.com/28/4728-050-C4ABAC20/activation-energy-equation-Arrhenius-reactants-products-reaction.jpg?w=400&h=300&c=crop.

4.2.3. Energy Profiles

Understanding energy profiles is relevant for the explanation of complex reaction mechanisms. An energy profile is a diagrammatic representation of the energy change that a system undergoes as it transforms reactants to products, with prominent features such as activation energy and intermediate steps being highlighted. The shape of the energy profile is influenced by a variety of factors, including the nature of the reactants involved as well as reaction conditions (e.g., pressure, temperature). Energy profiles are thus the most informative tools that chemists can have, as they can help guide reaction feasibility and rate. Through the analysis of the transition states and energy barriers of the reactions, researchers can better understand how reaction conditions can be managed to achieve the highest yield and selectivity in natural as well as engineered systems (Bar-Even et al., 2016). As reactions go on, they may involve short-lived species known as intermediates that may be observed in the energy profile. Intermediates are generally higher-energy states which have to be traversed in order for there to be a transformation from reactants to products. Their stability and presence may have a significant influence on the reaction pathway as a whole. In addition, thermodynamic driving forces that lead to assembly and disassembly of such intermediates can control reaction completeness and directionality. Clarifying these correlations helps us learn more about the control of enzymatic activity and metabolic fluxes in living systems. This kind of information is paramount, particularly when one works with bioengineering, where precision in metabolic pathways can lead to greater production of the target compound (Bar-Even et al., 2016). In complex reactions, energy profile control through manipulation of environmental conditions or the availability of a catalyst can lead to extreme changes in the kinetics of the reaction. The interaction of temperature and rate coefficients is the best illustration of this phenomenon, demonstrating how varying energy profiles can affect the rate of reaction. An increase in temperature will for instance lower the activation energy required for a reaction to occur, thus encouraging faster kinetics. Recent molecular interaction studies have shed fresh light on the nature of how temperature variations influence relaxation processes in molecular vibrational states, showing the central role played by energy transfer mechanisms in reaction dynamics. These findings underscore the intricately interdependent relationship between reaction pathways and energy profiles and provide critical insight into the further understanding of multistep reactions as well as the optimum design of more efficient chemical processes.

4.3. FACTORS INFLUENCING COMPLEX REACTIONS

Systematic responses are primarily governed by a set of intrinsic and extrinsic factors that decide their kinetics and thermodynamics. The concentration of reactants is a critical factor; elevated concentration has a tendency to accelerate the rate of reaction due to higher frequency of collision between particles of the reactants. Besides, the presence of catalysts can really change reaction pathways so that they can proceed via lower-energy state intermediates and thus enhance reaction rates without being consumed in the process. Temperature is another essential factor that affects kinetic energy in molecules, usually resulting in an increased rate of reaction as it enhances particle movement, leading to more efficient collisions. As a global perspective, knowing these parameters is critical for being able to forecast the behavior of complicated reactions under other circumstances, which is at the basis of areas of study like materials science and environmental chemistry. Besides the above factors, the state of reactants as solid, liquid, or gas can significantly influence the products of complex reactions. For instance, gaseous reactant reactions such as those in laser-induced breakdown spectroscopy (LIBS) tend to exhibit a typical plasma chemistry wherein the interactions between ion species and atmosphere gases produce multiple various oxides that yield valuable information on the initial samples. For LIBS specifically, the dynamic nature of the laser-induced plasma allows for extensive reactions among fragments resulting from organic compound dissociations. The resultant products ultimately come to play a significant role in characterizing the original compounds, illustrating how physical state variations can lead to complex reaction schemes that may complicate the spectroscopy but enhance the analytical capability of the method (Laserna-Vázquez et al., 2016). Furthermore, the interaction of metal ions with humic substances in heat reactions under oxidative conditions is a classic case of another prevailing factor in complex reactions. This interaction influences the structural stability and reactivity of the compounds being involved. For instance, the incorporation of metal ions into the structure of humic acids, studied by recent studies, creates new binding sites that alter thermal stability and reactivity. Thermogravimetric studies disclose that thermal evolution of the transformed compounds can significantly differ, manifesting the sensitive balance between chemical reactivity and structural self-assembly in organic matter (Nováčková et al., 2009). These findings indicate that various variables—ranging from chemical makeup to environmental conditions—play a key part in determining the trajectories and outcomes of complex reactions, proving the need for a qualitative appreciation in both experimental and theoretical domains.

4.3.1. Temperature Effects

The effect of temperature on complex reactions is important because it directly affects the kinetic activity of reactants and product formation. Raising the temperature increases the rate of molecular movement, thus adding to the frequency of collisions between reactants. This increase in activity not only accelerates reaction rates but can alter the very character of such processes. For instance, in thermochemical models employed to simulate atmospheres of celestial objects such as Brown Dwarfs and gas giants, changes in temperature can result in significant changes in molecular abundances. High temperatures would result in more formation of hydrocarbons and other carbon-bearing molecules, as observed in simulations that examine the carbon-to-oxygen ratio in different atmospheric conditions (Bilger et al., 2013). As temperature governs the dynamics of chemical species, the understanding of such effects is crucial to forecast behavior on Earth and beyond. Temperature is a critical variable in the area of catalytic reactions as well, governing reaction mechanisms and selectivity.

For example, for biphasic hydroformylation of 1-octene, temperature variation was observed to alter the activity of the rhodium sulfoxantphos catalyst. For elevated temperatures, the linear aldehyde selectivity remains to be favored, critical to the industry where output of the product needs to be exact. However, as temperature rises, the partial pressure dependence of carbon monoxide and hydrogen exhibits non-linear behavior, demonstrating more complicated underlying reaction mechanisms (Delmas et al., 2011). This subtlety demands ideal adjustment of reaction conditions with consideration for how temperature influences not just the rate, but selectivity and feasibility of desired products.

Also, the correlation between temperature and reaction dynamics is of utmost significance from the perspective of thermodynamics. An increase in temperature can cause shifting in the equilibrium composition of a complex reaction, potentially skewing the formation of certain products against others. This is particularly important in systems where there are competing reactions that can occur. In simulations used in complex reactions, e.g., ones investigating thermochemical processes in the atmospheres of celestial bodies, variations in temperature can lead to large variations in chemical relaxation timescales and stability of various molecular forms (Bilger et al., 2013). As researchers uncover these mechanisms, it is evident that temperature is a principal control parameter, ultimately dictating direction and efficiency of complex reactions. Knowledge about temperature effects therefore gives improved control over chemical processes for different applications, from astrobiological studies to industrial catalysis.

4.3.2. Concentration of Reactants

The concentration of reactants is a key element in the mechanism of complex reactions, particularly in the measure of reaction rates and position of equilibrium. Higher reactant concentrations usually bring about higher reaction rates due to higher opportunities for molecular collisions. This follows from the collision theory, which pronounces that on colliding with suitable energies and orientations, reactant molecules result in reactions. Additionally, the concentration can have a profound impact on the nature of products formed because variations can shift the position of the equilibrium in accordance with Le Chatelier's principle. In complex reactions involving multiple intermediates, as has been illustrated for the case of electrochemical reactions, such modifications can have permanent influences on resulting reaction products and kinetics. Besides, concentration of the reactant is vital in microelectrode electrochemical systems, as highlighted by recent studies. These systems allow precise current and potential measurements even at low supporting electrolyte concentrations to prevent competitive complexation or contamination. For instance, when studying complexation equilibria at low concentrations, it is vital to investigate how limiting currents can be affected by the concentration of electroinactive ionic species. Since these effects can alter migration effects and general behavior in voltametric studies, concentration changes must be controlled and observed carefully. This form of analysis delivers extremely valuable insight into the nature of complex reactions, particularly at electrode process interfaces where the elaboration of concentration effects is placed firmly in the spotlight (Bos et al., 1997).

The interaction between reaction kinetics and concentration becomes even stronger when considering complicated reaction mechanisms. When there are competitive processes or more than one reactant involved, the variance in their diffusion coefficients can produce considerable deviations from predicted behaviors purely based on simple models. For instance, if reactants' and products' diffusion coefficients are much different from each other, classical ideas of reaction mechanisms would not be applicable, and theoretical models more sophisticated will be needed to make reliable predictions. The simulation techniques revealed through the literature allow researchers to explore such complexities through modeling of concentration profiles developing during reaction medium, allowing a clear concept of their influences on both steady-state currents and overall reaction kinetics (Bos et al., 1995). The unmasking of such relationships continues to enhance our understanding of complex chemical behavior in various experimental conditions.

4.3.3. Presence of Catalysts

In the study of complex reactions, catalysts play a significant role in modifying the rate and yield of chemical transformations. Catalysts, as chemicals that speed up reactions but are not used up, make processes possible, which otherwise would proceed at rates too slow to be of practical importance. More importantly, catalysts regulate the direction of reaction through the formation of alternative mechanisms with lower activation energy. This idea is well illustrated in organocatalyst pathways that are able to catalyze a range of reactions such as the Diels-Alder reaction, and how these catalysts can achieve stereocontrol and selectivity. For instance, use of chiral α,β-unsaturated acyl ammonium salts as enantioselective Diels-Alder-lactonization organocascades illustrates how catalysts can direct complex reactions to yield desired stereochemical results, ultimately enabling valuable compounds to be produced under mild conditions (Abbasov et al., 2017).

Moreover, the intricate interaction between substrates and catalysts typically controls the energetics of a reaction and thereby the observed selectivity and result. For example, computer simulations in combination with experimental examination have illustrated how the transition state energetic barriers can vary with the presence of catalytic entities. Such types of insights prove to be most valuable in the investigation of stabilization effects that are contributed by catalysts; they can selectively stabilize specific transition states through specific interactions, such as hydrogen bonding. This aspect of catalysis not only accelerates reactions but also entails a high degree of control over product generation. This was typically observed in entropy-controlled diastereoselective reactions within Diels-Alder cycloadditions, where a catalyst dictates the formation of certain diastereomers in a predictable manner (Abbasov et al., 2017). The use of catalysts goes beyond conventional chemistry to incorporate biologically meaningful models, such as protocells, which focus on the complex dynamics of molecular interactions within the origin of life. Protocells are thought to have facilitated self-organization processes that were fundamental for the evolution of intricate biochemical networks.

By the mimicry of catalytic networks within lipid vesicles, researchers have begun to unravel how catalytic reactions could have operated within prebiotic environments. The semi-permeable nature of protocell membranes, for instance, allows selective uptake and containment of reactants, thereby regulating reaction dynamics in a way that optimizes the efficiency of catalysis. This perspective not only deepens the understanding of how catalytic processes could lead to life's origin but also opens up avenues for exploring how similar principles apply in modern complex reaction systems.

4.4. APPLICATIONS OF COMPLEX REACTIONS

In chemical contexts, complex interactions are particularly relevant in the context of azanone (HNO) reactions with metalloporphyrins. The reactivity of HNO towards biological molecules such as heme proteins and thiols is a reflection of its significance to both chemical and biological milieus. Transition metal porphyrins allow researchers to examine these types of interactions efficiently. Notably, studies have shown that porphyrins can serve as adequate models for interpreting the coordination chemistry of small molecules, including HNO. Kinetic and mechanistic insights gained from the studies offer a clearer view of the role of azanone in biological systems and enhancing the design of novel sensors employed in the detection of HNO, thus promoting the art of bioanalytical chemistry and drug discovery (Bikiel et al., 2014). Another impressive application of complex reactions is the use of geotechnical engineering technology, that of lime stabilization. This is the reaction of clay soils with lime which causes far-reaching physico-chemical transformations of the soil to enhance its properties.

Recent developments in recycling soil technology have made it enormously more efficient to blend fill materials with binders than in the past, making it possible to achieve more effective stabilization than the earlier practice. However, while technology allows for such applications, lime stabilization specifications lag behind innovations at times. This lack of compatibility causes challenges in optimizing application methods, particularly in the prevention of issues like excessive air voids and potential sulfate swelling that would compromise long-term stability and strength. This all the more underscores the need for tailored measures in project specification to meet varied soil conditions (Beetham et al., 2014). Furthermore, the use of complex reactions in various industrial processes is a sign of their general use in numerous industries. For instance, controlling complex reaction pathways makes it possible for chemists to create more effective synthesis pathways for drugs, leading to enhanced bioavailability and a reduction in side effects. Similarly, the design of catalysts capable of performing complex reactions can significantly impact energy usage in chemical production, allowing for more sustainable processes. With increasing environmental concerns in industries, the role of intricate reactions takes center stage while formulating processes minimizing waste and energy consumption. By extension, ongoing research into such reactions not only encourages technological innovation but also ensures that economic growth is concurrent with environmental stewardship, thereby providing a multi-dimensional avenue towards scientific investigation and innovation.

4.4.1. Industrial Processes

Industrial processes form the core of the transformation of raw materials into finished goods and bear the responsibility in many areas such as pharmaceuticals, manufacturing, and farming. Industrial processes typically involve complex reactions that require close control of temperature, pressure, and chemical composition to ensure maximum product yield and efficacy. For instance, aromatic compound nitration reactions are central in the production of important chemicals used in pharmaceuticals, pesticides, and defense. These reactions typically produce toxic waste, which is an illustration of the need for effective waste management practices to avoid negative impacts on the environment.

Figure 4.6. *Industrial processes.*

Source: https://www.mcgradyclarke.com/wp-content/uploads/2023/08/Industrial-Processes-1.png.

Techniques such as photocatalytic degradation assays have been evolved, revealing the presence of pollutants like RDX and HMX in wastewater, which emphasizes developing sustainable industrial methods that cause lesser environmental harm and maintain strict production (Hafidi et al., 2009). Chemical engineering involves reactor design and optimization, which is of utmost importance when handling the complex interaction of reaction and material flows. Newer reactors also feature rotating elements, which enable homogeneous mixing and reaction rate improvement. Simulation of such intricate geometrical arrangements poses problems in simulating their behavior, which leads to the need for sophisticated numerical techniques capable of estimating the material behavior under different conditions. One such technique involves dividing the computational domain into rotating and fixed parts so that fluid dynamics and reaction kinetics can be better modeled in industrial applications. With the application of a non-conformal mesh matching technique, scientists can study complex flow patterns in reactors, thereby enhancing performance and efficiency within various industrial processes (Baudry et al., 2017). As industries are more environmentally friendly, implementation of new technologies within industrial processes is even more important. Enhanced analysis techniques, e.g., solid-phase extraction coupled with HPLC, provide helpful tools to identify and quantify pollutants within effluent streams. This is especially necessary in hazardous chemicals handling businesses where environmental protection and regulatory compliance are non-negotiable. The ability to effectively treat harmful substances with high-tech treatments such as UV/TiO_2 degradation not only supports regulatory compliance but also enhances more environmentally friendly production loops. In addition, by increasingly improving understanding of complex reactions occurring in treatment systems and reactors, industries may set up mechanisms not only to improve product outputs but also to minimize the footprint of their operations on the environment, aligned with global efforts towards the adoption of sustainable manufacturing practices (Figure 4.6).

4.4.2. Biological Systems

Biological systems present a complicated pattern of interactions that may be considered intricate reactions with biochemical processes occupying the center stage powered by molecular interactions. Such systems are faced with an ever-present force from environmental factors, each of which adds variability and challenge to homeostasis. Variability in this direction tends to obscure complex dynamics, which in turn may conceal underlying regulatory mechanisms. One of the finest illustrations of the intricacy is the one in the interaction of ultraviolet radiation and DNA. The photochemical changes induced by the radiation directly addressed the necessity for an explanation of energetic interactions at the molecular level. As the authors Hadden et al. (2013) themselves described, knowing these interactions matters, particularly considering that they involve the chemical alteration of biological molecules. Also, reaction complexity in biomolecules demands a multi-level analysis that breaks down such systems into units of study from which they are reconstructed to reveal their interactions in their entirety. Between rising admiration of reaction complexity in biology is the need to assess the impact of noise on regulatory networks, which are the center of cellular operations. Biological networks are not only regulated by deterministic networks but also by stochastic fluctuations that have the capability of influencing cellular functions. Adding noise to such networks may lead to unscheduled oscillations, which can impair function. Against such noise, biological networks possess sophisticated feedback mechanisms, such as negative regulatory networks. As noted by Hinczewski et al. (2016), the biological networks are susceptible to analysis in terms of engineering principles such that researchers can use mathematical formalism to understand and measure noise control. Through theoretical models in application to biological systems, scientists can determine the inherent principles in how these complex networks remain stable even with fluctuations—a nature of biological resilience. Explaining our understanding of biological systems as complicated reactions not only offers insight into the processes of life at their essence but also unearths novel lines for biotechnological progress and treatment. Integrating experimental methods with theoretical considerations offers a new approach to studying such systems. Specifically, examining the ultrafast dynamics and photochemistry of smaller biomolecular species can illuminate the big picture of molecular interaction. While there are obstacles, such as the movement of large and delicate biomolecules into controlled environments for investigation, new advances provide optimistic responses to these questions. As researchers explore these complexities, they continue to unravel the details of biological reactions, expanding our understanding of life itself. From such questioning, it is evident that the richness of the interactions found in biological systems is fundamental not only to sustaining life but also to informing technology advances for controlling these systems toward health and environmental purposes (Figure 4.7).

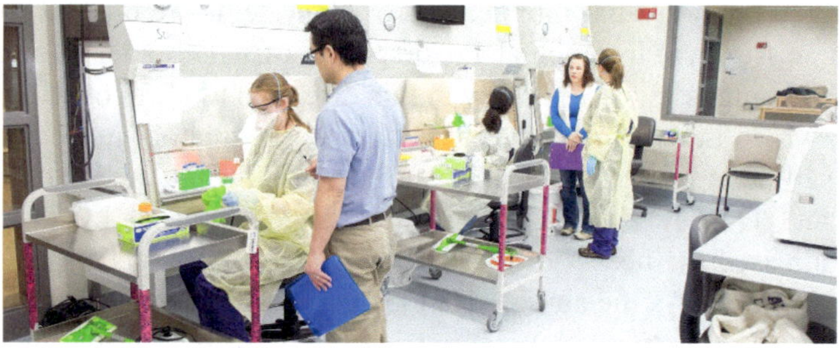

Figure 4.7. *Biological systems.*

Source: https://www.k-state.edu/academics/images/guides/2021/engineering/biological-systems-engineering.jpg.

4.4.3. Environmental Chemistry

The intricate topic of environmental chemistry entails the exploration of chemical processes and interactions in the environment, particularly those bearing human influence. Central to it is the nitrogen cycle, the conversion of nitrogenous compounds through a variety of biological and abiotic processes. Understanding such conversions is crucial since nitrogen compounds are determinative of both ecological balance and human health. For instance, denitrification, the process of converting nitrate to diatomic nitrogen (N_2), is essential for the limitation of reactive nitrogen species accumulation and consequently prevention of ecosystem eutrophication. New evidence, like the observation that denitrification processes are more complex than initially thought, highlights the need for a new comprehension of nitrogen cycling (Aller et al., 2005). This evolving knowledge of nitrogen dynamics underscores the profound intricacies beneath environmental chemistry, underlining the necessity for ongoing study to unravel such processes. In responding to complex reactions in environmental chemistry, it is necessary to consider the role of microbial communities and how they influence biogeochemical cycles. Microorganisms are centrally important in facilitating varied chemical conversions through metabolic processes that tend to be overlooked in conventional ecological models. For example, the inclusion of microbial activity in nitrogen cycling has revealed that anaerobic ammonium oxidation and other processes are significant players in the removal of nitrogen.

Figure 4.8. *Environmental chemistry.*

Source: https://cdn1.byjus.com/wp-content/uploads/2018/12/environmental-chemistry.png.

Furthermore, these hidden dynamics demonstrate a multifaceted hierarchy that is responsible for affecting ecosystem resilience and stability. As emphasized by research highlighting the need for enhancing our understanding of hierarchy in ecosystems, taking microbial interactions into consideration offers valuable information on ecological behavior, such as regime shifts and stability to environmental disturbances (Hosoda et al., 2016). Thus, the role played by microorganisms is not cumulative; they are at the heart of the rich tapestry of reactions which define environmental chemistry. Multidisciplinary understandings of the interaction between chemical processes and environmental systems underscore the necessity for a comprehensive approach in addressing contemporary environmental challenges. Through the synthesis of novel molecular and ecological data, researchers are in a better position to develop more dynamic models that capture the intricate interactions within ecosystems. Such a comprehensive approach lends itself to a deeper understanding of the intricate interactions among various chemical components and how such interactions govern ecosystem stability and health. Indeed, as depicted in advancements in the modeling approach, e.g., population–reaction model, addition of hierarchical dynamics allows for a more genuine imitation of ecological interactions and relationships (Hosoda et al., 2016). Identifying the character of environmental systems as complex, certainly, new research methods are needed to develop effective management plans towards ecological well-being and sustainability under chronic environmental change (Figure 4.8).

4.5. CASE STUDIES OF COMPLEX REACTIONS

The study of complex reactions encompasses a broad array of chemical processes, including ligand substitution reactions. These are basic in understanding how different compounds react within complex systems, particularly coordination chemistry. Recent developments in the field of rapid-reaction techniques have revolutionized the investigation of these reactions so that the kinetics thereof can be investigated in much greater detail. Although systematic procedures have been in large part established for cobalt(III) and platinum(II) complexes with monodentate ligands, the results of such experiments form the fundamental models used for extended application in the field. Investigation of such complex reactions not only goes to the heart of accounting for their convoluted mechanisms but also clarifies the key role molecular-orbital theory plays in relating rate data and structural data and providing access to a more complete representation of the process occurring in various chemical environments (Gray et al., 1966). Outside of ligand substitution reactions, metabolic networks are another area within the scope of complex reactions. Such biological networks interact in complex and interdependent manners necessary for cellular survival and activity regulation. The topological structure of such networks has been well appreciated; however, understanding the fundamental rules governing these dynamic interactions remains problematic. Importantly, the significance of cyclic forms lies at the heart of biological network resilience, enabling them to effectively adapt to environmental changes. There has been recent research suggesting ways of calculating fluxes for cyclic and acyclic systems, demonstrating how the fluxes are involved in performance and stability in metabolic activity. Such information is useful in elucidating the complex regulatory aspects governing biological processes and enhancing our understanding of biology, with research on various organisms such as *Escherichia coli* and *Methylobacterium extorquens* (Schwartz et al., 2009). The growing field of complex reactions highlights the requirement for an interdisciplinary approach that unites theoretical models and experimental findings. While scientists explore the foundations of complex reactions, they must consider the shifting paradigms of coordination chemistry and biochemistry. The interconnection between ligand substitution reactions and metabolism illustrates the broader implications of complex reactions across many areas of science. By combining the understanding of molecular-orbital theory and metabolic network analysis, scientists can construct better frameworks for examining complex interactions. Furthermore, establishing standard procedures allows one to test new reactions in a systematic manner, making generalization and predictions easier. At last, these case studies highlight the need for ongoing investigation into intricate responses, since they are both crucial to synthetic chemistry as well as biological function, setting the limits of our present understanding (Gray et al., 1966).

4.5.1. Haber Process for Ammonia Synthesis

The Haber process, a milestone reaction during the early 20th century, is an industrial method to synthesize ammonia from nitrogen and hydrogen gases. It is a challenging reaction to illustrate the marriage of thermodynamics, kinetics, and catalysis necessary for high yields and economic viability. Under high temperatures (around 450–500°C) and pressures (150–300 atm), atmospheric nitrogen reacts with hydrogen, typically derived from natural gas. The process employs an iron catalyst to accelerate the reaction so that it is brought to equilibrium effectively. Explanation of this process has not only transformed agriculture through more intensive production of nitrogenous fertilizers but has also placed in focus the need to develop new techniques to control nitrogen fixation and avoid environmental dangers due to production and use of ammonia. The intricacies of the Haber process are not only caused by operating conditions but also by the role of the catalyst and its profound effect on reaction kinetics. The iron-based catalysts have been maximized by surface area and access enhancement to active sites, which allow higher rates of reaction. Modern research into new methods of catalysis, such as the application of ruthenium materials, indicates promising lines towards greater efficiency and reduced energy consumption. Technological advancement notwithstanding, the process consists of complex interactions among chemically active substances, where electron configurations and bond energies dictate the feasibility of ammonia synthesis (Guo et al., 1999). Therefore, the search for alternative catalysts and reaction routes continues to be essential in the development of more sustainable processes for ammonia synthesis, reflecting the evolving nature of the Haber process in response to environmental pressures and economic considerations. As demand for ammonia increases worldwide, fueled mainly by agriculture, the Haber process comes under scrutiny for sustainability and environmental consideration. The traditional fossil-fuel usage to generate hydrogen is an environmental problem of significant proportions with the resultant carbon footprint. Researchers are now focusing on even cleaner means of generating ammonia, for instance, electrochemical nitrogen reduction—a new area of research seeking to harness renewable energy resources. Furthermore, inclusion of biotechnology-based approaches, i.e., microbial processes, is presenting the prospect of reducing energy signatures and non-renewable energy dependence. Biocatalysis research efforts point to the fact that advances in these areas will be capable of contributing significantly towards minimized energy consumption and facility costs (N/A, 2025). Thus, the future of the Haber process appears to be along the lines of carrying on with its fundamental role in ammonia synthesis along with the development into a greener mode of operation in line with global environmental objectives (Figure 4.9).

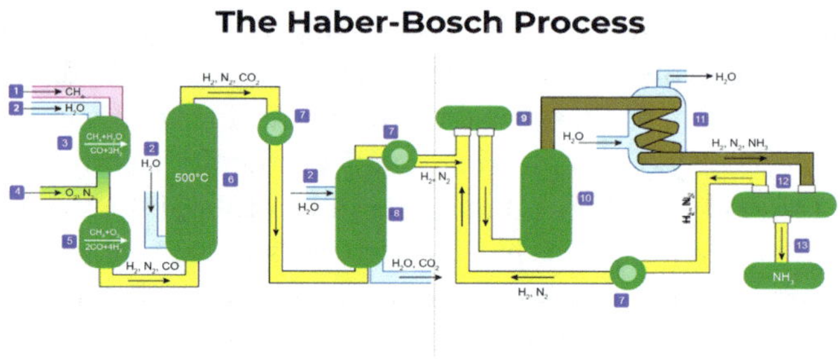

Figure 4.9. *Haber process.*

Source: https://media.geeksforgeeks.org/wp-content/uploads/20231213110847/Habers-Process.png.

4.5.2. Photosynthesis in Plants

Plants' photosynthesis is an advanced biochemical process where light energy gets trapped and stored in chemical form as mainly glucose. Photosynthesis takes place in plant cells in the form of chloroplasts, where chlorophyll captures sunlight, initiating a chain of reactions. Plants synthesize organic compounds for growth through two broad stages—light-dependent reactions and the Calvin cycle. The light reactions utilize the solar energy to produce NADPH and ATP, which are necessary for the subsequent process. Calvin cycle uses the energy carriers to fix carbon dioxide in the air to produce sugars ultimately. Photosynthesis plays an immense role in the growth of plants, productivity in an ecosystem, and the overall carbon cycle, thus demonstrating its importance from an ecological perspective. In particular, C_4 photosynthesis is a specialized accommodation that enhances carbon dioxide absorption efficiency, particularly in hot and high-intensity light environments. C_4 plants, such as maize, have acquired a unique anatomical organization which allows them to store carbon dioxide in special cells associated with their vascular tissues. This adaptation minimizes photorespiration, an energy-squandering process that may occur when oxygen competes with carbon dioxide as the target of fixation by the enzyme Rubisco. Recent advances in computer models depict that C_4 plants exhibit nonlinear relationships between carbon dioxide and oxygen concentration, which affect metabolic rates and responsiveness to environmental stress.

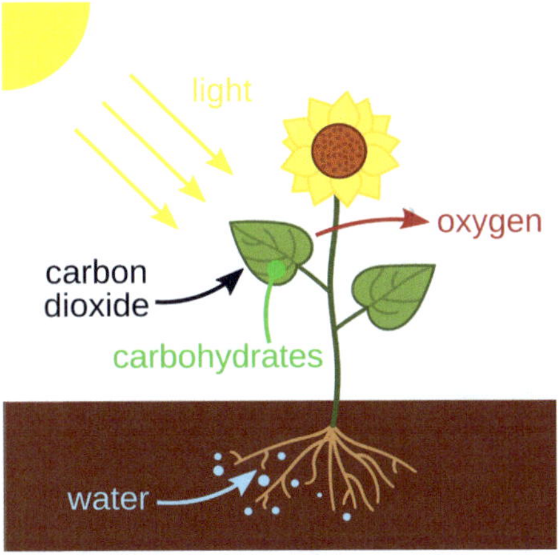

Figure 4.10. *Schematic of photosynthesis in plants. The carbohydrates produced are stored in or used by the plant.*

Source: https://upload.wikimedia.org/wikipedia/commons/5/55/Photosynthesis_en.svg.

Through examination of these nonlinear dynamics with advanced modeling techniques, researchers have obtained insight into the complex interactions governing C_4 plant metabolic states, detailing the complexity of their physiological adaptation (Bogart et al., 2015). Beyond plant biology, photosynthesis has significant relevance to sustainable energy technologies. As the world grapples with the problem of dependency on fossil fuels, there is an increasing need for the latest techniques of harnessing the power of the sun through the mechanisms of photosynthesis. Nanoscience and biotechnology are working in earnest to develop instruments mimicking nature to achieve greater efficiency in converting sunlight into usable forms of energy such as electricity and biofuels. These advances are meant to create efficient systems that emulate the competency of biological photosynthesis to address the critical issue of renewable energy resources. Ultimately, harnessing the power of photosynthesis is similar to attaining sustainable energy technologies in favor of worldwide initiatives to abate the impacts of climate change and promote ecological balance (George et al., 2007) (Figure 4.10).

4.5.3. Combustion Reactions in Engines

Understanding combustion reactions in an engine is relevant not just for performance improvement but also for minimizing harmful emissions. Combustion is a complex reaction and can be classified into two broad categories: complete combustion, occurring when there's sufficient oxygen, and incomplete combustion, due to oxygen deficiency. Complete combustion produces water and carbon dioxide with efficient release of energy, whereas incomplete combustion gives rise to the formation of unburnt hydrocarbons and carbon monoxide, which are harmful emissions.

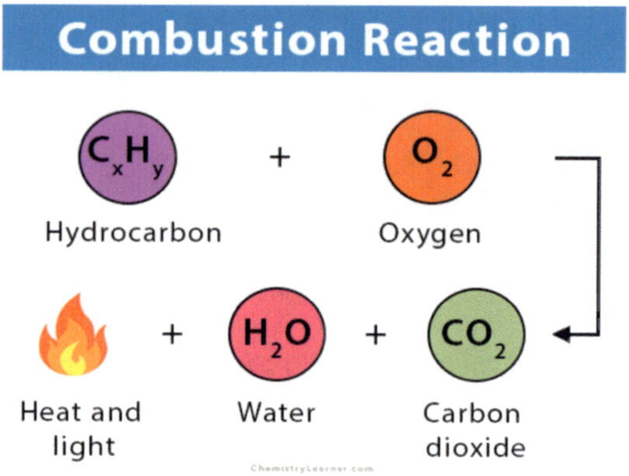

Figure 4.11. *Combustion reaction.*

Source: https://www.chemistrylearner.com/wp-content/uploads/2020/08/Combustion-Reaction.jpg.

Such reactions depend on the intricacy of the nature of fuel used, engine design, and operating conditions, necessitating further investigation of their mechanisms to enhance environmental sustainability. In recent years, alternative fuels such as biodiesel have gained acceptance because they are thought to produce fewer pollutants when burned compared to conventional diesel. Experiments have shown that the combustion quality of biodiesel relies significantly on its feedstocks, and its corresponding nitrogen oxides (NOx) and particulate matter (PM) emissions are varied under various combustion conditions. A study sheds light on oxygen content, fatty acid chain length, and the level of unsaturation in biodiesel being the determining factors to ascertain emission profiles (Kim et al., 2016). Through investigations of such correlations, engineers can design methods with which to utilize biodiesel more effectively, trading off performance against environmental impact. Such a focus on biodiesel introduces variability to the chemical composition of combustion reactions and combustion conditions, hence to combustion reactions themselves. Moreover, recent evidence from studies on butanol as a potential future biofuel has disclosed the complexities of butanol combustion in engine applications. Recent studies have developed skeletal mechanisms for butanol isomers, which allow for improved comprehension of butanol combustion under varying conditions, such as pressure and temperature (Brady et al., 2017). The findings indicate that butanol isomers have varying autoignition delays and fuel consumption patterns from those of typical gasoline surrogates. The elaborate comprehension reflects not only the complexity that comes with the prediction of combustion phenomena, but also the viability of incorporating cleaner alternative fuels within engine design. With increased research, the interface between combustion chemistry and engine efficiency will inevitably lead to evolutionary strategies for achieving improved emissions and performance with fuel (Figure 4.11).

4.6. CHALLENGES IN STUDYING COMPLEX REACTIONS

The study of complex reactions encompasses a multifold of challenges caused by the complex interaction between a multifold of chemical species and reaction channels. A major bottleneck arises from the fact that it becomes difficult to characterize the kinetic parameters governing such reactions. In media such as planetary atmospheres, where conditions can be very variable, available data on chemical kinetics are typically incomplete. For instance, while the chemistry of H, O, N, and Cl species is quite well documented in Earth atmospheres, sulfur species behavior is poorly defined. Moreover, the complexity of the higher hydrocarbons in Jovian planetary atmospheres highlights the shortage of existing models as researchers cannot correlate gas-phase reactions with aerosol and secondary product formation, hence complicating our understanding of these systems that are interdependent (Yung et al., 1990). The second challenge arises from the inconsistency of experimental approaches employed in studying complex reactions. Within education, in situations where inquiry teaching approaches are promoted, an understanding of the effectiveness of various teaching methods is vital in enabling students' learning motivation in science. Research has indicated significant correlations between students' intrinsic motivation and perceptions of laboratory experiences and of science-technology-society integration (Azevedo et al., 2011). But the teacher-facilitator image does not share a similar correlation, and so it can be argued that the learners may not have a full picture of the intricacies of scientific investigation taught in traditional forms. This reflects a vital disconnect although complicated reactions call for sophisticated pedagogical approaches, whether they deliver is left underrated, and therefore presents more challenges in disseminating information on complicated chemical process. Combination of theoretical models and experimental data is also an important challenge in research of complex reactions. While researchers attempt to build rigorous kinetic models that will reflect the performance of different chemical systems with precision, discrepancies between predicted results and experimental observations typically occur. These discrepancies could result from simplifications of models or uncontrolled variables in experimental protocols. Furthermore, chemical thought and development, particularly in light of new evidence relating to heterogeneous reactions, complicate matters even further as scientists grapple with how to synthesize gas-phase chemistry with solid and liquid phase interactions formation (Yung et al., 1990). Resolution of these issues involves not only greater methodological sophistication but also cooperation among disciplines of science, thus inspiring a more integrated appreciation of complex reactions in the overall context of chemical dynamics.

4.6.1. Difficulty in Measurement and Observation

The intricate nature of complex reactions often causes large-scale problems in measurement and observation. One of the primary difficulties arises because of dynamic interactions that occur simultaneously among such systems, and individual components become hard to isolate for study. For instance, in biochemical reactions, the existence of various substrates and products may obscure the specific pathways being under study. This complexity requires refined analytical methods which are able to provide accurate real-time output, but some of the more traditional methods are not sensitive enough to detect the nuance of such interactions. Therefore, scientists are usually left with no choice but to employ high-resolution methods that in some situations might be less than optimal and therefore leave it more challenging to achieve reliable measurements in the context of complex reactions. To further add complexity to the challenge, the time-evolving nature of complex reaction measurement complicates matters. The dynamic nature of such systems means reactions can evolve very rapidly, and so the measurements need to be fast and robust. In this regard, advanced modeling methods, such as the extended Kalman filter (EKF) algorithm, have been developed to support better analysis of the time-varying nature of complex biochemical reactions. By the application of a smaller number of observations, the EKF may sharpen the process of parameter estimation and enhance predictive capability of state-space models. The approach represents a significant achievement in the field of measurement technology but also illustrates the primary difficulties of monitoring transient behavior of complex reactions and provides a significant area of future research in the field (Du et al., 2011). The implication of measurement limitations extends beyond mere experimental complications; it influences underlying theoretical models driving scientific research into complex reactions. For example, in particle physics, observational complexity linked with neutrino detection highlights how measurement complications can hinder theory breakthroughs. Nuclear reactors, as significant neutrino research sources, demonstrate the complexities of obtaining correct measurements needed in approaching fundamental physics questions, such as neutrino mass hierarchy and the character of sterile neutrinos. This interplay of theory and measurement shows the way in which the observation difficulties can not only restrict the scope of the investigations but also the shifting scientific narrative of the complex reactions (Vogel et al., 2015). Measurement technologies may keep on advancing, yet there is a possibility to unveil such elusive interactions, the root problems remain the key area of concern for scientists trying to disentangle the complexity of the reaction dynamics.

4.6.2. Computational Modeling Limitations

The development of computational models in the understanding of complex reactions has uncovered gigantic discoveries; nevertheless, these models typically hit obstacles that can hinder their effectiveness. A major disappointment in computational modeling is the difficulty in understanding the multi-dimensional nature of biological processes. Biological systems naturally operate at various spatial and temporal scales, and therefore models must include all these different factors. As reported in the literature, complete understanding of the majority of biological processes is realized via multi-scale models that specifically model the dynamics in all involved scales. The synthesis of the scales complicates the modeling because errors in a scale may propagate to other scales, which leads to potential misinterpretation of the system-wide dynamics (Hasenauer et al., 2015). Thus, the complexity of these interactions accents the pressing need for formulation of robust modeling frameworks that can cater to and also forecast the complexity of reactions involved. Another significant limitation of computational modeling relates to the accessibility and quality of analysis tools required for effective model evaluation and data integration. During multi-scale reaction modeling, the ability to conduct model-based data integration is paramount in the sense that it makes it possible for scientists to juxtapose their findings with empirical realities. However, existing analysis tools are not well-placed to contend with the sheer volume of data generated by multi-scale models, and therefore inefficiencies in verification and enhancing such computation tools. Also, as highlighted in recent research, the interplay between spatial and temporal scales complicates the integration process, necessitating the development of novel modeling strategies that can efficiently bridge these features. These advances are needed to enhance the validity of computational models and ensure that they truly reflect the biological realities that they aim to model (Hasenauer et al., 2015). Lacking improved analytical techniques, the progress of computational modeling in the field of complex reactions might face significant setbacks. Lastly, the computational intensity of multi-scale modeling should not be overlooked. The more complex the models become in order to accommodate various scales, the computational requirements typically increase exponentially. This can increase the computation time, which can make it less practical to perform large-scale simulations or real-time analysis. The necessity of surrogate and low-order models has therefore developed as a feasible solution to the problem, enabling researchers to gain quicker estimates without sacrificing essential details of the biological processes being studied. Nevertheless, reliance upon these approximations may be of detriment in that they could simplify essential dynamics within the system into oversimplification. Therefore, while computational mitigation strategies are valuable, they must be weighed judiciously against the demands of accuracy and representativeness in modeling complex reactions. Finally, breaking these computational barriers will be the key to further advancing the field of computational modeling in biological systems.

4.6.3. Variability in Experimental Conditions

Variability in experimental conditions is the cornerstone of understanding complex reactions in scientific research. Each experiment is influenced by a humongous number of factors, ranging from temperature to pressure, concentration, and even catalysts' availability. Factors have the potential to influence reaction results and lead to different outcomes, which complicate data interpretation. For example, a rise in temperature can speed up a reaction, whereas a fall can slow it down and, therefore, introduce variability that scientists need to compensate for when analyzing. The way these conditions interact is central to establishing good models that can foretell behavior in intricate systems. Unless observed, these differences should be controlled, otherwise, researchers risk making false inferences that have the potential to mislead subsequent research on underlying chemical reaction and interaction laws. Additionally, inherent variation in experimental conditions will contribute to noise that obfuscates underlying signals researchers try to explain. When conducting experiments, researchers must document conditions carefully so that they can reproduce their outcome exactly. Variation need not only come from external environmental factors but also from the experimental design itself, say differences in instrumentation or human variation in sample preparation. This understanding works to support the necessity for employing robust methodological designs that can accommodate or circumvent variation. Advanced statistical methods can be applied to assess and interpret data set variation and enable scientists to learn useful information from seemingly chaotic results. By systematically accounting for variability, researchers are better able to understand complex reactions, eventually giving more precise and reliable results. Finally, acknowledgement of variability of experimental conditions transcends the simple matter of data collection; it also involves a broader philosophical interest in scientific practice. Variability in experimental systems reflects the complicated complexity of reality, which scientists try to model and understand. As highlighted in modern organizational research, appreciation for the impact that underlying principles such as attention and methodological discipline have on experimental outcomes within an organizational context has grown (Baer et al., 2015). It can enhance the robustness of scientific inquiry by encouraging researchers to embrace variability as part of their experimental design rather than something to be eliminated. This paradigm shift enables a more complex comprehension of complex reactions, the door to new theories and applications that are capable of developing from the subtle dance of variables that decide the outcomes of experiments (Glowacki et al., 2013).

SUMMARY

In brief, research into complex reactions not only amplifies their complicated nature but also the significance in a wide variety of scientific disciplines. Close inspection of such reactions enables progress in areas such as organic chemistry, where processes such as the Diels-Alder reaction are important. By delving into the dynamics over time and employing simulations to model such reactions, scientists are capable of penning complex stories that narrate process development through time. The reflections on forms of historical narrative, as considered in the simulation context, enable further insight into the intricacies of chemical reactions, contributing to a more sophisticated appreciation of actualized and potential effects in reactions (Wise et al., 2017). The implications of studying complex reactions extend beyond theoretical study; they yield practicalities that have impacts on technology and industry. For instance, research into physical and chemical changes in hard coal samples records the influence of low-intensity electrical fields on the organic matter in coal, demonstrating a new mechanism of destruction. Based on rigorous experimental methods and analysis, this information gives insight into upgrading coal conversion technology. The research identifies principal directions for the prevention of coal mining danger, thereby bringing together scientific research and practical use (Kamchatnyi et al., 2018). These inter-disciplinary connections make research in complex reactions timelier as a solution to problems of today. Ultimately, research in complex reactions is a confirmation of the complexity of the interactions that dictate chemical process. As scientists explore these processes further, the dance between applied science and theory will be the driving force in charting chemistry innovation as well as other cognate ones. The acknowledgment of both temporal change in response and historical narrative structure deepens our understanding so that there can be increased perception with synthesis of potentiality and reality. This collective pursuit for knowledge not just contributes to scientific advancement but also to deeper appreciation for the character of complexity (Wise et al., 2017). In the future, the study of intricate reactions will no doubt be a lodestar in the quest for more scientific truths and innovation.

Appreciation of the complex interactions of complex reactions requires an overall re-cap of main points identifying their multi-dimensional nature. A central aspect is appreciating the interplay between biological systems and geochemical environments, particularly in settings such as deep geological repositories for radioactive wastes. The role of microbial activities in such systems is influential, as they can affect both the stability and safety of radioactive waste storage techniques. This places emphasis on the imperative of well-educated integration of microbiology into safety assessment and performance testing, as stressed by Humphreys et al. (2010). A realization of these interactions not only improves our theoretical understanding; it also has advantages in practical applications in waste treatment, showing the efficacy of combined study methodologies in solving problematic reactions. Furthermore, the spatial and temporal scales at which biological processes are active are of central importance to comprehending problematic reactions. Accurate modeling of such processes requires multi-scale models that can respond to the diversity of properties at different observation scales. According to Hasenauer et al. (2015), integrative models must cross scale boundaries so that they can identify concealed biological processes. The challenges involved in constructing such integrative models cannot be overstated because they require advanced mathematical techniques and innovative computational approaches. Techniques of model-based data integration and hypothesis testing come into play here to enable synthesis of information across scales and verifying the robustness of performance predictions in intricate systems. Generally, the subtle nature of complex reactions consists of interactions of bio-, geochemical, and mathematical components. To be able to fully appreciate these interactions, account must be taken of the specific contexts in which reactions occur as well as the broader context for applications such as waste management and the environmental sciences. The integration of microbiology into safety cases, in addition to ongoing development in multi-scale modeling, represents a major research area that can help to further our knowledge of complex reactions. Through investigating these areas of significance, researchers and professionals can gain a better comprehension of the issues and possibilities of complex reactions and ultimately design better approaches to addressing their consequences in diverse settings.

REFERENCES

1. Abbasov, M. E., Hudson, B. M., Romo, D., Tantillo, D. J., et al. (2017). Stereodivergent, Diels–Alder-initiated organocascades employing α,β-unsaturated acylammonium salts: Scope, mechanism, and application. https://core.ac.uk/download/323070907.pdf (accessed on 10th September 2025).
2. Ablikim, U., Ahmed, M., Azyazov, V. N., Bashkirov, E., et al. (2018). Low-temperature formation of polycyclic aromatic hydrocarbons in Titan's atmosphere. https://core.ac.uk/download/294625779.pdf (accessed on 10th September 2025).
3. Agliari, E., Barra, A., Landolfi, G., Murciano, S., et al. (2018). Complex reaction kinetics in chemistry: A unified picture suggested by mechanics in physics. https://core.ac.uk/download/154948478.pdf (accessed on 10th September 2025).
4. Agúndez, M., Cuppen, H. M., Geppert, W., Herbst, E., et al. (2010). Reaction networks for interstellar chemical modelling: Improvements and challenges. http://arxiv.org/abs/1011.1184 (accessed on 10th September 2025).
5. Albéniz Jiménez, A. C., & García Loma, R. (2019). Vinylic addition polynorbornene in catalysis. https://core.ac.uk/download/228074028.pdf (accessed on 10th September 2025).
6. Alberts, B., Andreini, C., Balche, W. E., Barker, H. A., Benner, S. A., Berge, C., Braakman, R., et al. (2012). The compositional and evolutionary logic of metabolism. http://arxiv.org/abs/1207.5532 (accessed on 10th September 2025).
7. Aller, R. C., Canfield, D. E., Dalsgaard, T., Engström, P., et al. (2005). Nitrogen removal in marine environments: Recent findings and future research challenges. https://core.ac.uk/download/12043452.pdf (accessed on 10th September 2025).
8. Alvarado, J., Doeff, M. M., Kroll, T., Liu, Y., et al. (2020). Thermal stress-induced charge and structure heterogeneity in emerging cathode materials. https://core.ac.uk/download/286728976.pdf (accessed on 10th September 2025).
9. Andrianantoandro, E., Arkin, A., Ashkenasy, G., Bath, J., Benner, S. A., Soloveichik, D., Winfree, E., et al. (2009). DNA as a universal substrate for chemical kinetics. https://core.ac.uk/download/4884366.pdf (accessed on 10th September 2025).

10. Azevedo, M., Baptista, M., Carvalho, C., Conboy, J., et al. (2011). Student perceptions of secondary science teachers' practices following curricular change. https://core.ac.uk/download/12424895.pdf (accessed on 10th September 2025).

11. Baer, R. A., Bono, J. E., Brewer, J. A., Brown, K. W., et al. (2015). Contemplating mindfulness at work: An integrative review. https://core.ac.uk/download/80496021.pdf (accessed on 10th September 2025).

12. Baig, M. W. (2017). On the theory of time dilation in chemical kinetics. http://arxiv.org/abs/1203.2510 (accessed on 10th September 2025).

13. Banerjee, R., Barham, J., Beilstein, G., Cahard, D., Cossi, M., Doni, E., et al. (2017). Electron transfer and hydride transfer pathways in the Stoltz–Grubbs reducing system (KOtBu–Et_3SiH). https://core.ac.uk/download/110689724.pdf (accessed on 10th September 2025).

14. Bar-Even, A., Davidi, D., Flamholz, A., Liebermeister, W., et al. (2016). The protein cost of metabolic fluxes: Prediction from enzymatic rate laws and cost minimization. http://arxiv.org/abs/1604.00167 (accessed on 10th September 2025).

15. Baudry, C., Bennani, L., Fede, P., Laviéville, J., et al. (2017). Numerical simulation of unsteady dense granular flows with rotating geometries. https://core.ac.uk/download/83534150.pdf (accessed on 10th September 2025).

16. Beetham, P., Bell, F. G., Bergado, D. T., Bin, S., Bohn, H. L., BSI, et al. (2014). Lime stabilization for earthworks: A UK perspective. https://core.ac.uk/download/30640618.pdf (accessed on 10th September 2025).

17. Bikiel, D. E., Doctorovich, F., Marti, M. A., Pellegrino, J., et al. (2014). Reactions of HNO with metal porphyrins: Underscoring the biological relevance of HNO. https://core.ac.uk/download/159297833.pdf (accessed on 10th September 2025).

18. Bilger, C., Helling, C., & Rimmer, P. (2013). Small hydrocarbon molecules in cloud-forming brown dwarf and giant gas planet atmospheres. http://arxiv.org/abs/1307.2565 (accessed on 10th September 2025).

19. Bogart, E., & Myers, C. R. (2015). Multiscale metabolic modeling of C_4 plants: Connecting nonlinear genome-scale models to leaf-scale metabolism in developing maize leaves. http://arxiv.org/abs/1502.07969 (accessed on 10th September 2025).

20. Bos, M., Palys, M. J., Stojek, Z., et al. (1997). Voltammetric investigation of the complexation equilibria in the presence of a low level of supporting electrolyte. Part 1: Steady-state current–potential curves for inert complexes. https://core.ac.uk/download/pdf/11453121.pdf (accessed on 10th September 2025).

21. Bos, M., van der Linden, W. E., Palys, M. J., Stojek, Z., et al. (1995). Digital simulation of chronopotentiometry and steady-state voltammetric curves at microelectrodes in the presence of a low concentration of supporting electrolyte. https://core.ac.uk/download/pdf/11453115.pdf (accessed on 10th September 2025).

22. Brady, K. B., Hui, X., Niemeyer, K. E., Sung, C. J., et al. (2017). Reduced chemistry for butanol isomers at engine-relevant conditions. http://arxiv.org/abs/1706.02043 (accessed on 10th September 2025).

23. Branco Pinto, D., Constales, D., Marin, G., Yablonsky, G. S., et al. (2017). Kinetic intersections as a source of information on rate coefficients. https://core.ac.uk/download/91288153.pdf (accessed on 10th September 2025).

24. Crabtree, G. W., Green, M. A., Grätzel, M., Hamakawa, Y., Lewis, N. S., O'Regan, B., et al. (2007). Solar energy conversion. https://core.ac.uk/download/4880865.pdf (accessed on 10th September 2025).

25. Damiani, C., Filisetti, A., Graudenzi, A., Panini, T., et al. (2014). A stochastic model of catalytic reaction networks in protocells. http://arxiv.org/abs/1407.8508 (accessed on 10th September 2025).

26. Delmas, H., Deshpande, R. M., Julcour-Lebigue, C., Kelkar, M., et al. (2011). Kinetics of hydroformylation of 1-octene in ionic liquid–organic biphasic media using rhodium sulfoxantphos catalyst. https://core.ac.uk/download/12043525.pdf (accessed on 10th September 2025).

27. Du, M., Li, Y., Liu, X., Wang, J., et al. (2011). Inference of nonlinear state-space models for sandwich-type lateral flow immunoassay using extended Kalman filtering. https://core.ac.uk/download/337165.pdf (accessed on 10th September 2025).

28. Glowacki, D. R., Harvey, J. N., & Lightfoot, R. (2013). Non-equilibrium phenomena and molecular reaction dynamics: Mode space, energy space and conformer space. https://core.ac.uk/download/29026941.pdf (accessed on 10th September 2025).

29. Gray, H. B., & Langford, C. H. (1966). *Ligand Substitution Processes.* https://core.ac.uk/download/4890295.pdf (accessed on 10th September 2025).

30. Guo, L. (1999). *The Synthesis and Characterization of New Higher Nuclearity Arene-Ruthenium-Sulfur Clusters: A Thesis Presented in Partial Fulfillment of The Requirements for The Degree of Master of Science in Chemistry at Massey University, New Zealand.* https://core.ac.uk/download/162617233.pdf (accessed on 10th September 2025).

31. Hadden, D. J., Lodish, H., Rossle, S., Satzger, H., Scoles, G., Vastola, F. J., Yang, X., et al. (2013). Recent advances in experimental techniques

to probe fast excited-state dynamics in biological molecules in the gas phase: Dynamics in nucleotides, amino acids and beyond. https://core.ac.uk/download/18327700.pdf (accessed on 10th September 2025).

32. Hafidi, M., Merlina, G., Perchet, G. T., Pinelli, E., et al. (2009). Evaluation of a TiO_2 photocatalysis treatment on nitrophenols and nitramines-contaminated plant wastewaters by solid-phase extraction coupled with ESI HPLC–MS. https://core.ac.uk/download/12041818.pdf (accessed on 10th September 2025).

33. Hangos, K. M., Szederkényi, G., & Tuza, Z. (2011). Finding weakly reversible realizations of chemical reaction networks using optimization. http://arxiv.org/abs/1103.4741 (accessed on 10th September 2025).

34. Hasenauer, J., Hross, S., Jagiella, N., Theis, F. J., et al. (2015). Data-driven modelling of biological multi-scale processes. http://arxiv.org/abs/1506.06392 (accessed on 10th September 2025).

35. Hinczewski, M., & Thirumalai, D. (2016). Noise control in gene regulatory networks with negative feedback. http://arxiv.org/abs/1603.02620 (accessed on 10th September 2025).

36. Hosoda, K., Kadowaki, K., Nakamura, Y., Tsuda, Y., et al. (2016). Population–reaction model and microbial experimental ecosystems for understanding hierarchical dynamics of ecosystems. https://core.ac.uk/download/42368857.pdf (accessed on 10th September 2025).

37. Humphreys, P., Metcalfe, R., & West, J. M. (2010). Microbial effects on repository performance. https://core.ac.uk/download/55755.pdf (accessed on 10th September 2025).

38. Kamchatnyi, O., Molchanov, O., Rudakov, D., Soboliev, V., et al. (2018). Destabilization of the hard coal microstructure by a weak electric field. https://core.ac.uk/download/168413992.pdf (accessed on 10th September 2025).

39. Kim, D. S., Kumar, A., Kuppili, S. K., Omidvarborna, H., et al. (2016). Experimental modeling of NO_x and PM generation from combustion of various biodiesel blends for urban transport buses. https://core.ac.uk/download/70427234.pdf (accessed on 10th September 2025).

40. Laserna-Vázquez, J. J. (2016). *Chemistry in the laser-induced plasma.* https://core.ac.uk/download/62909677.pdf (accessed on 10th September 2025).

41. Luise, M., Monti, M., Sanguinetti, L., Tschudin, C., et al. (2013). A chemistry-inspired framework for achieving consensus in wireless sensor

networks. http://arxiv.org/abs/1311.1084 (accessed on 10th September 2025).

42. Marley, M. S., & Zahnle, K. J. (2014). Methane, carbon monoxide, and ammonia in brown dwarfs and self-luminous giant planets. https://core.ac.uk/download/pdf/189597261.pdf (accessed on 10th September 2025).
43. Marshall, P., McCabe, D. C., Rajakumar, B., Ravishankara, A. R., et al. (2006). The relaxation of OH (v = 1) and OD (v = 1) by H_2O and D_2O at temperatures from 251 to 390 K. https://core.ac.uk/download/4873815.pdf (accessed on 10th September 2025).
44. N/A. (2025). *ECUT (Energy Conversion and Utilization Technologies Program). Biocatalysis Project.* https://core.ac.uk/download/pdf/42839622.pdf (accessed on 10th September 2025).
45. Nováčková, K. (2009). Stability of salts and complexes of humic acids. https://core.ac.uk/download/30278457.pdf (accessed on 10th September 2025).
46. Proppe, J., & Reiher, M. (2018). Mechanism deduction from noisy chemical reaction networks. http://arxiv.org/abs/1803.09346 (accessed on 10th September 2025).
47. Raynal, L., Van Elk, E. P., Van Loo, S., et al. (2007). Mass transfer characteristics in structured packing for CO_2 emission reduction processes. https://core.ac.uk/download/pdf/12937217.pdf (accessed on 10th September 2025).
48. Regev, C., Priami, C., Versari, C., Hoare, C. A. R., Gillespie, D., Ciocchetta, F., Hillston, J., et al. (2009). Process algebra modelling styles for biomolecular processes. https://core.ac.uk/download/81097.pdf (accessed on 10th September 2025).
49. Schwartz, J. M., dos Santos, M. T., & Maurí, S. (2009). Organizing metabolic networks: Cycles in flux distributions. https://core.ac.uk/download/pdf/288898.pdf (accessed on 10th September 2025).
50. Shillady, D. D. (2002). *Mathematics for Non-Science Majors Chemistry Course.* https://core.ac.uk/download/215487690.pdf (accessed on 10th September 2025).
51. Vogel, P., Wen, L., & Zhang, C. (2015). Neutrino oscillation studies with reactors. http://arxiv.org/abs/1503.01059 (accessed on 10th September 2025).

CHAPTER 5

POLYMERIZATION KINETICS

Contents

Introduction 163

5.1. Types of Polymerizations 167

5.2. Mechanisms of Polymerization 172

5.3. Factors Affecting Polymerization Rates 177

5.4. Kinetic Models in Polymerization 181

5.5. Experimental Techniques for Studying Polymerization 185

5.6. Applications of Polymerization Kinetics 190

Summary 196

References 198

INTRODUCTION

Polymerization kinetics is a basic field of the knowledge of how macromolecules obtain their characteristics through chemical reactions. Polymerization, or the production of the larger, more intricate structures from the smaller units named monomers, is a key aspect of materials science, biochemistry, and engineering. The kinetics of the reactions influence the properties of the generated polymers to a great extent, such as molecular weight, thermal stability, and mechanical strength. This body of knowledge not only provides answers to simple questions about reaction rate and mechanism but also has industrial application to those industries that focus on the production of man-made materials. Thus, the description of the polymerization dynamics allows scientists to be in control of the process to yield specialty polymers intended for specific purpose applications, making this discipline a dynamic interface between theory and application. Kinetics of polymerization can be classified into two broad categories: chain-growth and step-growth polymerizations. Both processes exhibit distinct kinetic behaviors that define polymerization rates and mechanisms.

Figure 5.1. *Polymerization kinetics.*

Source:https://images.unsplash.com/photo-1743056311244-5a4b8f195c26?crop=entropy&cs=tinysrgb&fit=max&fm=jpg&ixid=M3w3MDgyMDR8MHwxfHNlYXJjaHwxfHxwb2x5bWVyaXphdGlvbiUyMGtpbmV0aWN-zJTJDJTIwdGhlcm1vZHluYW1pY3MlMkMlMjByZWFjdGlvbiUyMG-1lY2hhbmlzbXMlMkMlMjBwb2x5bWVyJTIwc3ludGhlc2lzfGVufDB8M-Hx8fDE3NDkwNzk2MDF8MA&ixlib=rb-4.1.0&q=80&w=500&?utm_source=numberanalytics&utm_medium=referral.

Chain-growth polymerization features fast growth of the polymer chain through repeated addition reactions due to unsaturation or reactive groups present in the monomers. This is generally accompanied by substantial molecular weight development over time, which makes it possible to form high-performance materials. Step growth polymerization, however, involves a gradual reaction between functional groups such that polymers develop much more gradually but have polymers of convergent structures and molecular weights. It is critical to grasp these mechanisms so that polymerization processes can be synthesized effectively and product performance optimized across various applications.

Additionally, advancements in polymerization kinetics during the past several years have led to new methods and materials with better properties. As an instance, research has discovered new reaction conditions and catalysts that can influence the rate and yield of polymerization substantially. Performance of various engineered materials, including conductive polymers and oxygen vacancy metal oxides, indicates that organizing building blocks can provide astounding gains in energy consumption (Hao et al., 2022; Zhang et al., 2017). These developments not only reinforce the theoretical foundations of polymerization kinetics but also create prospects for practical uses in technologies such as batteries and sensors. As the field continues to evolve, continued study of polymerization kinetics remains a focus in enhancing both scientific understanding and industrial application (Figure 5.1).

Definition of Polymerization

The concept of polymerization is central to the formation of macromolecules, where smaller molecules, or monomers, chemically unite to give large and complex molecules known as polymers. Polymerization occurs through various mechanisms, which are largely addition (chain-growth) and condensation (step-growth) polymerization. In addition, polymerization, monomers containing unsaturated bonds (usually double bonds) unite to produce long chains, often initiated by free radicals, heat, or light. Alternatively, condensation polymerization involves the monomer reaction with functional groups to produce the evolution of small molecules, such as water, in the process. Such elementary difference is very significant when it comes to investigating polymerization kinetics since the mode of polymerization and its rate have a direct implication on the resultant polymers' properties and applications. Further, the polymerization kinetics entails the measurement of the rate of these processes and the influencing factors like temperature, pressure, concentration of reactants, and catalysts. Each mechanism of polymerization has its own typical kinetic patterns due to the differences in the mechanisms of monomer addition and the kinds of bonds formed. For instance, chain-growth polymerization will typically show a rapid increase in molecular weight initially, followed by levelling, which is characteristic of the exhaustion of the available monomers and retardation of the reaction. Steep-growth polymerization, by contrast, will show a gradual increase in molecular weight until all the monomers have been consumed. Understandings of these kinetic behaviors are important not only for theoretical comprehension but also for uses, such as designing new materials and optimizing industrial processes related to polymer manufacturing. Moreover, a growing significance of polymerization is its environmental dimension, namely polymer degradation and microplastic generation. When polymers degrade in their natural environment, they form smaller particles that negatively affect ecosystems, as the case of the dispersion of nanoplastics illustrates. This is one aspect that ties in with recent research that highlights the role of polymerization kinetics not only in a laboratory setting but also in environmental and health settings. For instance, polymers that break down at low rates are responsible for long-term pollution, and understanding the mechanism of their breakdown would result in more sustainable materials. The study of polymerization kinetics thus has a twofold mandate; it includes fostering scientific knowledge and addressing current worldwide concerns over polymer waste, like material consequences advanced in studies in regard to environmental degradation (Olubukola et al., 2017; Patra et al., 2018) (Figure 5.2).

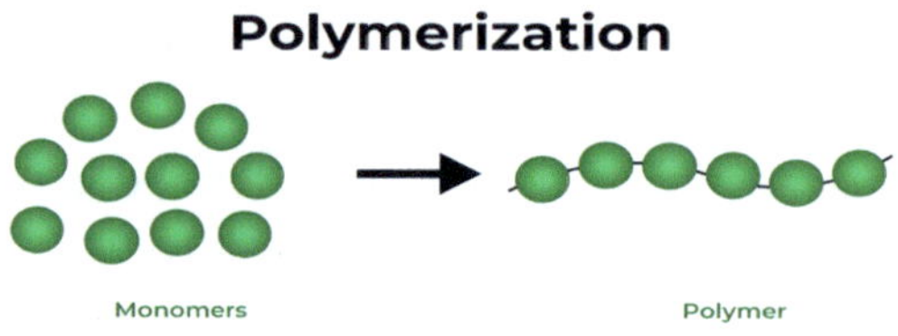

Figure 5.2. *Polymerization.*

Source: https://media.geeksforgeeks.org/wp-content/uploads/20231019182925/Polymerization.png.

Importance of Polymerization Kinetics

The applicability of polymerization kinetics extends beyond simple chemical reactions to affect many everyday applications across industries from pharmaceuticals to materials science to biotechnology. Chemists are able to engineer polymer properties through the control of critical variables such as reaction rates, temperature, and concentration using the understanding of polymer formation kinetics. For instance, kinetics of swelling of polyelectrolyte gels, which are typical in biosystems as well as in many types of industrial applications, can be regulated through control of ionic strength and counter-ion species in the solution. This level of control is required in applications that require specific mechanical or chemical properties from the resulting polymers, thereby underscoring the significance of polymerization kinetics to studies and practical use (Horkay, 2021). In advanced manufacturing technologies, particularly in three-dimensional (3D) printing, polymerization kinetics are crucial aspects for achieving desired precision and functionality in printed materials. Ultrafast laser 3D lithography illustrates the manner in which photopolymerization kinetics affects spatial resolution and object accuracy of printed items. Such parameters as light exposure and material response under consideration can be a determinant of the success of printing. Scientists can further evolve methods encompassing photoexcitation and chemical modification based on better understanding of these kinetic processes. In so doing, ongoing research on the intricate mechanism of polymerization kinetics provides access to material property innovation and manufacturing process developments, precipitating breakthroughs in various fields of technology (Skliutas et al., 2021). Furthermore, the importance of polymerization kinetics also arises from the development of new materials with desired properties intended for various purposes. The coupling of polymerization kinetics with nanotechnology, to use one illustration, enables the creation of nanoscale materials that in turn enable enhanced strength, flexibility, and thermal stability. Such blending, as well as enlarging the scope for polymer use, also focuses attention on extremely high requirements for interdisciplinary work that incorporates concepts in chemistry, physics, and engineering. Thus, an obsession with polymerization kinetics can lead to new materials and processes that address contemporary problems in sustainability, medicine, and other fields, making it pivotal for the construction of contemporary science and technology.

5.1. TYPES OF POLYMERIZATIONS

The field of polymerization is replete with various techniques through which polymers may be synthesized by distinct mechanisms. Of these, chain-growth polymerization and step-growth polymerization are the two principal classes. Chain-growth polymerization, especially characterized by sequential monomer addition for the accumulation of a growing chain via mechanisms such as radical, cationic, or anionic mechanisms, facilitates high rates of molecular weight growth. This method is particularly useful in the synthesis of polymers of highly characterized structures and low polydispersity required for application in materials science. Step-growth polymerization, on the other hand, is the reaction process between multifunctional or bifunctional monomers, where the polymer Matrices form as a result of the formation of covalent bonds along the molecule, which typically forms higher molecular weights at lower conversion levels. This disparity in mechanisms emphasizes the diverse approaches involved in polymerization reactions that influence the resultant material properties of the final materials. Among the several methods used in chain-growth polymerization, atom transfer radical polymerization (ATRP) is one significant breakthrough. ATRP goes by a reversible deactivation mechanism in which it is possible to have a fine control over molecular weight and polymer architecture of the resulting polymers. This technique is particularly suitable for use in the environment and utilizes infinitesimal quantities of catalysts, which allow direct polymer synthesis on surfaces, such as when grafting polymer brushes onto substrates using ppm-level catalysts and allowing the kinetics of polymerization to occur under optimal conditions (Guo et al., 2024). Moreover, the versatility of ATRP has been translated into copolymerization techniques by which a number of monomers may be added sequentially to form tailored properties for specific applications. Thus, ATRP not only enlarges the polymer chemist's repertoire but also accentuates the impact of polymer structure on dictating the materials' function and performance. In step-growth polymerization, advances have enabled the synthesis of complex copolymers with targeted properties. Vinyl addition copolymerization comes to mind perfectly here; recent studies have demonstrated its capacity for making particular thermal stability and crystalline polymers by copolymerizing monomers like 4-vinyl-cyclohexene and ethylene with the help of specific catalysts (Ali et al., 2022). This methodology shows the control that polymerization kinetics has over the final product, illustrating how polymerization rates and molecular weight distribution of the resultant polymers are functions of catalyst choice and reaction conditions. These kinetics are essential to comprehend in order to maximize polymer synthesis and advance the boundaries of application in each area, from medicine to packaging. Control of kinetic parameters then allows the tailoring of polymer properties to the needs of the most advanced technological applications, confirming the value of integrative studies of polymerization processes (Figure 5.3).

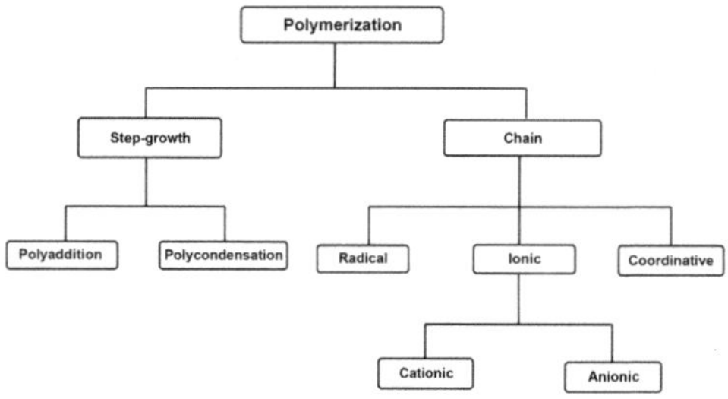

Figure 5.3. *Types of polymerizations.*

Source: https://cms-media.bartleby.com/wp-content/uploads/sites/2/2021/10/11090615/Polymerization_kinetics_jpg-1.png.

5.1.1. Addition Polymerization

Addition polymerization, which is an important process in polymer chemistry, is the study that deals with the process by which successive monomer linkages form long-chain polymers from unsaturated monomers. It is characterized by its chain-growth process, as opposed to condensation polymerization where the monomers release small molecules upon reaction. Addition polymerization, which is in most instances activated through free radicals, cationic, or anionic systems, follows a series of stages: initiation, propagation, and termination. Of special interest is the observation that the reactivity of different monomers can vary substantially, influencing the kinetic rates of polymerization. Understanding these kinetics informs us on how polymers are generated and what characteristics they possess, of absolute relevance in applications ranging from plastics to elastomers.

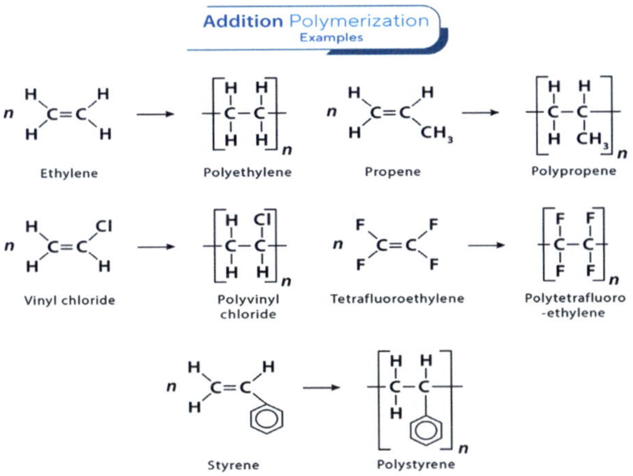

Figure 5.4. *Addition polymerization.*

Source: https://www.chemistrylearner.com/wp-content/uploads/2023/07/Addition-Polymerization.jpg.

In addition, the addition polymerization kinetics are critical in designing well-defined copolymers with targeted properties. Research proves that with manipulation of reaction conditions—temperature and monomer feed ratios—the researchers can obtain targeted molecular weights and distributions. For example, new research with statistical and block copolymers prepared from n-hexyl isocyanate and 2-chloroethyl isocyanate reported how the presence of a titanic complex brought about catalysis of the polymerization to optically active products with prescriptive structural characteristics (Mantzara et al., 2024). In addition, mechanical and thermal properties of the polymers were extensively analyzed through techniques like Differential Scanning Calorimetry (DSC) and Thermogravimetric Analysis (TGA). Not only has such research enhanced our understanding of the behavior of polymers, but it has also promoted innovation in material science. In addition, the use of sophisticated technologies like Reversible Addition-Fragmentation chain Transfer (RAFT) polymerization has made us better at controlling the reactivity ratios of various monomers when copolymerizing. For instance, RAFT copolymerization of 2-chloroethyl vinyl ether and N-vinyl pyrrolidone allowed for precise property control through control of the concentration of the chain transfer agent (Plachouras et al., 2023). By employing varying techniques to determine monomer reactivity ratios and dyad sequence fractions, researchers can effectively control the structural properties of the resultant polymers. These findings concerning addition polymerization kinetics not only illuminate inherent aspects of polymerization but also pave the way toward the development of new materials with defined functionality for various industrial applications (Figure 5.4).

5.1.2. Condensation Polymerization

Condensation polymerization is a significant process during the polymer synthesis of most polymers, particularly those made up of enormous macromolecular architectures required in industry and research. In contrast to addition polymerization, condensation polymerization involves stepwise coupling of monomers with at least two reactive functional groups during which there is loss of low molecular weight species such as water or methanol in polymer formation. This property has a tendency to give rise to high molecular weight polymers through sequential building of chains, which is essentially under kinetic control. Understanding of the inherent kinetics of such reactions is of importance because it governs rates and yields of reactions and is therefore vital for optimizing conditions in an industrial setup. The intricate relationship among pressure, temperature, and the presence of catalytic influences affects the efficiency and rate of condensation polymerization, pointing to its applicability in practical use. Kinetic investigation of condensation polymerization reveals that it is not only dictated by the reactivity of involved functional groups but also by the molecular weight of forming polymer chains. When polymer chains become longer, termination reactions become dominant, influencing the overall rate of polymerization. It is this balance between reaction kinetics and chain length that is central to how polymers can be engineered for specific end-use properties. Molecular diffusion and heat motion also take fundamental roles in determining the microstructures of polymers that ultimately contribute to the final material's mechanical properties. Intermolecular associations in condensation polymerization lead to new properties of the final products, highlighting that the reaction conditions need to be controlled in order to generate required polymer properties. Supermolecular structures in condensation polymerization have also recently been the focus of interest, and their strong impact on materials science and regenerative design. Importantly, the mechanics and polymer network dynamics can be explored through advanced modeling approaches, taking into account the role of thermal and mechanical stresses throughout the process of polymerization. For instance, the formation of complex architectures in polymer networks can mimic biological processes such as tissue growth, where the mechanical signals govern molecular organization. This correlation of polymerization kinetics with biological models is an indication of the growing knowledge of material sciences, where not just polymer properties are subject to kinetics but also the designs of materials in the future. As research increasingly closes the gap between classical polymer chemistry and bioactive processes, the understanding of condensation polymerization will undoubtedly influence modern applications in fields ranging from drug delivery to biofabrication, as illustrated in research on tissue architecture (He et al., 2025) and combustion chemistry impacting soot formation (Wen, 2025) (Figure 5.5).

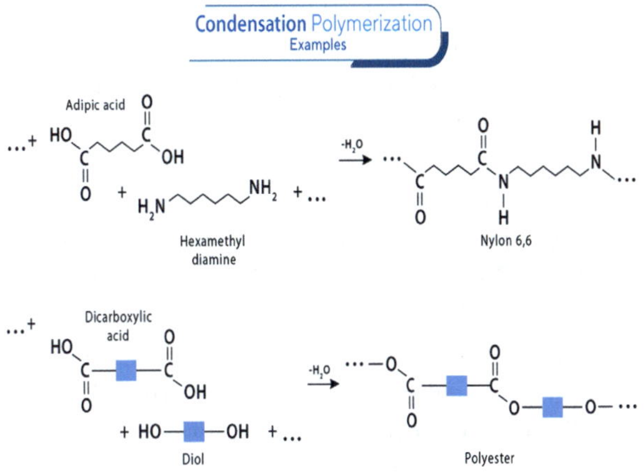

Figure 5.5. *Condensation polymerization.*

Source: https://www.chemistrylearner.com/wp-content/uploads/2023/07/Condensation-Polymerization.jpg.

5.1.3. Copolymerization

The copolymerization process gives valuable insight into polymerization kinetics, particularly on how materials of particular properties are synthesized. Copolymerization involves simultaneous or successive polymerization of two or more dissimilar monomers for the synthesis of polymers exhibiting rich diversity in physical and chemical characters. The process is useful in synthesizing advanced materials that meet specific requirements for performance, such as increased strength or improved biocompatibility. The kinetics of copolymerization is governed by parameters like the reactivity ratio between the monomers, which determine the composition and the sequence distribution of the resulting copolymer. Such interactions lead to block, random, or alternating copolymers that possess characteristic mechanical properties and applications, particularly in fields like biomedical engineering and materials science, where specialty properties are required. In biomedicine, synthetic or natural copolymers have found much promise, especially in tissue engineering. Copolymerization allows multiple monomeric units to be integrated, and thus physical properties such as elasticity, degradation, and biocompatibility can be modulated. For instance, hydrogels created through copolymerization can function as scaffolds for cellular activity due to their hydrophilic, three-dimensional matrices that can house biological fluids and nutrients. Their ability to respond to environmental stimuli, such as pH or temperature, also underscores the relevance of copolymerization kinetics to the formation of therapeutic applications (Chai et al., 2017). Therefore, understanding the dynamics of copolymer formation is important in designing optimum biomaterials that operate under physiological conditions and facilitate tissue or organ regeneration. In addition, the role of copolymerization in the modification of hydrogels illustrates the functional applications of the kinetic process. By utilizing the varied swelling behaviors and mechanical properties achieved through copolymerization of various monomers, scientists are able to engineer responsive hydrogels for application in specific areas, such as drug delivery systems or biosensors. By applying the principles of polymerization kinetics, researchers can predict the effect of changes in monomer concentration or type on the macro and micro material properties of the final product (Reddy et al., 2021). The development of such sophisticated hydrogels, which could find applications in the controlled release of medication or cellular growth support, is the merging of chemistry and biology that points to the redemptive power of copolymerization in contemporary material science. With continued research, emphasis on copolymerization kinetics will always be at the center of creating innovative responses to intricate biomedical issues.

5.2. MECHANISMS OF POLYMERIZATION

The mechanisms of polymerization play a key role in the understanding of polymerization kinetics and eventually influence material properties and function. Various polymerization processes like addition, condensation, and copolymerization differ in the mechanisms involved for the formation of polymers from minute monomer building blocks. Addition polymerization, for instance, typically involves free radicals or ions added sequentially to unsaturated monomers and leads to long-chain molecules. This mechanism is characterized by its ability to yield high molecular weight polymers both quickly and efficiently under specific conditions, thus providing a foundation for most industrial applications. Further, temperature, solvent, as well as initiator concentration, significantly determine the kinetics of such reactions, including the molecular weight as well as the polymer distribution. Hence, detailed studies of polymerization mechanisms enable strategic control as well as optimization of polymers towards applications. In radiation-induced polymerization, various mechanisms describe how ionizing radiation can change polymer structures effectively.

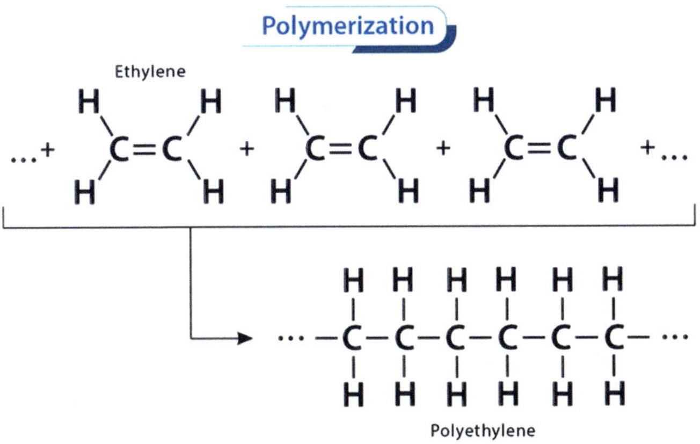

Figure 5.6. *Mechanisms of polymerization.*

Ionizing radiation generates free radicals and excited states, enhancing the complexity of polymerization processes. LLET ionizing radiation, for instance, initiates physical and chemical property changes due to crosslinking or chain scission depending on the intrinsic property of the polymer and dose. Competition among oxidative degradation and crosslinking further indicates a unique feature of radiation chemistry, where various conditions in the environment such as oxygen content alter the outcome of the polymerization reaction. The knowledge of these interactions enhances the capability to tailor polymer properties for multiple uses, from healthcare devices to environmental applications, demonstrating the significance of appreciating mechanisms in radiation-induced processes (Ashfaq et al., 2020). The alteration of polymer chains through procedures such as dopant-induced modifications or grafting procedures also encompasses the complexities inherent in polymerization procedures. For example, advancements in photocatalysis illustrate how dopants influence charge separation and reactivity improvement, expanding the application of polymeric materials (Zhang et al., 2020). This highlights that polymerization reactions are not limited to the initial formation of the polymer backbone but also encompass additional changes that can significantly enhance material properties. Additionally, the inclusion of methods like radiation-induced grafting presents new opportunities for the achievement of properties that meet modern society demands. Thus, a profound knowledge of polymerization processes allows scientists to produce and design polymers that satisfy specific functional requirements in a wide range of applications, thus delineating the significance of this data in polymerization kinetics (Figure 5.6).Source: https://www.chemistrylearner.com/wp-content/uploads/2023/07/Polymerization.jpg.

5.2.1. Free Radical Mechanism

Free radical mechanism is the core in the explanation of polymerization kinetics because it initiates the process of monomers transformation into complex polymer structures by way of a process of reaction steps. Free radical mechanism begins with the generation of free radicals that initiate the polymerization process. The free radicals are formed through thermal, photochemical, or chemical activities resulting in chain reactions that are quick in their progression. This heightened reactivity plays a critical role in determining the rate and timing of polymerization, which has a distinct influence on the properties of the resulting end product. Deep insight into these kinetics allows scientists to regulate numerous conditions, such as temperature and light exposure, to achieve optimal polymerization outcomes, therefore the significance of detailed kinetic modeling in order to predict and regulate radical species behavior throughout the process. The mechanism of free radical polymerization is also demonstrated by the action of stimuli-responsive biomaterials, which capitalize on particular environmental cues to enhance targeted drug delivery systems. Native free radicals such as reactive oxygen species (ROS) are particularly involved in crosslinking and immobilizing polymer networks like acrylated polyethylene glycol diacrylate (PEGDA). As evident from prior studies, ROS can act to mediate the formation of stable high molecular-weight polymer networks that are crucial for viable biomaterials applications. The dynamic capacity of these systems to respond to pathological conditions indicates the significance of having knowledge about free radical mechanisms in basic and applied applications. Consequently, this convergence of polymer chemistry and biomedical engineering is transforming therapeutic modalities through the precise timing and dosing of drug delivery through innovative two-phase targeting strategies, as evidenced in Emily et al. (2023). The complexity of polymerization kinetics is also enhanced by the advancements in photopolymerization technologies, which employ light to initiate free radical reactions. This method has a number of significant advantages, including fast reaction kinetics and regulation of the space and time of material formation. Polymerization of liquid monomers to solid polymers takes place at astonishing rates, demanding acute understanding of the curing kinetics that occurs. Studies indicate that such photopolymerizations not only exhibit high efficiency but are also flexible in many applications, providing room for improved quality and efficiency in production processes. In addition, probing the theoretical models encompassing diffusion-controlled effects and oxygen inhibition is crucial for maximizing these systems. By elucidating the intricate kinetics of free-radical and cationic photopolymerizations, researchers can maximize the reliability and performance of industrially produced materials, thus accelerating the field as exemplified by Lang et al. (2022).

5.2.2. Ionic Mechanism

The ionic mechanism plays a critical role in polymerization kinetics, particularly for reactions where oxidation processes influence polymer behavior. Support is garnered from studies on the kinetics of various oxidation reactions, for example, the oxidation of 3-Methyl indole by potassium bromates under aqueous conditions. The second-order reaction kinetics with respect to reactants confirms the role of ionic interaction throughout the course of the reaction. Most importantly, the constancy of the rate of reaction with ionic strength points to charged intermediates not playing a dominant role. Instead, the polarity of solvents is seen to be the strongest factor in rate control of the reaction, bringing into view the intricate relationship between ionic mechanisms and polymerization kinetics as a response to the solvent media (Periyasami et al., 2025). In addition to the information derived through oxidation studies, reaction mechanisms with imidazolium fluorochromate demonstrate similar trends in their kinetic features. The first-order dependency demonstrated with respect to both oxidant and D-mannose further emphasizes the contribution of substrate concentration in defining the reaction velocity. Finally, the enhancement of reaction rates under the condition of perchloric acid presence also demonstrates that ionic species possess the potential in changing the kinetics of oxidation reactions. This is most pertinent when taken into account with regard to the absence of polymerization in such intermediate reactions and thus it is inferred that ionic mechanisms are the ones that control the pathways to product formation without being founded on radical propagation processes (Tomar et al., 2024). The influence of ionic strength and also solvent characteristics also signifies the broader implications of ionic mechanisms in polymerization kinetics. For instance, the retardation of reaction rates with solvents of higher dielectric constant violates conventional understanding of how reaction kinetics is affected by solvent polarity. Such findings illustrate how an understanding of ionic interactions can provide valuable insights into the mechanisms of polymerization, particularly under those circumstances where polymerization is suppressed or does not proceed. With scientists continuing to study the roles played by ionic species and solvent media, their combination will further enrich the theoretical frame under polymerization kinetics as well as assist in developing more effective synthetic processes (Periyasami et al., 2025; Tomar et al., 2024).

5.2.3. Coordination Mechanism

The coordination mechanism is at the core of polymerization reaction kinetics, particularly in coordination polymers (CPs). CPs are characterized by their unique structural organization, which has the potential to significantly influence the rate of chemical reactions within them. The study of kinetics involving CPs allows one to understand how structural features, for example, ligand electronegativity, can direct reaction rates. To give an example, in the investigation of [2 + 2] photocycloaddition reactions, researchers have established that the electronegativity of carboxylate ligand substituents influences the kinetics, showing how coordination structure directly affects molecular interactions and reactivity. This sets the role of ligand architecture in regulating the pathways and efficiencies of polymerization reactions and therefore coordinates mechanisms as a focal point of investigation in polymerization kinetics studies (Wang et al., 2025). In exploring the coordination mechanism, its significance in catalyzed depolymerization processes should be considered. Studies of metal-catalyzed melt-phase depolymerization techniques provide information concerning kinetic mechanisms involved in polymer degradation and reformation. For polyesters, studies have used techniques such as thermogravimetric analysis to study the effect of certain parameters, including molar mass and end-group chemistry, on net reaction rates. The systematic kinetic studies of model polyesters like poly(δ-valerolactone) (PVL) set the coordination mechanism as facilitating a ring-closing depolymerization route under catalysis, which is vital in switching between polymer and back to monomer. Not only does this mechanism illuminate the action of polymer under different conditions but also helps in designing more efficient recycling technologies for circular polymers (Wilfred et al., 2024). Overall, the inclusion of coordination mechanisms in the comprehension of polymerization kinetics provides an integrated view of how molecular interactions govern polymer system behavior. The comprehension of the structural and electronic implications on the reaction rate has long-term repercussions, particularly in the creation of sustainable materials and new applications of polymers. With a growing demand for materials with properties that are tailored to specific requirements, understanding the mechanisms of coordination is critical in unlocking the doors to next-generation polymers. By having a better appreciation of the mechanism present, researchers are able to plan for improvements in ways to enhance efficiencies of reactions and optimize the performance of materials, further paving the way for polymer science (Wang et al., 2025; Wilfred et al., 2024).

5.3. FACTORS AFFECTING POLYMERIZATION RATES

Understanding the factors that determine polymerization rates is crucial in optimizing industrial processes and enhancing the performance of polymer products. Among the key factors is monomer concentration, which directly influences the reaction kinetics. From aqueous precipitation polymerization, variations in monomer concentration can radically alter the polymerization rate, as evidenced by the kinetic equations derived for different initiation systems. For instance, the rate of polymerization (Rp) established for acrylonitrile (AN) indicates it is under a non-linear dependence with monomer concentration, mirroring on a dependence that demands great control under controlled circumstances. Such knowledge on monomer dynamics provides precious guidance for the synthesis of preferred polymers and desired weights, thus their application in various technological uses (Yan et al., 2024). Another significant factor influencing polymerization rates is the mode of initiation. Different initiators may lead to large variations in rates of reaction because their efficiency, mechanism of action, and their interaction with the monomer play a pivotal role in the rate of polymer formation. The ammonium persulfate-ammonium sulfite monohydrate (APS-AS) system provides an example of how one can manipulate the conditions of initiation in order to create specific polymer characteristics. The comparative investigation of initiation systems shows that temperature and solvent type play equally significant roles, with their interaction affecting the nucleation sites and thus the dynamics of polymerization. Understanding these initiation factors is at the heart of chemists seeking to create polymers with specific industrial specifications, as the initiation process can decide the stability and quality of the ultimate polymer chain (Yan et al., 2024). Polymerization rates are also influenced by the environmental conditions, including temperature and pressure. Temperature variations may lead to changes in molecular mobility and collision frequency among reacting moieties, consequently changing the rate of polymer chain formation. Higher temperatures can speed up reaction, but they can also potentially create undesirable by-products or alter polymer properties. In addition, extrinsic conditions such as pressure affect reactant solubility and thus change the effective concentration of interacting molecules. An example of starch digestion in maize is quite relevant and illustrates the role of extrinsic conditions in controlling rates of reaction, which also reflects how intrinsic material properties are able to interact with extrinsic environments to control digestion kinetics as well. By keeping a keen eye on these environmental factors, manufacturers can make polymerization processes more efficient and provide better product results.

5.3.1. Temperature

Temperature is a key parameter of polymerization kinetics that influences rates of reaction, construction of polymer structures, and ultimate properties of the ensuing material. As temperature rises, kinetic energy of molecules also rises, which can lead to the acceleration of reactive species in a polymerization system. This phenomenon is well pronounced in free radical polymerization, where increasing temperatures will speed up the rate of radical generation. Nonetheless, temperature effects are not uniform across all polymer systems as various polymers will react differently to thermal stability. The complex interplay of factors like activation energy and thermal history significantly affects the behavior of polymerization reactions, testifying to the imperative to utilize precise control of temperature during synthesis activities to realize the desired material properties (Li et al., 2021). Furthermore, temperature's influence does not end with regulating reaction rates; it can also dictate dominant reaction pathways in polymerization. For instance, at elevated temperatures, reactions may favor chain scission against crosslinking in particular polymers, with consequential changes in molecular weight and structural integrity. This equilibrium between the two processes is of greatest importance since it corresponds to the mechanical, thermal, and chemical properties of the resulting polymer. The use of advanced simulation techniques such as ReaxFF molecular dynamics has provided researchers with an understanding of such temperature-dependent processes, and it is used to explain the variations in reaction outcomes due to differences in temperature. Studies have shown that, while higher temperatures can increase reaction rates, they can also be challenging to control and reproduce, particularly in mass-scale processes (Li et al., 2021). In addition to affecting the kinetic of polymerization, temperature also has a vital function in radiation-induced changes in polymers. Ionizing radiation, employed extensively to graft and crosslink polymer chains, possesses different efficiencies depending on exposure temperature. Under low temperatures, radicals generated can recombine more readily and thus decrease the rate of polymerization, whereas at high temperatures, the mobility of polymer chains can be enhanced, which may lead to improved radical interactions and thus alter the yield of the desired products ultimately. Moreover, the degree of crystallinity as well as the use of additives also possess the potential to alter the way temperature influences radiation chemistry. Understanding these interactions is important to provide an improved design insight for creating successful polymerization protocols, so that the thermal regimes employed maximize the desired properties of the end materials without compromising on safety or efficacy.

5.3.2. Concentration of Reactants

Reactant concentration is extremely critical in polymerization kinetics, with the rate of polymer formation and the molecular weight of the polymers produced being strongly influenced by it. It is usually the case that increased reactant concentrations bring about increased rates of reaction through the optimization of collision frequency between molecules, according to collision theory. In the case of radical polymerization, monomer initial concentration directly affects the generation of active centers and, in turn, the generation of polymer chains. The differences in reactant concentrations, particularly in heterogeneous systems, cause the dynamic structural transition of active sites to be key. Atomic level modifications can design the reaction pathways and enhance the activity of the catalysts and, in turn, produce a more efficient polymerization process. These interactions highlight that an understanding of concentration dynamics is vital in developing methods for maximizing polymer yield and controlling polymer properties. In catalyst design usage, reactant and catalyst species concentrations are correlated with one another and can affect the effectiveness and selectivity of the polymerization reaction. In heterogeneous catalysts, as outlined within metal catalyst research, the concentration of active surface sites can vary with reaction advancement. It impacts overall process kinetics, affecting polymerization rates and product quality. As noted, under reaction conditions, catalyst structure is dictated by interaction with reactant concentrations, hence resulting in liquid-phase reactions having the ability to alter active sites due to increased atomic mobility. Hence, the concentration of reactants needs to be traced and controlled so that catalysts are able to function optimally during polymerization reaction to stabilize active sites for improved outcomes (Liu et al., 2020). Moreover, solubility of the reactants in solvents becomes a determining factor in polymerization kinetics. Slightly soluble reactants can limit the overall concentration of active species and therefore the reaction efficiency, leading to low conversions or less-than-ideal polymer characteristics. The choice of solvent not only affects the solubilization of reactants but also is able to affect the reaction temperature and the diffusion of particles in the reaction medium. Dilution of the reactants typically yields improved rates of polymerization, but temperature and solvent viscosity must be considered carefully to avoid problems such as precipitation or gelation. As such, optimization of reactant concentration as a function of solvent characteristics becomes a prerequisite for achieving desirable polymerization kinetics and product uniformity (Song et al., 2020).

5.3.3. Presence of Catalysts

The function of catalysts is at the heart of the efficacy and selectivity of polymerization reactions. Catalysts not only enhance polymerization rates but allow the control of molecular weight and polydispersity, important facets in the achievement of desired polymer properties. Controlled radical polymerization (CRP) techniques like photoinduced electron transfer-reversible addition–fragmentation chain transfer polymerization (SI-PET-RAFT) demonstrate the controlling role of catalysts in the kinetics of polymerization. In SI-PET-RAFT, a photoredox catalyst is employed to enable polymerization under mild conditions with high control over the polymerization process. The flexibility of this method—enabling a wide range of monomers and control of chain end functionality—emphasizes the importance of catalysts in the development of commercially significant polymerization techniques that are versatile and efficient (Li et al., 2019). In addition, low concentrations of metal catalysts have been shown to yield good outcomes in polymerization kinetics. In mechanically controlled atom transfer radical polymerization (mechanoATRP), the use of ultrasound as a stimulation factor with a low catalyst concentration of copper has been a remarkable achievement. Such a method not only facilitates observation of the first-order kinetics of methyl acrylate polymerization but also demonstrates that control of environment is critical for the desired molecular weight distribution. Such new approaches demonstrate how catalytic existence and operating conditions intricate with each other to produce polymers of specified properties, thus broadening the frontiers of polymer science and application (Wang et al., 2017). Moreover, judicious selection of the catalysts and the polymerization environment can dramatically influence the morphological and functional properties of polymers formed. For instance, by using specific catalysts in conjunction with ultrasound, as proven by mechanoATRP, scientists can engineer reaction times as well as effective molecular weights. The changes dictated by catalysts advantageously alter polydispersity indices to yield products that are closer to those demanded by the industry. Ultimately, not only do catalysts simplify complex polymerizations, but they also pave the way for advanced materials of properties engineered to satisfy certain requirements, demonstrating the pivotal role of knowledge of kinetics of polymerization since the impact of a catalyst has a cascading ramification on ultimate material properties. Synthesis of advanced polymers is thus inextricably linked with the catalytic choices made within such advanced kinetic schemes.

5.4. KINETIC MODELS IN POLYMERIZATION

Understanding the kinetic models of polymerization is crucial in the optimization of the synthesis of polymers with desired properties. The most rudimentary role played by these models is that they describe how reaction rates respond to the reactant concentration, temperature, and other environmental conditions. A number of kinetic models, such as rate equations of the mass action theory, provide a vehicle for polymerization reaction behavior anticipation. Such models allow for the examination of chain initiation, propagation, and termination reactions, all of which are central to designing efficient polymerization techniques. By fully modeling these reactions, researchers can gain insight into the conditions for optimum yield and molecular weight control and ultimately design specific polymers for specific applications. The kinetic model analysis also allows for major differences in reaction pathways and product formation. For example, in organosolv hemicellulose research, different models were found to vary in their predictability based on conversion rates and product distributions. In a study, the model that employed a xylose intermediate had greater correspondence with experimental data, highlighting the complexity involved in properly simulating polymerization pathways. These differences in model behavior are frequently caused by the influence of oligomeric species that have been formed as a result of the reaction, and these complicate the kinetics of polymerization and degradation. By fully appreciating such nuances, chemists can refine their practice in polymer synthesis and ensure the achievement of targeted molecular architectures with reduced unwanted side reactions. In addition to being applied in process synthesis, kinetic models of polymerization can also be highly valuable tools for the functional testing of the resulting products. For instance, the use of reversible addition-fragmentation chain transfer precipitation polymerization could lead to new materials with selectivity recognition properties, as observed in work for the removal of environmental contaminants. These models encompass adsorption kinetics, which describe how rates both are a function of concentration and of the chemical nature of the binding sites. Not only do these findings help to understand polymer behavior in engineering use, but they also demonstrate the cohesion of kinetic theories in all areas of science. The implications of these studies are to work towards designing efficient separation techniques and environmental remediation methods, thus reiterating the need for accurate kinetic modeling in polymerization processes.

5.4.1. Rate Equations

In polymerization kinetics studies, rate equations are simple tools employed to describe the speed at which polymerization reactions occur. Rate equations typically consist of parameters such as monomer concentration, temperature, and reaction time. The most common forms of rate equations employed in polymerization are zero-order, first-order, and second-order reactions, usually derived from empirical information from observations of experimental data. Understanding these correlations is vital because they guide chemists on how to control reaction conditions to achieve desired polymer properties, such as molecular weight and mechanical response. A good example of this control is in the synthesis of polyacrylic elastomer latex, where the concentration of cellulose nanocrystal (CNC) varied to affect crystallization rate and mechanical performance of blends, putting into perspective the importance of optimizing rate equations in complex polymer systems (Meng et al., 2023). In photopolymerization, particularly in holographic data storage devices, rate equations play a crucial role in shedding light on the kinetics of dual monomer systems. Nonlocal extended polymerization-driven diffusion considers both temporal and spatial effects in the medium to account for a better understanding of how the growth of polymer chains occurs under various conditions. Empirical measurements of values such as diffusion coefficient and polymerization rate allow chemists to fine-tune the reaction for enhanced material performance. Using equations for refractive index modulation in relation to diffraction efficiency, scientists are able to maximize photopolymer formulation, thereby achieving high optical clarity needed for effective holographic storage devices (Liu et al., 2021). This interdependence between kinetics and material properties continues to emphasize the intricate relationship between kinetics and resultant polymer performance. The consequences of rate equations extend far beyond the immediate understanding of polymerization kinetics; they significantly influence industrial process and application. For instance, for the production of high-performance materials like PBMG/CNC dispersions, rate equations can be employed to predict the influence of composition or conditions on crystallization rates and consequently on mechanical performance. By exploring crystallization halftime under varying CNC content, researchers can identify ideal conditions that lead to enhanced mechanical properties of the PLA matrix. This predictive capacity allows material scientists to create next-generation polymers tailored to application, advancing the science of polymers and its applications. Thus, rate equations are not abstract concepts but essential parts driving innovation in polymer technologies.

5.4.2. Chain Growth versus Step Growth Kinetics

The kinetics of polymerization may be grouped under two general classes of the mechanism of polymerization: chain growth and step growth mechanisms. The polymerization of chain growth is related to the rapid addition of monomers onto an active center, giving rise to a growing polymer chain. The mechanism typically involves initiation, propagation, and termination processes, with the polymerization rate mainly dependent on the concentration of reactive species. Conversely, step growth polymerization occurs as a series of reactions between functional groups to allow the incremental build-up of the polymer chains. The process requires at least two distinct functional groups and has the potential to result in the formation of oligomers before reaching high molecular weights. Understanding these mechanisms is essential because the kinetics not only dictate the molecular weight distribution of polymers formed but also their chemical and physical nature and thus their applications in various industrial processes. Living polymerization is one of the special characteristics of chain growth kinetics.

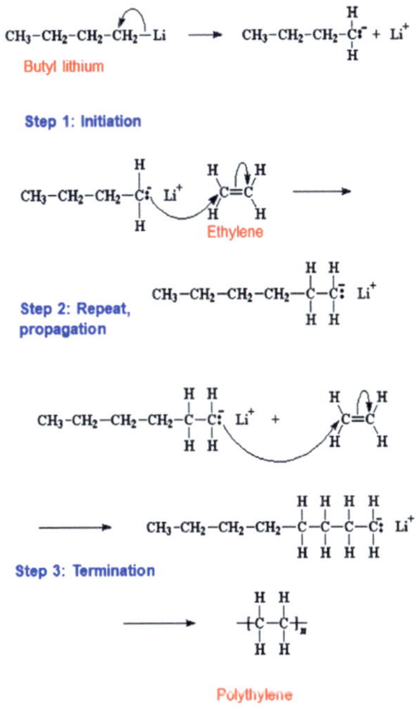

Figure 5.7. *Steps of chain-growth polymerization of ethylene to form polyethylene.*

Source: https://ars.els-cdn.com/content/image/3-s2.0-B9780128132944000042-f03-22-9780128132944.jpg.

Chain growth systems, once a polymer chain is established, can be made to grow indefinitely as long as monomers are present. This provides a tight molecular weight distribution, which is advantageous for applications where precise material properties are desired. For instance, the field of peptoid assembly research indicates that the sequence of building blocks can significantly impact the growth kinetics, with effects on properties such as stability and self-repairability (Schmid et al., 2024). Surprisingly, it is reported that there are certain sequences with extremely low detachment rates once incorporated into the chain, which emphasizes the control and effectiveness of chain growth kinetics in polymerizing. Step growth polymerization generates broader distributions of molecular weight, which merely reflect the random nature of reactions being carried out. On the other hand, step growth kinetics mechanisms provide unusual advantages, particularly in the synthesis of high molecular weight polymers without the need to require a living mechanism. The reaction is typically the coupling between two reactive species to give polymer growth as evidenced in various resin formulations (Karnes et al., 2022). It offers the flexibility to modify reaction conditions and functional group concentrations, resulting in tailored properties for a specific end application. For the investigation of trifunctional isocyanurate resins, various polymerization pathways result in varying mechanical properties which relate to the underlying mechanism of reaction, underscoring the importance of understanding both kinetics to improve polymer design. Thus chain and step growth mechanisms both offer useful ways to design polymers with diverse properties for a range of technological applications (Figure 5.7).

5.4.3. Mathematical Modeling Approaches

Mathematical modeling approaches are basic frameworks within the research of polymerization kinetics that allow researchers to model and predict the performance of various polymerization reactions. Through the development of mathematical models, researchers can formulate differential equations representing the kinetics of polymer buildup and the relationship between monomer concentration, reaction rate, and temperature. This predictive power is particularly valuable in condition optimization for polymer synthesis, where one can adjust parameters such as catalyst loading and reaction time to achieve desired properties in the end product. Various modeling approaches, including empirical and mechanistic methods, offer benefits depending on the specific polymerization system being studied. Recent advances in computational techniques further enhance the validity of such models because they are capable of dealing with advanced variables and interactions that occur during polymerization reactions. In addition to traditional approaches, recent models utilize machine learning algorithms to provide more accurate predictions and improve kinetics analysis accuracy for polymerization. By feeding prior experiments' big data sets into machine learning models, researchers are able to identify patterns and correlations that more traditional models might not.

Through process-based modeling on data, real-time control over polymerization processes is guaranteed, hence optimizing efficiency and reducing waste. Furthermore, the advances in multiphysics modeling emphasize the necessity of including many factors, such as solvent effects and heat dynamics, which have the potential to make a great difference in polymerization kinetics. Utilizing these sophisticated mathematical modeling techniques, scientists are equipped to address the complexity in polymer production and thus achieve material science breakthroughs. Fuzzy logic integration into mathematical modeling is another avenue that can be utilized in exploring the uncertainty associated with polymerization kinetics. For instance, as mentioned in the case of pesticide concentration kinetics, fuzzy models are quite capable of dealing with variability and imprecision that may be found in experimental data. These kinds of models would make strong prediction development possible, particularly where precise measurements of reaction conditions are hard to obtain. With the use of fuzzy inference systems along with conventional mathematical tools, researchers can develop more robust and flexible models to analyze polymerization reactions under diverse situations. This flexibility not only enhances the understanding of kinetics but also provides avenues towards improved formulation strategies. By basing itself on these inter-disciplinary modeling paradigms, polymerization kinetics as a field continues evolving, underlining the importance of mathematical modeling in building theoretical and applied applications within material science (Chen et al., 2023; Leidy et al., 2024).

5.5. EXPERIMENTALTECHNIQUESFORSTUDYING POLYMERIZATION

The versatility of experimental approaches to the study of polymerization lies in their ability to provide insight into the kinetics of various polymerization reactions. Experimental techniques such as linear sweep voltammetry (LSV) have been particularly useful in the study of heterogeneous radical polymerization mechanisms. For instance, the control of the thickness of the diffusion layer through techniques such as stirring and ultrasonic waves can increase the rate of polymerization without compromising the quality of polymer produced. As described in comparative research, the utilization of these external mass transfer enhancement strategies allows scientists to provide insight into the kinetics of reactions under variable conditions, further advancing radical polymerization knowledge. Such practices allow space for theoretical simulations to be cross validated with experimental observation, which constitutes an integrated method toward investigating polymerization kinetics taking into consideration practical as well as theoretical perspectives (Tang et al., 2025). Apart from electrochemically mediated techniques, solid-state techniques also offer stable structures for the study of polymerization kinetics. Advanced characterization techniques like Scanning Electron Microscopy (SEM), Fourier Transform Infrared Spectroscopy (FTIR), and X-ray Diffraction (XRD) define the compositional and structural details of formed polymers. The methods enable assessment of various parameters that affect polymerization, such as temperature, pH, and monomer concentration. The interaction between these variables can have strong impacts on the reaction kinetics of the polymerization. For example, studies on adsorption efficacy of single polymer particles show that varying environmental conditions can lead to varying efficiencies and polymerization rates, validating the vital importance of precise control in experimental conditions. These complete analyses ensure that product polymers have desired properties and functionalities (Şeyma et al., 2024). In addition, the integration of kinetic modeling and experimentation increases total polymer behavior knowledge. The application of isothermal models like the Langmuir isotherm and kinetic models like the second-order reaction model enables researchers to correlate experimental data quantitatively with theoretical estimates. This not only validates experiment but also optimizes synthesis conditions for polymers. By comparing these kinetic models with experimental observations side-by-side, disparities are uncovered and avenues for methodological refinement are uncovered. The findings are critical feedback to refine experimental design and theoretical models and eventually lead to more effective means of doing polymer research. This is a constant loop required to enhance the kinetics of polymerization and ensure that novel materials of polymers meet the evolving demands of many applications.

5.5.1. Differential Scanning Calorimetry (DSC)

Differential Scanning Calorimetry (DSC) is a powerful thermal analysis method applied to examine the heat behavior and crystallization tendencies of polymers throughout various phases of polymerization. By determining the heat flow involved in material transitions, such as melting or crystallization, DSC facilitates researchers to gain significant information on the polymer formation mechanism and kinetics. For instance, DSC can examine nonisothermal crystallization behavior of different samples of polylactic acid in a most efficient manner, which reveals the influence of chain length on crystallinity. Based on this data, polymer performance can be predicted, thus making DSC a valuable tool for optimizing conditions applied during the synthesis and processing of polymers. DSC's ability to interpret crystallization kinetics has been demonstrated through the investigation of three-arm polylactic acid (TA-PLA) samples of varying chain lengths. Surprisingly, it was revealed through analysis that there was a high inverse relationship of chain length to crystallization ability; for example,

TA-PLA12.1 with the highest chain length only reached a crystallinity of 38.21% when cooled at a rate of 1°C/min, whereas TA-PLA3.9 reached a high crystallinity of 70.75%. This variation highlights the significance of molecular structure on thermal properties, since longer chain length not only influences the rates of crystallization but also influences crystalline morphology as observed using polarized optical microscopy. This kind of crystallization morphology highlights the complexity of polymer physics, underscoring the excellent applicability of DSC in exploring polymerization kinetics. Further, DSC is not merely a simple thermal change; it is also an important source of information for the kinetic and structural investigation of copolymers obtained via various process schemes, such as coordination polymerization. In one such significant investigation, the thermal stability of block copolymers obtained from n-hexyl isocyanate and 2-chloroethyl isocyanate was evaluated by means of DSC as well as other techniques. This approach provided valuable insight regarding the reactivity ratios of the monomers and their influence on the architecture of the resulting copolymer. The strong characterization provided by DSC, including the calculation of activation energies for thermal degradation, is imperative in describing the manner in which these materials respond under thermal stress and are durable. Hence, DSC not only gives information about kinetic parameters but also contributes to the broader knowledge of polymer structure effect on thermal properties in the field of polymerization dynamics (Figure 5.8).

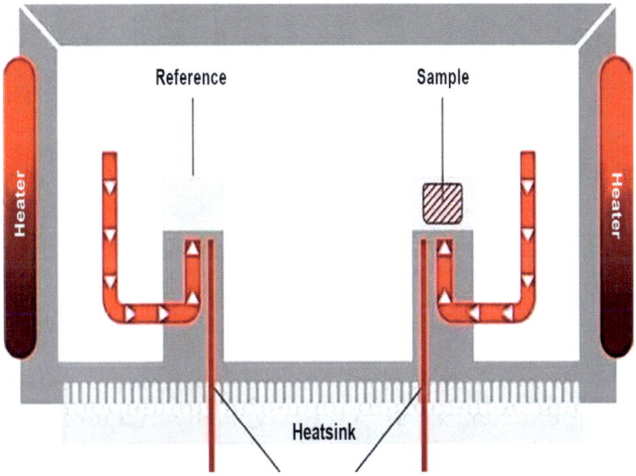

Figure 5.8. *Exploring differential scanning calorimetry (DSC).*

Source: https://www.azom.com/image-handler/ts/20230306023612/ri/600/src/images/Article_Images/ImageForArticle_22521_16780879668181326.jpg.

5.5.2. Gel Permeation Chromatography (GPC)

Gel Permeation Chromatography (GPC) is a critical analysis technique in polymer science, particularly in the case of polymerization kinetics. Using this process, scientists can differentiate between polymers based on their molecular weight distribution, which is critical in assessing the impact of molecular weight on the physical characteristics of polymers. Through calibration of GPC with standards of known molecular weights, it is possible to accurately determine the molecular weights of freshly prepared polymers, as evidenced in the preparation of the monomers 1-(isopropylamino)-3-(1-naphthyloxy)-2-propanoacrylate (IANOPA) and 1-(isopropylamino)-3-(1-naphthyloxy)-2-propanomethacrylate (IANOPMA). Polymer characterization through GPC is crucial in the evaluation of their potential use, e.g., effectiveness as antibacterial agents, thereby linking molecular features to functional effects in utilitarian applications (Kumar et al., 2024).

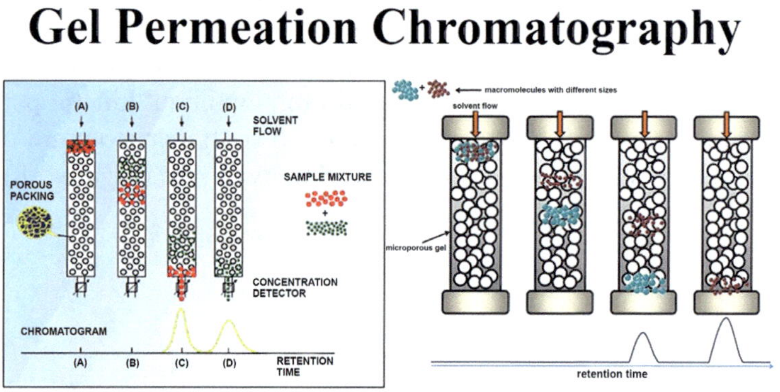

Figure 5.9. *Gel permeation chromatography (GPC).*

Source: https://microbenotes.com/wp-content/uploads/2019/04/Gel-Permeation-Chromatography.jpg.

Additionally, the deficiency of GPC typically requires supplementary methods for improved copolymer analysis. While GPC effectively determines molecular weight, it does not necessarily report copolymer composition. These deficiencies were addressed in recent years by applying artificial neural networks (ANNs) to the interpretation of GPC data, which allow scientists to not only detect molecular weights but also predict copolymer chemical compositions. For instance, GPC-ANN method has exhibited high accuracy of correlation between experimental and calculated values of an *r*-value of over 0.98 (Nagy et al., 2023). The combination of GPC with advanced computational methods is a perfect example of a breakthrough in polymer characterization, where the understanding of polymerization kinetics is enhanced by the insight into molecular architectures. Overall, the inclusion of GPC in the analysis of polymers significantly improves the study of the kinetics of polymerization. As this method and technique are further refined by researchers, the knowledge generated from the analysis using GPC can significantly influence the design of polymers for specific applications. Synthesis and characterization of molecules such as poly-IANOPA and poly-IANOPMA provide the practical relevance of the method. By revealing molecular weight distributions and characteristics of newly prepared materials, GPC not only assists in the optimization of polymer characteristics but also assists in filling the gap between theoretical studies and applications, thereby facilitating innovation in a wide range of industrial sectors (Kumar et al., 2024) (Figure 5.9).

5.5.3. Nuclear Magnetic Resonance (NMR) Spectroscopy

Nuclear magnetic resonance (NMR) spectroscopy has emerged as an important analysis tool in the study of polymerization kinetics. The technique allows researchers to ask questions regarding the molecular structure, motion, and interactions of polymer systems and thereby gain a deeper understanding of how materials evolve over time during polymerization reactions. It is possible with the help of NMR spectroscopy for scientists to monitor the course of the reaction, determine the degree of polymerization, and assess the insertion of various monomers into the polymer chain. Because NMR responds to the variability in chemical environment, it is particularly useful for the observation of varying behavior of polymers that is traceable to particular catalyst systems or reaction conditions. This capability is central to optimizing polymerization strategies and improving the qualities of the subsequent materials. The incorporation of NMR spectroscopy into the discipline of polymer science presages a more profound comprehension of polymer behavior as a function of various conditions. Within the realm of polymerization kinetics, researchers utilize numerous NMR techniques, including proton NMR and diffusion-ordered NMR, to gather data on molecular weight distribution and mobility of chains. These measurements are central to the establishment of the kinetics of reactions, for instance, initiation, propagation, and termination steps that establish the resulting properties of the polymer. Besides, advances in NMR technology, for instance, high-resolution technique and multidimensional NMR, further enhance the ability to investigate complicated polymer systems. By knowing the influence of different factors on end product properties and on reaction rates, researchers can design polymers for targeted applications, thus bridging the gap between applied and basic research. NMR spectroscopy is also vital in polymer matrix characterization, particularly when it comes to drug delivery via polymer nanocarriers, as explained in recent literature. For instance, citation shows application in explaining pH-induced drug release kinetics of drugs from polymeric drug delivery systems as it indicates how NMR facilitates the discovery of dynamic drug molecule interactions and interactions with drug carriers. Such is essential in drug release studies where drug release kinetics guides designing effective nanomedicines with regulated drug release behaviors for minimizing side effects. Thus, the inclusion of NMR as a common analysis tool can contribute towards polymer design innovations where the polymerization kinetics control is accurate and performance in real applications within biomedicine and materials science improves. Finally, the inclusion of NMR spectroscopy into polymerization studies is a key drive towards developing new polymeric materials and uses (Figure 5.10).

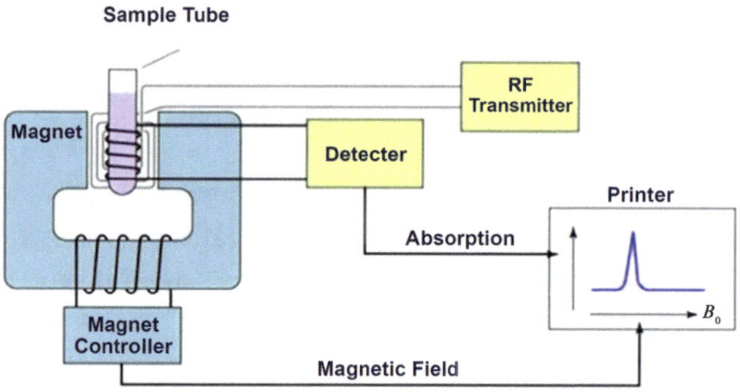

Figure 5.10. *Nuclear magnetic resonance (NMR) spectroscopy.*

Source: https://microbenotes.com/wp-content/uploads/2018/11/Nuclear-Magnetic-Resonance-NMR-Spectroscopy.jpg.

5.6. APPLICATIONS OF POLYMERIZATION KINETICS

Knowledge of polymerization kinetics is a critical foundation stone for the behavior and properties of the majority of polymers under conditions of practical reality. In the production of ultra-high-molecular-weight polyethylene (UHMWPE), for instance, the intricate interaction between conditions of polymerization and properties of the final material is of overriding significance. Recent advances in catalytic synthesis have led to the achievement of disentangled UHMWPE, with exceptional mechanical robustness and enhanced processability. This is partly because of optimization of polymerization conditions that limit entanglements of chains, thus allowing better processing technologies through novel heterogeneous catalytic systems. The implications of these developments ripple across a wide range of engineering and biomedical applications, and this highlights the importance of kinetics in informing material performance and design. By the utilization of knowledge gained through polymerization kinetics, researchers have been able to develop polymers of enhanced performance features ideal for specific industrial demands (Li, 2025). Polymerization kinetics is also one of its important uses in the application of 3D printing, using biodegradable materials such as polylactide (PLA). The kinetic degradation of PLA 3D-printed structures by hydrolysis is an important aspect in determining the durability and environmental applicability of the material. Recent studies have employed advanced techniques like micro-computed tomography and X-ray diffraction to analyze structural and molecular changes that occur during degradation. By understanding the kinetics involved, researchers can optimize the additive manufacturing process to achieve the best material performance and degradation rate to make the process more environmentally sustainable. The application of polymerization kinetics, in this case, enhances the quality of the printed materials in addition to contributing to pushing the debate on sustainable manufacturing processes (Bartłomiej et al., 2024). Furthermore, the integration of polymerization kinetics into the field of material science has potential revolutionizing capabilities in the production of new polymers with engineered properties. Enhanced catalyst engineering and control over polymerization have enabled possibilities for designing innovative materials designed for particular industrial applications faster. As an example, the use of engineered catalysts can control chain propagation and crystallization kinetics, thereby affecting polymer properties like tensile strength and flexibility. With the progression of science, an understanding of the underlying kinetics of polymerization will remain paramount to both scientists and engineers. It enables innovation, synthesizing polymers that can address today's problems in a variety of industries, from automobiles to biomedical. Thus, the strategic deployment of polymerization kinetics remains the foundation of material innovation and industrial heritage and still prescribes the boundaries of polymer science.

5.6.1. Material Science

The study of the kinetics of polymerization gives profound insights into material science, particularly where polymers are formed and remain stable. The understanding of polymer reaction kinetics is critical in terms of optimizing processing conditions and altering material properties. In recent developments, focus has been at the interface between polymerization rate and structural integrity because both contribute significantly to the overall performance of the final material. This junction is particularly important in advanced applications, such as 3D printing, where one has such precise control over polymerization kinetics that it can create enhanced material properties. With the use of techniques such as laser confocal microscopy, researchers can investigate the photopolymerization kinetics in real-time, providing an extended investigation into how various process conditions affect the properties of printed materials and thereby bridge the gap between theoretical models and actual applications (Arinstein et al., 2018). Additionally, thermooxidative stability of polymer composites such as poly(methyl methacrylate) (PMMA) has attracted keen interest in material science.

Figure 5.11. *Material science.*

Source: https://astiacademy.ac.ae/wp-content/uploads/2023/08/Why-Materials-Science-is-Important.jpg.

The addition of silica nanoparticles and micro into PMMA demonstrates how variations in filler size and filler content are capable of dramatically altering thermal degradation behaviors. Through the use of thermogravimetry, researchers are able to simulate thermooxidative degradation conditions and examine the role of these additives on the overall stability of the polymer matrix. The results indicate that larger silica particles can destabilize PMMA composites, and smaller nanoparticles exert very little effect at low concentrations. This knowledge of the composites performance under thermal stress elucidates the critical balance between reinforcement and stability, emphasizing the need for accurate material design in functionally demanding applications that involve resilience and endurance (Čechová et al., 2012). The integration of polymerization kinetics with advancements in material manufacturing leads to state-of-the-art methods for designing functional materials. This can particularly be observed for optical devices, in which 3D printed optical components with controllable and consistent properties are very important. There are issues still, though, since structural defects in the printed structure can get in the way of propagating light, thereby affecting performance. The knowledge gained from the study of photopolymerization kinetics enables researchers to make method designs that can enhance the clarity and functionality of printed optical devices. By rendering the factors that limit light performance rational, researchers can propose experimental approaches to engineer the best optical properties, thus opening the door to innovations in applications from telecommunications to biomedical devices, where precision and material integrity are paramount (Arinstein et al., 2018) (Figure 5.11).

5.6.2. Pharmaceutical Development

In drug development, the use of polymers in drug formulation increases in importance. The intricate relationships between the kinetics of polymerization and drug biocompatibility have opened up avenues for the design of more effective drug delivery systems. Such breakthroughs like molecularly imprinted polymers (MIPs) have emerged as crucial tools in active pharmaceutical ingredient extraction and purification. The synthesis techniques, including bulk polymerization and surface precipitation methods, provide special advantages of specificity and binding capacity. For instance, the utilization of D-panthenol as a template for imprinting in MIPs has been successful in providing selective adsorption properties that lead to increased recovery yields of the target compounds from complex matrices. Hence, polymerization kinetics has a direct influence on the efficacy of MIPs, ensuring the stability and safety of drugs (Andrea et al., 2025). The application of hydrogels to drug development provides a perfect example of how development has been advanced through studies of polymerization kinetics. Polyvinyl alcohol (PVA)-methylcellulose (MC) blends were established to work efficiently as far as their swelling characteristics and rates of drug release are concerned. Experiments have also indicated that the increase in volumetric content of PVA increases the swelling capacity within artificial saliva, a key factor in developing effective drug delivery systems. Moreover, the sorption and release kinetics of active pharmaceutical ingredients from these hydrogels have established that while an increase in content of PVA contributes to retention capacity, it tends to leach out release rates. This becomes crucial in cases of controlled drug delivery where both dosage and time duration are of utmost concern for therapeutic success (Zarina et al., 2025). The application of polymer-based systems in drug delivery initiates a study of their mucoadhesive nature, and this can have a critical impact on patient compliance and efficiency. The ability of PVA-MC hydrogels to be adhered to oral mucosa for an extended period of time signifies their potential for buccal drug delivery devices. Investigations showed that mucoadhesive properties are influenced by ratios of polymers in hydrogels, whereby taking into account the polymerization kinetic parameters can lead to tailored formulation to achieve optimal adhesion. These advancements not just enhance the therapeutic value of the medicine but also ensure stability and lifespan in providing prolonged release. Their continuous research into the intricate relationship between polymerization procedures and drug formulation can make pharmaceutical development more sophisticated in meeting clinical needs (Zarina et al., 2025) (Figure 5.12).

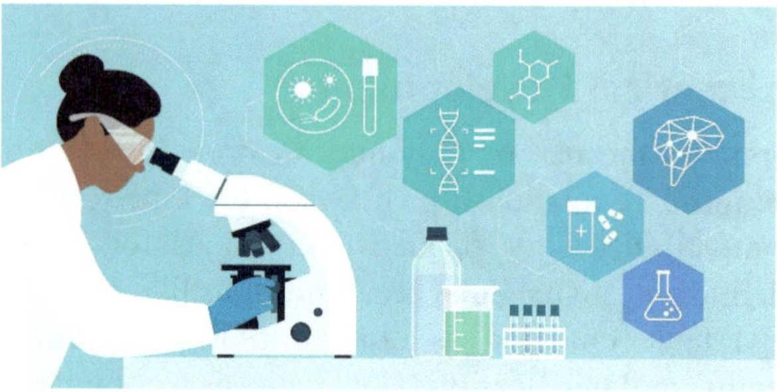

Figure 5.12. *Pharmaceutical development.*

Source: https://www.news-medical.net/images/Article_Images/ImageForArticle_25145_17234677524466225.jpg.

5.6.3. Environmental Impact and Biodegradable Polymers

The environmental impact of traditional polymers, particularly plastics, has been a matter of concern over the past few decades. The traditional plastics are highly durable and resistant to degradability, leading to massive buildup in landfill sites and nature. Their long lifespan generates severe environmental impacts, particularly in aquatic environments where plastic wastes are the causative agents of pollution and damage to wildlife. The pressing need for environment-friendly alternatives has spurred the development of biodegradable polymers, which hold promise for the reduction of plastic wastes and attendant environmental problems. By harnessing the natural degradation systems catalyzed by microorganisms, biodegradable polymers can reduce pollution if integrated into waste management systems. The capacity of such materials to enhance environmental sustainability, as pointed out by other studies, highlights their role in addressing the problem of plastic waste through more eco-friendly means (Yao et al., 2024). Biodegradable polymers such as polylactic acid (PLA) and starch-based items are founded on typical chemical frameworks to guarantee their degradation within natural environments. Degradation can either be enzymatic or abiotic, thus allowing polymer chains to be broken down into smaller fragments that are susceptible to microbial metabolism. For PLA, research has illustrated how catalysts, for instance, $SnCl_2$, play vital roles both in the polymerization kinetics as well as the degree of resulting biodegradability of the polymer. In addition, modifications of the physical and chemical properties of starch polymers also influence their kinetics of degradation, which are adapted to specific applications.

Figure 5.13. *Environmental problems caused by polymers.*

Source: https://www.azocleantech.com/images/Article_Images/ImageForArticle_1189_16161632736447928.jpg.

The interaction of environmental conditions, such as temperature and water content, and the intrinsic property of the polymer plays an essential role in governing the efficiency of the biodegradation process. Because of this, it is important to appreciate the underlying kinetics for more effective biodegradable products (Yao et al., 2024). The transition to biodegradable polymers also entails a wider consideration of their life cycle and overall environmental impact. Although these materials present a solution to plastic pollution, their production and end-of-life scenarios also need to be taken into account. Production of biodegradable polymers remains rooted in petrochemical processing, which can negate some of their positive environmental benefit. It is crucial to assess the lifecycle energy inputs and emissions of traditional and biodegradable polymers in establishing a scientifically sound basis for their use. Moreover, current research is creating new techniques for the stimulation of polymer biodegradation, including the addition of additives and modifying the processing routes. Thus, knowledge regarding the polymerization kinetics of such biodegradable plastics is not only necessary for their enhanced functional characteristics but also for enabling their sustainable consumption behaviors and conserving the environment (Gonçalo et al., 2023) (Figure 5.13).

SUMMARY

Polymerization kinetics is a significant topic in understanding the more general uses of polymer science, particularly in environmental and health problems. As the use of synthetic polymers continues to rise, their impact on ecosystems and human health has become a significant issue. Microplastics, created through the degradation of larger plastic pieces, can now be detected in nearly every habitat and are categorized as a significant pollutant. Furthermore, some of these microplastics also contain toxic additives that were inherent to their original polymer structure. It is worth adding that in its lifecycle, a polymer changes, leading to chemicals that have grave health and environmental implications. Thus, the effects of polymerization are not mere chemical reactions; they are intertwined with social responsibility and the environment, warranting an in-depth exploration of prevention and sustainability measures. Moreover, extending the foundation of polymerization kinetics can uncover vital insights into the design and utilization of future materials that break microplastic-related risks. In recent studies, it was mentioned that multiple physical–chemical properties of thermoplastics make them as well-complexed stressors in nature, impacting biodiversity as well as ecosystem balance. Therefore, this complexity necessitates an overhaul of the utilization and production of polymers in consumer products. There is a necessity to offer a rigorous review of polymer formulations, in this instance, the possibly poisonous additives, to inform regulatory measures and public health approaches. The case is that examining catalyst design in conjunction with the oxygen evolution reaction within polymers can create innovative modifications that enhance performance but reduce environmental impacts. The integration of these technological advancements strongly keeps the focus on future efforts in polymerization kinetics. In short, the understanding in polymerization kinetic research is key to resolving the intricate problems generated by synthetic polymers. The need for comprehensive research on microplastics' environmental and health effects represents a critical juncture for the field. Current literature indicates a deficiency in the comprehension of the fate of various polymer additives and their long-term effects, which remain insufficiently covered. Closing these gaps in knowledge ensures that not only are future polymer science breakthroughs effective, but they will also be environmentally friendly. As we move forward, it is important that there be a concerted push from researchers, policymakers, and industry stakeholders to make sure that polymerization kinetic breakthroughs result in practices that secure human health and the environment. This widespread approach not only increases scientific understanding but also opens the way for sustainable polymer chemistry in the future.

The polymerization kinetics is a topic that provides essential data regarding the mechanisms of polymer formation, the required macromolecules for biological systems. Another essential aspect of polymerization is the character of polymerases, the enzymes responsible for catalyzing the nucleic acid synthesis of molecules like DNA and RNA. Through a complex kinetic comparison, researchers have demonstrated that Therminator DNA polymerase is very efficient in catalyzing the addition of both DNA and threosyl nucleic acid (TNA) derivatives. Comparing the rate of single-nucleotide incorporations, they noted that the incorporation rate of TNA did not decrease drastically when using a DNA-TNA chimeric primer, demonstrating the polymerase flexibility. These findings underscore the enzyme's ability to shuttle the gap between the various forms of nucleic acids, suggesting that TNA may be capable of potentially assuming a more central role in life's early evolutionary stages of molecular machinery, in support of current hypotheses on life origins (Chaput et al., 2005). Apart from single polymerase efficiency, even the collective behavior of the molecular machines offers intriguing insights into polymerization kinetics. Polymerases' ability to tap chemical energy for processive movement along a template is suggestive of the involved and advanced nature of these processes, essentially equivalent to tape-copying Turing machines. Biological polymerization relies on such natural nanomachines that, processively moving along linear templates, sequentially add monomers and ultimately lead to the synthesis of critical macromolecules like DNA, RNA, and proteins. Because these interactions are susceptible to analysis within the framework of statistical physics, they have common operating modes, despite the fact that each one of these polymerases performs a distinct function. This observation not only sheds light on the fundamental nature of life but also provokes the development of novel kinetic models, which can potentially forecast the behavior of these systems in a variety of settings (Chowdhury et al., 2012). The implications of these kinetic studies are extensive, extending to both theoretical and practical applications in fields like molecular biology and synthetic chemistry. For instance, the identification of the minimum difference in the catalytic activity between TNA and DNA synthesis opens up our appreciation of polymerization dynamics, suggesting potential avenues for genetic material engineering. Also, the understanding gained from the research on the operation of polymerizing machines can be utilized for creating new methods of synthesizing synthetic polymer systems. Through the understanding of how these biological machines operate and the kinetic frameworks behind their operation, scientists can innovate across fields from biotechnology to materials science. Hence, the richness of polymerization kinetics not only exposes the molecular system development history but also offers a doorway to future technological advancements in polymeric synthesis and design.

REFERENCES

1. Ali, A., Naveed, A., Rasheed, T., Aziz, T., Imran, M., Zhang, Z., Ullah, M., et al. (2022). Methods for predicting ethylene/cyclic olefin copolymerization rates promoted by single-site metallocene: Kinetics is the key. *Polymers, 14*. https://www.semanticscholar.org/paper/7eb9834a84ce1da0cd82901aac1d6a59df6029e2 (accessed on 10th September 2025).
2. Alimi, O. S., Farner, J. M., Hernandez, L. M., & Tufenkji, N. (2017). Microplastics and nanoplastics in aquatic environments: Aggregation, deposition, and enhanced contaminant transport. *Environmental Science & Technology, 52*, 1704–1724. https://doi.org/10.1021/acs.est.7b05559 (accessed on 10th September 2025).
3. Arinstein, A., Camposeo, A., Cardarelli, F., Persano, L., et al. (2018). 3D printing of optical materials: An investigation of the microscopic properties. https://core.ac.uk/download/587855339.pdf (accessed on 10th September 2025).
4. Ashfaq, A., Clochard, M. C., Coqueret, X., Dispenza, C., Driscoll, M., Ulański, P., & Al-Sheikhly, M. (2020). Polymerization reactions and modifications of polymers by ionizing radiation. *Polymers, 12*, 2877–2877. https://doi.org/10.3390/polym12122877 (accessed on 10th September 2025).
5. Campanale, C., Massarelli, C., Savino, I., Locaputo, V., & Uricchio, V. F. (2020). A detailed review study on potential effects of microplastics and additives of concern on human health. *International Journal of Environmental Research and Public Health, 17*, 1212–1212. https://doi.org/10.3390/ijerph17041212 (accessed on 10th September 2025).
6. Čechová, E. (2012). Thermooxidative stability of PMMA composites. https://core.ac.uk/download/30296004.pdf (accessed on 10th September 2025).
7. Chai, Q., Jiao, Y., & Yu, X. (2017). Hydrogels for biomedical applications: Their characteristics and the mechanisms behind them. *Gels, 3*, 6–6. https://doi.org/10.3390/gels3010006 (accessed on 10th September 2025).
8. Chaput, J. C., Horhota, A., Ichida, J. K., McLaughlin, L. W., et al. (2005). Kinetic analysis of an efficient DNA-dependent TNA polymerase. https://core.ac.uk/download/323075466.pdf (accessed on 10th September 2025).
9. Chen, C., & Pan, Z. (2023). An overview of progress, challenges, needs and trends in mathematical modeling approaches in food drying. *Drying Technology, 41*, 2586–2605. https://www.semanticscholar.org/paper/0c5165ab28ab84ee1c301ef5651c1cf91a21f50a (accessed on 10th September 2025).

10. Chowdhury, D., & Sharma, A. K. (2012). Template-directed bio polymerization: Tape-copying Turing machines. http://arxiv.org/abs/1212.6827 (accessed on 10th September 2025).
11. DiMartini, E. T., Lowe, C. J., & Shreiber, D. (2023). Alternative chemistries for free radical-initiated targeting and immobilization. *Journal of Functional Biomaterials, 14*. https://www.semanticscholar.org/paper/c0134339468b2a69c4484c194a1d50c2d84918a0 (accessed on 10th September 2025).
12. Diment, W. T., Gowda, R. R., & Chen, E. Y. X. (2024). Unraveling the mechanism of catalyzed melt-phase polyester depolymerization via studies of kinetics and model reactions. *Journal of the American Chemical Society*. https://www.semanticscholar.org/paper/c8d64c6eb78b742cc2e88ba058421f3e7c3e8e73 (accessed on 10th September 2025).
13. Espitia-Cruz, L. V., & Villamizar-Roa, É. J. (2024). Comparison of two approaches for solving a fuzzy system describing the degradation kinetics of pesticides in plant uptake models. *Mathematical Modelling of Natural Phenomena*. https://www.semanticscholar.org/paper/55ca6a39f201adf1f62f62e2dea76508ab127eab (accessed on 10th September 2025).
14. Guo, T., He, B., Mu, R., Li, J., Sun, C., Wang, R., Zhang, G., et al. (2024). Electrochemically mediated surface-initiated atom transfers radical polymerization by ppm of CuII/Tris(2-pyridylmethyl) amine. *Langmuir: The ACS Journal of Surfaces and Colloids*. https://www.semanticscholar.org/paper/dc2888d6837b88058456b3baec641c18923dbf43 (accessed on 10th September 2025).
15. Hao, H., Hutter, T., Boyce, B., Watt, J., Liu, P., & Mitlin, D. (2022). Review of multifunctional separators: Stabilizing the cathode and the anode for alkali (Li, Na, and K) metal–sulfur and selenium batteries. *Chemical Reviews, 122*, 8053–8125. https://doi.org/10.1021/acs.chemrev.1c00838 (accessed on 10th September 2025).
16. He, J. X., Sui, B., Jin, Y., Zheng, C. X., & Jin, F. (2025). Cell condensation initiates organogenesis: The role of actin dynamics in supracellular self-organizing process. *Cell & Bioscience, 15*. https://www.semanticscholar.org/paper/961b9d8186cdbc147814248f307c3057a2e02174 (accessed on 10th September 2025).
17. Horkay, F. (2021). Polyelectrolyte gels: A unique class of soft materials. *Gels, 7*, 102–102. https://doi.org/10.3390/gels7030102 (accessed on 10th September 2025).

18. Karnes, J., Weisgraber, T., Cook, C. C., Wang, D. N., Crowhurst, J., Fox, C., Harris, B. S., et al. (2022). Isolating chemical reaction mechanism as a variable with reactive coarse-grained molecular dynamics: Step-growth versus chain-growth polymerization. *Macromolecules*. https://www.semanticscholar.org/paper/4a19c3c5f2081ddd75a01fff4519a2dac0dac7bb (accessed on 10th September 2025).
19. Kenessova, Z. A., Mun, G. A., Urkimbayeva, P. I., Toktabayeva, A. K., Rakhmetullayeva, R., Yermukhambetova, B., Kenzhebai, Z., et al. (2025). Development and characterization of polymer blends based on polyvinyl alcohol for application as pharmaceutical dosage form. *Polymers*. https://www.semanticscholar.org/paper/3717c368a30e894c1d6301f97f4ac5c7a1a03eb4 (accessed on 10th September 2025).
20. Klimczuk, B., Rudnicka, A., Owczarek, O., Puszkarz, A., Szparaga, G., & Puchalski, M. (2024). Investigation of the hydrolytic degradation kinetics of 3D-printed PLA structures under a thermally accelerated regime. *Materials, 17*. https://www.semanticscholar.org/paper/d0226d531e3a9af195550dc48955cfd14aa18c7e (accessed on 10th September 2025).
21. Köchermann, J., Mühlenberg, J., & Klemm, M. (2018). Kinetics of hydrothermal furfural production from organosolv hemicellulose and *d*-xylose. *Industrial & Engineering Chemistry Research*. https://www.semanticscholar.org/paper/074bf0263ac0b4c65b3e61a30ddb857924d87f2e (accessed on 10th September 2025).
22. Kumar, J. S., Kamaraj, P., Vivekanand, P. A., Periyasami, G., Rahaman, M., Karthikeyan, P., Uthappa, U. T., et al. (2024). Design of novel poly(propranolol) acrylate and methacrylate polymers through radical polymerization for antibacterial activity and metal ion absorption. *International Journal of Polymer Science*. https://www.semanticscholar.org/paper/2004f90abd304aacbf5aca26df079cc2ad5c7f2b (accessed on 10th September 2025).
23. Lang, M., Hirner, S., Wiesbrock, F., & Fuchs, P. (2022). A review on modeling cure kinetics and mechanisms of photopolymerization. *Polymers, 14*. https://www.semanticscholar.org/paper/caaa658456a7710465bb32ba840b578295555a1a (accessed on 10th September 2025).
24. Li, J., Dong, R., Wang, X., Xiong, H., Xu, S., Shen, D., Song, X., et al. (2015). One-pot synthesis of magnetic molecularly imprinted microspheres by RAFT precipitation polymerization for the fast and selective removal of 17β-estradiol. *RSC Advances,*

5, 10611–10618. https://www.semanticscholar.org/paper/b9f14af3ec9e0304f46098fc111644c7764646d2 (accessed on 10th September 2025).

25. Li, L. (2025). Engineering nascent disentangled ultra-high-molecular-weight polyethylene based on heterogeneous catalytic polymerization. *Organics*. https://www.semanticscholar.org/paper/c850f7dbf3b91ae71126cfd7ae15c0aacd81018c (accessed on 10th September 2025).
26. Li, M., Fromel, M., Ranaweera, D., Rocha, S. P. D., Boyer, C., & Pester, C. W. (2019). SI-PET-RAFT: Surface-initiated photoinduced electron transfer–reversible addition–fragmentation chain transfer polymerization. *ACS Macro Letters, 8*, 374–380. https://doi.org/10.1021/acsmacrolett.9b00089 (accessed on 10th September 2025).
27. Li, X., Zheng, M., Ren, C., & Guo, L. (2021). ReaxFF molecular dynamics simulations of thermal reactivity of various fuels in pyrolysis and combustion. *Energy & Fuels, 35*, 11707–11739. https://doi.org/10.1021/acs.energyfuels.1c01266 (accessed on 10th September 2025).
28. Libánská, T., Špringer, L., Pestova, K., Kotalík, R., Konefał, A., Šimonová, T., Křížek, T., et al. (2023). Using surface plasmon resonance, capillary electrophoresis and diffusion-ordered NMR spectroscopy to study drug release kinetics. *Communications Chemistry, 6*. https://www.semanticscholar.org/paper/957ec58efe9e30f19bef830041cf8f7a6ece2fff (accessed on 10th September 2025).
29. Liu, L., & Corma, A. (2020). Evolution of isolated atoms and clusters in catalysis. *Trends in Chemistry, 2*, 383–400. https://doi.org/10.1016/j.trechm.2020.02.003 (accessed on 10th September 2025).
30. Liu, Y., Lv, H., Kang, B., Li, S., Zhen, Z., & Zheng, J. (2021). Diffusion kinetics investigations of the dual monomer system in the photopolymer holographic material. *Proceedings of SPIE, 12066*, 120661N–120661N-7. https://www.semanticscholar.org/paper/76cc511124d9a57d3ff65212a8e16c5ceded3b43 (accessed on 10th September 2025).
31. Mantzara, D., Katara, A., Panteli, M., Stavrakaki, I. G., Plachouras, N., Choinopoulos, I., & Pitsikalis, M. (2024). Synthesis and characterization of statistical and block copolymers of *n*-hexyl isocyanate and 2-chloroethyl isocyanate via coordination polymerization. *Journal of Polymer Science*. https://www.semanticscholar.org/paper/67d916184385821ae70316b15242d63cdae12944 (accessed on 10th September 2025).
32. Meng, W. H., Zhang, X., Hu, X., Liu, Y., Zhang, J., Qu, X., & Abdel-Magid, B. (2023). Mechanical properties and non-isothermal crystallization kinetics of

polylactic acid modified by polyacrylic elastomers and cellulose nanocrystals. *Polymers, 15*. https://www.semanticscholar.org/paper/3271fbb39d8f7ae904b038953127117081b9b5fc (accessed on 10th September 2025).

33. Nagy, T., Róth, G., Benedek, M., Kuki, A., Timári, I., Zsuga, M., & Kéki, S. (2023). Enhanced copolymer characterization for polyethers using gel permeation chromatography combined with artificial neural networks. *Analytical Chemistry*. https://www.semanticscholar.org/paper/aa6e4af5f8ec9a401753bceaf2278a48e6fd39dd (accessed on 10th September 2025).
34. Patra, J. K., Das, G., Fraceto, L. F., Ramos Campos, E. V., Rodriguez-Torres, M. del P., Acosta-Torres, L. S., Díaz-Torres, L. A., et al. (2018). Nano based drug delivery systems: Recent developments and future prospects. *Journal of Nanobiotechnology, 16*. https://doi.org/10.1186/s12951-018-0392-8 (accessed on 10th September 2025).
35. Periyasami, N., & Kumaraguru, K. (2025). Comparative and kinetic investigation of oxidation of 3-methylindole by potassium bromate in two different solvents. *International Journal of Environmental Sciences*. https://www.semanticscholar.org/paper/530292c8839cb74d6dc0162d7969b3017cb949e4 (accessed on 10th September 2025).
36. Plachouras, N., & Pitsikalis, M. (2023). Statistical copolymers of N–vinylpyrrolidone and 2–chloroethyl vinyl ether via radical RAFT polymerization: Monomer reactivity ratios, thermal properties, and kinetics of thermal decomposition of the statistical copolymers. *Polymers, 15*. https://www.semanticscholar.org/paper/03b84613ed49a6c610b388659f6252a5151a2dc4 (accessed on 10th September 2025).
37. Reddy, M. S. B., Ponnamma, D., Choudhary, R., & Sadasivuni, K. K. (2021). A comparative review of natural and synthetic biopolymer composite scaffolds. *Polymers, 13*, 1105–1105. https://doi.org/10.3390/polym13071105 (accessed on 10th September 2025).
38. Shaikhah, D., Rossi, C. O., De Luca, G., Angelico, R., Calandra, P., & Caputo, P. (2024). The use of nuclear magnetic resonance spectroscopy (NMR) to characterize bitumen used in the road pavements industry: A review. *Molecules, 29*. https://www.semanticscholar.org/paper/54dd564d20f1a5f95989aca2e8ea7b4dca5ef862 (accessed on 10th September 2025).
39. Skliutas, E., Lebedevaite, M., Kabouraki, E., Baldacchini, T., Ostrauskaitė, J., Vamvakaki, M., Farsari, M., et al. (2021). Polymerization mechanisms initiated by spatio-temporally confined light. *Nanophotonics, 10*, 1211–1242. https://doi.org/10.1515/nanoph-2020-0551 (accessed on 10th September 2025).

40. Song, J., Wei, C., Huang, Z. F., Liu, C., Zeng, L., Wang, X., & Xu, Z. J. (2020). A review on fundamentals for designing oxygen evolution electrocatalysts. *Chemical Society Reviews, 49*, 2196–2214. https://doi.org/10.1039/c9cs00607a (accessed on 10th September 2025).

41. Špačková, A., Solčianska-Schwarz, S., Kövérová, M., Lukačovič Vajová, I., Tiňo, R., & Hroboňová, K. (2025). Molecularly imprinted adsorbent and its magnetic version for solid phase extraction of D-panthenol: Comparison of synthesis approaches, properties, and applicability for different real samples extraction. *Journal of Separation Science, 48*(5), e70161. https://www.semanticscholar.org/paper/370595bcd48da2d7cc05a720cf9b474cd0ffbeb5 (accessed on 10th September 2025).

42. Tang, L., Jin, J., Xiong, J., Guo, J. K., Huang, X. Y., Zhou, Y. N., Luo, Z., et al. (2025). The crucial role of the diffusion layer in externally controlled heterogeneous reversible deactivation radical polymerization systems: A case study of eATRP. *AIChE Journal*. https://www.semanticscholar.org/paper/e6e7a3864eb56106ef9b6bdeeb5b002b21009fcb (accessed on 10th September 2025).

43. Tiago, G. A. O., Mariquito, A., Martins-Dias, S., & Marques, A. C. (2023). The problem of polyethylene waste–Recent attempts for its mitigation. *The Science of The Total Environment, 892*, 164629–164629. https://doi.org/10.1016/j.scitotenv.2023.164629 (accessed on 10th September 2025).

44. Tomar, A., Arora, U., & Singh, J. (2024). The study of the kinetics of D-mannose oxidation with imidazolium fluorochromate in aqueous acetic acid medium through spectrophotometric analysis. *French-Ukrainian Journal of Chemistry*. https://www.semanticscholar.org/paper/fa89fc31db399189727c0556ced229152a0452e5 (accessed on 10th September 2025).

45. Wang, Y., Zhang, Q., Huang, X. Y., Liu, Q., & Lang, J. (2025). Modulating solid-state photocycloaddition kinetics via ligand substituents and crystal structures in one-dimensional coordination polymers. *Journal of the American Chemical Society*. https://www.semanticscholar.org/paper/9f831d576dde8b2a704b1a057853bd6caca0af6b (accessed on 10th September 2025).

46. Wang, Z., Pan, X., Yan, J., Dadashi-Silab, S., Xie, G., Zhang, J., Wang, Z., et al. (2017). Temporal control in mechanically controlled atom transfers radical polymerization using low ppm of Cu catalyst. *ACS Macro Letters, 6*, 546–549. https://doi.org/10.1021/acsmacrolett.7b00152 (accessed on 10th September 2025).

47. Wen, X. (2025). Flamelet/quadrature-based moment method for modeling laminar shooting pulverized solid fuel flames. *Physics of Fluids*. https://www.semanticscholar.org/paper/28f827044e4b333e8f475cff1521c2413ee71550 (accessed on 10th September 2025).

48. Yadav Schmid, S., Ma, X., Hammons, J., Mergelsberg, S., Harris, B. S., Ferron, T. J., Yang, W., et al. (2024). Influence of peptoid sequence on the mechanisms and kinetics of 2D assembly. *ACS Nano, 18*, 3497–3508. https://www.semanticscholar.org/paper/8ae8f09112caa5df097e6e23da90c5f6e0b2e34a (accessed on 10th September 2025).

49. Yalçın Turan, Ş., Kara, A., & Tekin, N. (2024). Removal of acid violet 7 from aqueous solution with polymer matrix composite particles by adsorption and photocatalytic decolorization methods: Isotherms, kinetics, and thermodynamic studies. *International Journal of Environmental Analytical Chemistry, 105*, 1996–2021. https://www.semanticscholar.org/paper/9a6a543720058867c876c3da0b6a8f466332be2b (accessed on 10th September 2025).

50. Yan, H., Jiao, Y., Jin, B., Zhang, H., Fu, Z., Jia, S., & Deng, Y. (2024). Insights of mechanism and kinetics of acrylonitrile aqueous precipitation polymerization. *Journal of Macromolecular Science, Part A, 61*, 805–821. https://www.semanticscholar.org/paper/0f14d7217ecb215a51bc9909f51e7265c60b3c2a (accessed on 10th September 2025).

51. Yang, A., Yue, L., Shao, Y., Yang, Y., Wang, H., & Zou, G. (2025). Nonisothermal crystallization behavior and crystallization morphology of three-arm polylactic acid with different chain lengths. *Journal of Applied Polymer Science*. https://www.semanticscholar.org/paper/dee3f254500e8ebb6962975fd2bf1ccb48cff8cc (accessed on 10th September 2025).

52. Yao, X., Yang, X., Lu, Y., Qiu, Y., & Zeng, Q. (2024). Review of the synthesis and degradation mechanisms of some biodegradable polymers in natural environments. *Polymers, 17*, 66–66. https://doi.org/10.3390/polym17010066 (accessed on 10th September 2025).

53. Zhang, P., Tong, Y., Liu, Y., Vequizo, J. J. M., Sun, H., Yang, C., Yamakata, A., et al. (2020). Heteroatom dopants promote two-electron O_2 reduction for photocatalytic production of H_2O_2 on polymeric carbon nitride. *Angewandte Chemie International Edition, 59*, 16209–16217. https://doi.org/10.1002/anie.202006747 (accessed on 10th September 2025).

54. Zhang, Y., Ding, Z., Foster, C. W., Banks, C. E., Qiu, X., & Ji, X. (2017). Oxygen vacancies evoked blue TiO_2(B) nanobelts with efficiency

enhancement in sodium storage behaviors. *Advanced Functional Materials, 27*. https://doi.org/10.1002/adfm.201700856 (accessed on 10th September 2025).

55. Zurak, D., Vlajsović, D., Duvnjak, M., Salajpal, K., & Kljak, K. (2023). Factors affecting starch digestibility rate of maize grain in poultry. *World's Poultry Science Journal, 79*, 43–68. https://www.semanticscholar.org/paper/e55693eb59fd51d03378287186c3eb48e44f9aa5 (accessed on 10th September 2025).

CHAPTER 6

CATALYSIS

Contents

Introduction 207

6.1. Types of Catalysts 212

6.2. Mechanisms of Catalysis 218

6.3. Factors Affecting Catalytic Activity 223

6.4. Kinetics of Catalytic Reactions 227

6.5. Applications of Catalysis 232

6.6. Challenges in Catalysis 237

Summary 241

References 244

INTRODUCTION

Catalysis is a chemical fundamental principle which highly increases the rate of reactions without being degraded in the reaction process. Essentially, catalysis is a question of interaction between a catalyst and reactants and results in lowering activation energy barriers, thereby making it simpler for substrates to be transformed into products. Catalysis is crucial not just in scientific research but also in industrial operations in a broad array of sectors from pharmaceuticals, petrochemicals, through environmental science. Respected for its application in increasing reaction rates, catalysis often leads to enhanced yield and selectivity of target products. As such, understanding the essential principles of catalytic activity and the determinant of catalytic activity is crucial to chemists who want to optimize reaction conditions as well as design novel catalytic systems. In-depth interrogation of catalytic mechanisms reveals intricate information about the functioning of different catalysts. For example, the selectivity of a catalyst has a key role to play in defining the character of chemical conversions. Active site geometric orientation, substrate orientation, and molecular interaction type are some of the aspects that define the effectiveness of a catalyst. Advanced approaches like molecular dynamics simulations and spectroscopic methods have maximized our understanding of the mechanisms. A practical example is in investigations of human platelet ALOX12, where site-directed mutagenesis has identified important residues that influence substrate binding and product formation, and single mutations resulting in dramatic product specificity changes.

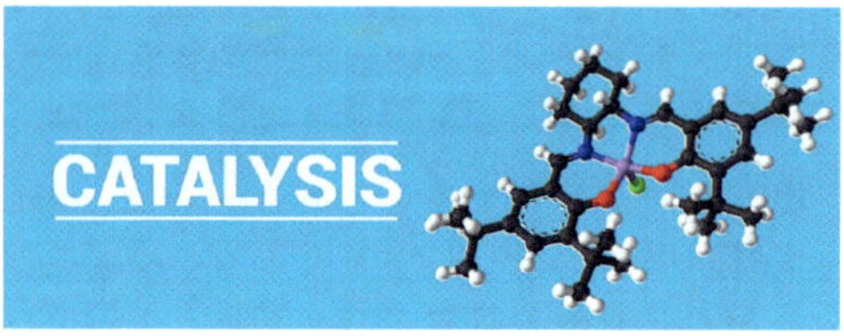

Figure 6.1. *Catalysis.*

Source: https://cdn1.byjus.com/wp-content/uploads/2018/12/catalysis.png.

Such insights are instrumental to the designing of catalysts for maximum reactivity and selectivity in processes of synthesis, thus bridging theory and application (Aleem et al., 2019). The implications of catalysis are not merely of interest to academics. Zeolites, for instance, have been instrumental in catalytic industrial processes due to their unique structural characteristics and ability to act as mediators for chemical changes under varied conditions. Zeolite catalysis studies have shown how the acidity and topology of the materials can significantly influence reaction rates and mechanisms. As demonstrated in cutting-edge modeling studies, the presence of guest molecules and temperature oscillations may influence zeolite-catalyzed reaction kinetic processes, for example, methanol-to-hydrocarbon conversion. Based on the understanding of such dynamics, chemists can leverage the special characteristics of catalysts to optimize efficiency and make the process compatible with other feedstocks, thereby facilitating sustainable operation in the chemical industry (Bailleul et al.,

2016). This underscores the pivotal position of catalysis not just in the evolution of chemical science but also in the resolution of this century's challenge in materials and energy (Figure 6.1).

Definition of Catalysis

The definition of catalysis involves a key aspect of chemical kinetics that has a significant impact on the rates of reaction. Catalysis is a reaction that involves a substance, or a catalyst, which accelerates a chemical reaction without being permanently changed in the process. This acceleration occurs through the provision of a different reaction pathway with lower activation energy than the uncatalyzed route. In this process, the catalyst interacts with reactants to bring about the transition state and thus an increased rate of formation of products. The application of catalysis in kinetics cannot be overstated since it facilitates reactions to occur under milder conditions and with greater efficiency, which is extremely important in industrial and biological processes. Enzyme catalysis in biochemistry is a particularly intriguing subdiscipline of catalysis that illustrates the character of the process.

Enzymes are biological catalysts that have gained recognition for being specific and effective in triggering biochemical processes. They lower the activation energy by stabilizing the transition state and typically require to be supplied with special conditions such as temperature and pH in order to perform optimally. As observed in many experiments, such as multisite phosphorylation mechanisms, the cycles of regulation present in biochemistry pathways are capable of generating bistable behavior that helps to further accentuate the catalytic role in dynamic systems (Krakauer et al., 2007). These results serve to further develop our understanding of the intricate relationship between catalytic mechanisms and the kinetics that govern complex biological processes. The importance of catalysis extends to the non-biological world to industrial processes, in which it is employed to initiate productivity and selectivity in chemical synthesis. For instance, catalysis is a key factor in the synthesis of pharmaceuticals and agrochemicals, in which selective reactions are employed to produce desired products with few by-products. Moreover, the progress made in the understanding of catalytic processes and those explored in studies of surfactants' effects on enzymatic function supports the dynamism of interactions between catalysts and their environment (Bančáková et al., 2014). As reliance on catalytic processes keeps growing with the current demands for increasingly sustainable chemical processes, identification and understanding of catalysis remain a core area of study in chemical kinetics and scientific studies in general (Figure 6.2).

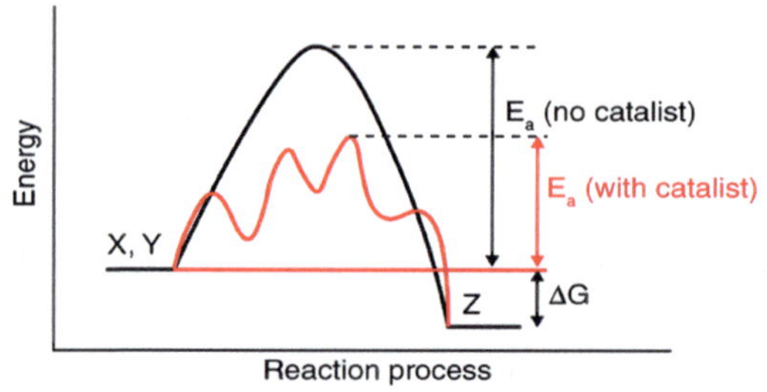

Figure 6.2. *How does a catalyst work.*

Source: https://cdn1.byjus.com/wp-content/uploads/2017/08/Catalysis-How-does-a-catalyst-work-1.png.

Importance in Chemical Reactions

Chemical reactions are the fundamental processes that drive the changes of matter. In these processes, kinetics is central since it determines the rate at which reactants transform to products. An understanding of the importance of reaction kinetics is not an academic endeavor only; it holds critical importance to industries ranging from pharmaceutical to petrochemical. Accelerating the rates of reactions through catalytic processes demonstrates the way kinetic parameters can promote efficiency in production. Catalysts, which lower the activation energy required for reactions, induce faster and selective conversions and thus significantly impact overall reaction kinetics. Creating a comprehensive understanding of these kinetic parameters, particularly when working with complex reaction networks comprising many pathways and products, is important. The fine balance between thermodynamics and kinetics also helps to further underscore the importance of chemical reactions. Thermodynamics provides a general account of the feasibility and extent of reaction, whereas kinetics provides a report on reaction speed and mechanism. The degree of rate control is particularly significant, illustrating how varying kinetic factors will influence selectivity and rates of conversion in chemical reaction. As described by Andreasen et al. (2011), the extension of Campbell's degree of rate control concept brings with it a collection of variables allowing for extensive study of kinetic and thermodynamic factors. Improved insight aids chemists and reaction engineers in optimization of reaction conditions and selectivity improvement, thereby optimizing yields in synthesis reactions. Consequently, research into these kinetic dimensions informs the design of more effective catalytic strategies. In practice, the study of kinetics often involves complex techniques that are designed for serious study of reaction mechanisms. Biphasic hydroformylation reactions, say—based on modern catalytic systems—reflect the sophisticated interplay between kinetics, reaction conditions, and selectivity to products. As defined by Delmas et al. (2011), the behavior of the system is characterized by first-order dependence for certain reactants but increasing complexity with others, accentuating the complexity of the system in catalyst interactions and substrate dynamics. By employing mathematical modeling and statistical approaches to simulate experimental data, researchers are made more predictive of the outcomes of reactions. This convergence of theory with laboratory practice not only emphasizes the significance of kinetics in catalysis but also results in trail-blazing enhancements in the technologies of chemical production (Figure 6.3).

Figure 6.3. *Chemical reactions.*

Source: https://www.vlaby.com/public/lfm/photos/1/BLOG/images/img_3902.jpg.

Overview of Catalytic Processes

Catalytic reactions are at the core of speeding up the rate of chemical reactions without being consumed in the process. It is essential to understand such mechanisms, particularly under kinetics where the speed of reaction matters a great deal in the results in the industry and lab. Catalysts work by providing a different reaction pathway that has a lower activation energy than the reaction without the catalyst. This concept lies at the heart of both heterogeneous and homogeneous catalysis in which the catalyst is in a different phase from the reactant. Catalysis is widely applied in a wide variety of fields like organic synthesis, petrochemical processing, and pollution treatment, as indicative of their function in advancing chemical engineering and industrial chemistry. Computational sciences have substantially encouraged the research of catalytic processes. It is now possible for scientists to employ sophisticated models in order to explore mechanisms of catalytic action on a molecular level. A case in point is the discovery of biocatalysts, which has shed great detail into catalytic mechanisms. The use of these with coarse-graining procedures produces thermodynamically correct reaction network models. These advances recognize stoichiometries and reaction fluxes but also forces acting on these biochemical systems. Understanding the connection between reaction forces and fluxes at fine-grained and coarse-grained levels helps explain previous presumptions in the literature, ultimately enhancing our comprehension of the underlying principles of catalysis and kinetics in complex biological systems (Esposito et al., 2018). Beyond biochemical catalysis, zeolites are irreplaceable substances for industrial catalytic uses. Their unique structural features render them as effective catalysts for converting various feedstocks. In the latest research involving current ab initio molecular dynamics approaches, it has become feasible to comprehend the role of catalyst topology, acidity, and external parameters in choosing reaction pathways. As a case in point, research studies involving methanol to hydrocarbon conversion depict a dualistic nature for the reaction mechanisms—concerted or stepwise. The ability to closely simulate reaction conditions enables researchers to learn about the conditions that favor a particular mechanism over the other, hence offering essential insights for optimizing zeolite-catalyzed reactions and the design of sustainable chemical technologies (Bailleul et al., 2016).

6.1. TYPES OF CATALYSTS

Categorization of catalysts is a fundamental assumption for studying their use in chemical reactions, particularly chemical kinetics. Catalysts are broadly categorized into two big groups: homogeneous and heterogeneous catalysts. Homogeneous catalysts act in the same phase as that of reactants, which has the tendency to deliver equal reaction conditions. Their effectiveness is usually attributed to their ability to form intermediate complexes with reactants, thereby lowering the activation energy for a reaction. Heterogeneous catalysts, however, are in an alternate phase, typically used in solid state form in gaseous or liquid reactant reactions. This phase difference can facilitate distinct kinetic behaviors and mechanisms, influencing the net rates of reaction. It is necessary to understand the distinction between them, as the choice of type of catalyst has a big influence on the selectivity and efficiency of industrial processes, such as fatty acid methyl esters from waste cooking oil production, whose functioning relies directly on catalyst choice (Aubin et al., 2013). Heterogeneous catalysts, specifically, have received much interest because of their ease of separation from reaction mixtures in a short time and possibility of reuse, rendering them economically feasible for volumetric applications. Such catalysts tend to work by promoting reactions on the surface in which the active sites are typically located on the surface of the catalyst. Their activity can be altered by pressure, temperature, and reactant coverage and result in an intricate connection between kinetic performance and physical properties of the catalyst. Recent studies have elucidated some of the phenomena, including compensation effects, where the measured activation energy is observed to linearly correlate with surface reactant coverage (Farra et al., 2012). This interaction shows an interplay of multi-layered nature between kinetic rates and thermodynamic parameters, calling for an in-depth description of reaction environment and catalyst surface. Catalyst design and innovation continue to advance with advances in engineering and materials science, enabling more efficient catalytic systems to be created. Researchers are now exploring new catalyst formulations, such as bifunctional catalysts that combine various active sites in order to promote tandem reactions, thereby maximizing processes. In addition, integration with innovative reactor technologies, including microstructured reactors and reactive distillation systems, has been discovered to maximize catalytic efficiency, especially in biodiesel production where continuous flow processes enhance yield and reduce reaction time (Aubin et al., 2013). With the expanded basic understanding of various catalysts, so does the potential for optimizing industrial applications, which is precisely most critical in overcoming the needs of sustainable chemistry and energy conversion processes (Figure 6.4).

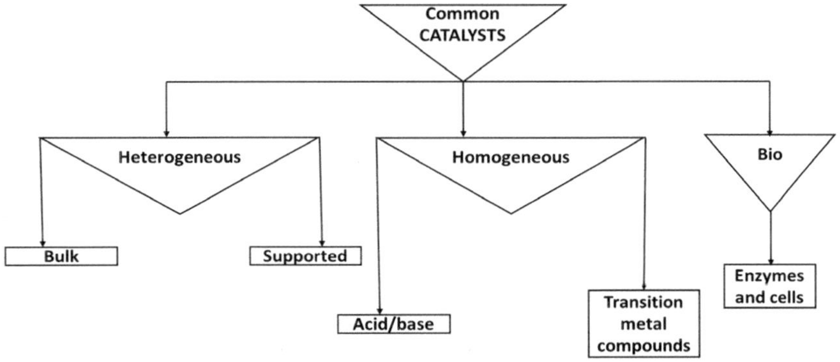

Figure 6.4. *Common catalyst types with emphasis on heterogeneous and homogeneous catalysts.*

Source: https://www.researchgate.net/publication/365421649/figure/fig2/AS:1143 1281097401741@1668578746706/Common-catalyst-types-with-emphasis-on-heterogeneous-and-homogeneous-catalysts.ppm.

6.6.1. Homogeneous Catalysts

Homogeneous catalyst research has become increasingly important in modern chemical kinetics because homogeneous catalysts can increase the rates of reactions with high selectivity. In contrast to heterogeneous catalysts that work at the boundary surface between two different phases, homogeneous catalysts are dissolved in the same phase as the reactants, which favors equal distribution and interaction between the reactants and catalyst. This condition allows for more efficient collision and reaction between the molecules, which ultimately leads to lower activation energy requirements for various processes. Moreover, homogeneous catalysts are characterized with well-defined structures, and thus scientists can play with their properties at the molecular level and regulate their catalytic activity based on a specific reaction.

Therefore, such versatility has led to improvements in many industrial operations, from pharmaceuticals to petrochemicals, showcasing the leading role of homogeneous catalysts in optimizing chemical transformation efficiencies in kinetic studies. Experiments on homogeneous catalysts' reaction kinetics studies provide interesting information regarding their mechanism and effectiveness. For example, a comprehensive kinetic model was developed to investigate the liquid-phase transesterification of ethyl acetate and methanol over a variety of ion-exchange resins. The model accounted for some of the important reaction pathways that entailed occupation of active sites by methanol, allowing for successful proton exchange, which was one of the keys to the transesterification reaction. The results indicated that the ability of the polymer resins, which were predominantly characterized in terms of their cross-linking character, to swell had a profound influence on active site accessibility. Such findings highlight that kinetics is deeply linked with the structural characteristics of catalysts and offer optimization pathways wherein increased understanding of catalyst design can enhance the efficiency of reactions in homogeneous catalytic processes (Clercq et al., 2012).

The payoffs of such research are greater than theoretical modeling, with attention directed toward the fundamental interface between catalyst design and performance during operation. In optimizing homogeneous catalysts, selectivity, solubility, and environmental conditions all directly influence the resulting reaction kinetics. Improved catalytic performance often demands stringent control of reaction factors such as temperature and concentration, as has been demonstrated in experiments where adjustments to these variables brought about astonishing difference in reaction dynamics. Such parameters as reactant ratio and activation energy are also critical because they can readily influence the rate coefficients of many catalytic systems. The general information from these kinetic evaluations not only aids in advancing basic chemical knowledge but is also crucial in pushing inventive catalytic applications into multiple industrial environments (Pinto et al., 2017) (Figure 6.5).

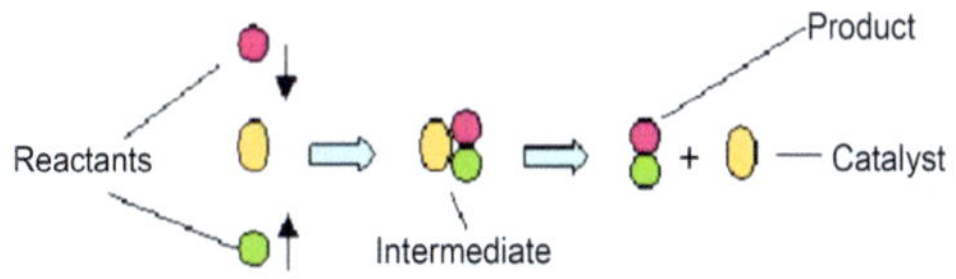

Figure 6.5. *Homogeneous catalysts.*

Source: https://alevelchemistry.co.uk/wp-content/uploads/2019/02/Catalyst_3.png.

6.6.2. Heterogeneous Catalysts

Heterogeneous catalysts play a crucial role in enhancing chemical reactions through the provision of a surface where reactants can interact more efficiently. Unlike similar-phase reactant homogeneous catalysts, heterogeneous catalysts are typically solids that catalyze gaseous or liquid phase reaction. This positioning allows for easier separation of the catalyst from the products and thus facilitates purification. The power of heterogeneous catalysts is most evident in a variety of industrial applications, where they not only speed reactions but also promote selectivity for desired products. Their physical structure, such as pore diameter and surface area, are key to catalytic activity, providing a structure that can be tailored to refine specific reactions. A fine example of heterogeneous catalysis is the propane oxidative dehydrogenation. In this reaction, the catalyst enhances the conversion of propane to desired products and regulates the reaction conditions, such as pressure and temperature. From research, increased pressures have been shown to lead to improved yields and selectivity for catalyzed and non-catalyzed reactions and thus optimizing reaction parameters is important. Specifically, it is shown that maximum product yields are obtainable by adjusting partial pressures of the reactants without altering their proportions. Such findings validate the technology potential of heterogeneous catalysts in finely tuning reaction kinetics to obtain the best results, a principle that lies at the basis of contemporary industrial chemistry (Kotanjac et al., 2007). Moreover, heterogeneous catalysis advancements are specially groundbreaking concerning the production of fatty acid methyl esters (FAME) from waste cooking oil (WCO). With various innovative process equipment, such as microwave reactors and reactive distillation, transesterification and esterification processes can be significantly enhanced in terms of efficiency and effectiveness. In particular, the application of heterogeneous catalysts in the said processes has been proven to have positive effects in reaction efficiency, a case of how catalyst selection can influence overall operational efficiency. Further research on investigating continuous flow equipment that integrates both reaction and separation processes is further indicative of the potential in optimizing FAME production through heterogeneous catalysis. With a rising need for sustainable and efficient biofuels, the importance of these catalytic processes cannot be overstressed, acting as a remarkable impact on the environmental as well as economic aspects of energy generation (Aubin et al., 2013) (Figure 6.6).

Figure 6.6. *Heterogeneous catalysts.*

Source: https://www.hidenanalytical.com/wp-content/uploads/2022/05/Heterogeneous-Catalysts-scaled.jpg.

6.6.3. Biocatalysts

The application of biocatalysts, specifically enzymes, has gained great attention in catalysis due to their efficiency and specificity. Biocatalysts are biological molecules that catalyze chemical reactions, which is a process of natural metabolic reactions and industrial processes. Enzymes, as biocatalysts, possess remarkable catalytic capabilities that even outperform traditional chemical catalysts by offering milder reaction conditions, thus decreasing energy consumption and by-product formation. Biocatalysis research has progressed from understanding simple biochemical reactions to coming up with useful applications in the guise of drugs, biofuels, and waste management, making them unavoidable for sustainability efforts. Their use in catalysis, therefore, is a testimony to how biological processes can spearhead technological advancement in kinetic processes, hence illustrating the link between biology and applied chemistry. While investigating the kinetics of biocatalysis, a number of key factors need to be taken into account, such as substrate concentration, enzyme activity, and conditions. Kinetic studies on biocatalysts tend to portray intricate patterns that regulate reaction rates. For example, an in-depth study of enzyme kinetics can yield information regarding parameters such as k_{cat} and K_M values, which are essential in determining enzyme efficiency as well as substrate affinity. Furthermore, new experimental techniques, such as highly parallelized measurements of biomolecular function, have enhanced our understanding of enzyme behaviors for different sequences, facilitating more optimized design of biosystems.

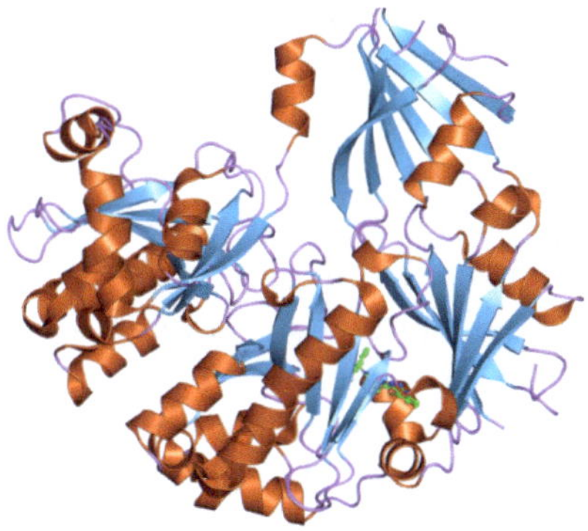

Figure 6.7. *Three-dimensional structure of an enzyme. Biocatalysis utilizes these biological macromolecules to catalyze small-molecule transformations.*

Source: https://upload.wikimedia.org/wikipedia/commons/9/9b/PDB_3bl7_EBI.png.

The advancements display the dynamic interplay of enzyme structure and function and therefore open the door for enhanced biocatalytic application that is capable of radically altering industrial processes. Research into these interactions not only drives technological progress but also addresses today's demands for efficient, eco-friendly chemical processes. The impact of biocatalysts on economic and energy efficiency ensures the argument for greater use is persuasive. Recent intensive work has focused on the optimization of bioprocess design for less energy consumption and reduced operational costs. Particularly, encapsulation and immobilized biocatalyst system development have facilitated improved enzyme reusability and stability with cost-effective use in industry. By the application of such principles, industries can facilitate possible reduction of energy expenses while achieving product yield maximization. The scientific research endeavors into many aspects of biocatalysis accentuate the twin imperative towards scientific consciousness and application and depict the role of the biocatalysts as a cornerstone in the new paradigm of catalysis. Behind this setting lie economic viability and ecological sustainability as the pillars of future chemical manufacturing (Figure 6.7).

6.2. MECHANISMS OF CATALYSIS

The intricate mechanisms of catalysis occupy a central position in the field of kinetics, with a profound influence on reaction rates and mechanism. Catalytic reactions involve a number of steps that have the ability to change the energy landscape of a reaction, hence making it occur faster than otherwise under the same circumstances in the absence of the catalyst. Through effective lowering of the activation energy barrier, catalysts facilitate the transition state and intermediate species formation that would otherwise be hindered in non-catalytic reactions. Surprisingly, these processes are dynamic, dependent on parameters such as catalyst composition, surface properties, and reaction conditions, thus exhibiting the dynamic character of catalytic action. Explanation of these mechanisms is of extreme importance, particularly as researchers develop new catalytic systems that can operate on environmentally friendly terms with maximum reaction efficiencies and sustainability of chemical reactions. Microscopic analyses of catalytic mechanisms have the potential to reveal the underlying rules governing the behavior of various catalytic materials, as in recent studies on zeolites. Modern ab initio molecular dynamics techniques test the role of topology, acidity, and guest molecules in influencing reaction mechanisms and kinetics for zeolite-catalyzed reactions. For example, a detailed modeling of crucial reaction steps in methanol to hydrocarbon conversion can decide whether reactions occur via concerted or stepwise mechanisms. From this view, researchers can manipulate reaction conditions (such as temperature and acidity) to modulate the behavior of the catalytic material and reveal classical as well as emerging reaction mechanisms for the transformation of reactants. These findings reemphasize the necessity of further understanding catalytic behavior at the molecular level in order to advance catalytic technology and maximize the chemical process of interest (Bailleul et al., 2016). In the context of microbial catalysis, the complexity of the reaction mechanism may also be observed, particularly in bioelectrochemical system applications such as microbial fuel cells (MFCs). Oxygen reduction reaction (ORR) is a nice example where microbial catalysis serves as a reformatory factor that optimizes efficiency under some working conditions. Despite various hypothesized mechanisms for microbial ORR, ongoing studies strive to explain such pathways more precisely, particularly regarding parameters like pH and mass transfer rates. A clear understanding of the ecology of electroactive biofilms is crucial in developing an efficient microbial oxygen-reducing cathode and hence highlights the interaction between electrochemical and biological parameters in catalysis. This research not only strengthens catalysis theory but also has a utilitarian application for enhancing the energy production sustainability by microbial systems (Bergel et al., 2012).

6.2.1. Transition State Theory

Transition state theory (TST) is a key mechanism for catalyzed reaction kinetics, particularly in outlining how the rate of reaction relates to the transition complex encountered in a reaction. According to TST, the reaction passes through a high-energy transition state with maximum potential energy of the system before it can be converted into product. The theory allows one to calculate the activation energies and provides data concerning factors influencing these energy barriers, i.e., temperature and pressure. Of special interest is that the stability of the transition state is influenced by specific conditions of a reaction that lead to equilibrium shift toward reactants or products, which underlines the importance of TST in predicting the outcome of catalytic reactions. In the laboratory, TST has been applied to describe catalytic reaction mechanisms with complex catalysis.

For instance, recent studies of the water oxidation reaction on the *n*-$SrTiO_3$ surface have shown the dynamics of transition states under various conditions of pH, temperature, and ionic strength. The demonstration of two simultaneous transition state paths for charge-trapping intermediates demonstrates how the states influence the activation barriers in the process of catalysis. This research is a model for application of transition state theory to the isolation of key activation barriers needed to optimize catalytic efficiency in significant processes like water oxidation.

The results not only improve theoretical understanding but also are of relevance to the development of next-generation catalysts. Moreover, the union of TST with statistical mechanics provides a comprehensive method to the simulation of heterogeneous catalysis. Through the decoupling of density-functional theory from kinetic Monte Carlo simulations, researchers have been able to examine the steady-state dynamics of catalytic surfaces under real operating conditions. It has already been successfully used in the oxidation of CO over RuO_2(110) to offer insight into surface structures and the factors that determine turnover frequencies. The findings in accordance with experimental data indicate the utility of TST to outline complex surface interactions and verify an enhanced understanding of catalyst activity under real conditions. Ultimately, TST remains vital in an effort to improve theoretical models and applicability in catalysis and kinetics research towards more effective catalytic systems (Figure 6.8).

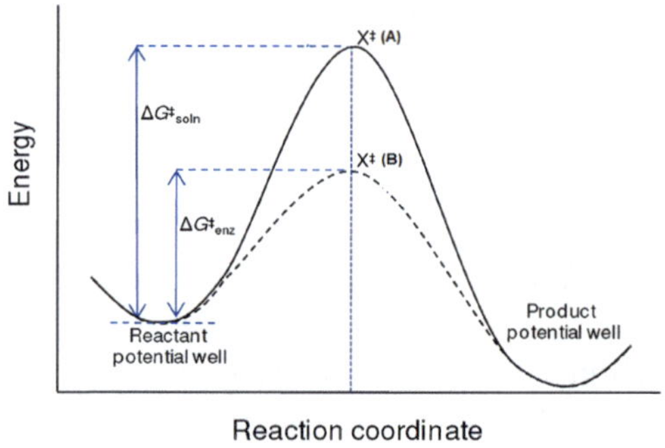

Figure 6.8. *Transition state theory model of catalysis. X‡ (A) is the transition state for a reaction in solution. X‡ (B) is the transition state for the enzyme-mediated reaction.*

Source: https://www.researchgate.net/profile/Osbourne-Quaye/publication/268304262/figure/fig4/AS:669529814335502@1536639755774/Transition-state-theory-model-of-catalysis-X-A-is-the-transition-state-for-a.ppm.

6.2.2. Reaction Coordinate Diagrams

The importance of reaction coordinate diagrams in catalysis is extensive since reaction coordinate diagrams provide a visual explanation for the energy changes in a chemical reaction. By using the demonstration of how the correlation of reactants' energy, the transition states, and products' energy is depicted, reaction coordinate diagrams enable chemists to describe the mechanism through which reactants turn into products. When undergoing catalytic reactions, the activation energy of the reaction is lowered, hence lowering the rate of reaction. This is particularly observed in enzyme-catalyzed reactions where the enzyme provides an alternative reaction pathway with a low transition state. Understanding of such profiles of energy is essential to reaction kinetics prediction as well as optimization of conditions for optimum catalytic activity, and therefore reaction coordinate diagrams are an essential tool in the study of catalysis and kinetics of biochemical reactions (Scholey et al., 2013). In addition to facilitating the study of enzyme-catalyzed reactions, reaction coordinate diagrams also play a very important role in explaining the kinetics of chemical reactions under multiple external parameters. Specifically, the diagrams would be able to illustrate how temperature, concentration, and pressure changes affect the energy landscape of the reactions. For instance, increasing the temperature has a tendency to decrease the activation energy and enable more molecules of the reactants to reach the transition state, hence enhancing the reaction rate. This is especially relevant in heterogeneous catalysis, where what occurs between catalyst surface and reactants determines the efficiency of the catalytic process. With reaction coordinate diagrams, researchers are able to identify optimal conditions that will maximize the rate of reactions based on what they already understand from both thermodynamics and kinetics when designing their experiments (Bazant et al., 2020). In addition, reaction coordinate diagrams are not only applied in traditional chemical reactions but also in complex cellular processes such as intracellular transport and motor protein action. The utility of being able to analyze the energy processes at work in these biological systems sheds light on the mechanisms behind cellular motility and force generation. Reaction coordinate diagrams not only allow qualitative assessments of energy levels but also relate macroscopic properties of cell behavior to the microscopic origins of chemical kinetics. By integrating these diagrams in a structured manner into catalysis and kinetics study, researchers are able to gain a comprehensive understanding of engineered and natural catalytic systems, informing innovations in drug delivery, biomimetic engineering, and the development of efficient energy systems (Scholey et al., 2013) (Figure 6.9).

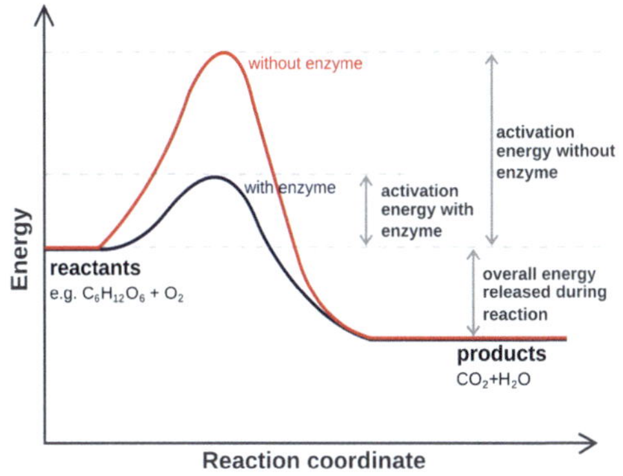

Figure 6.9. *Diagram of a catalytic reaction, showing the energy level as a function of the reaction coordinate. For a catalyzed reaction, the activation energy is lower.*

Source: https://upload.wikimedia.org/wikipedia/commons/e/e3/Activation2_updated.svg.

6.2.3. Role of Activation Energy

The idea of activation energy is the backbone of catalytic reaction kinetics. Activation energy refers to the minimum energy barrier that must be overcome in order to convert reactants to products. By catalysis, a catalyst lowers this energy barrier and hence increases the reaction rate without permanently changing itself. This catalytic activity is often described using the Arrhenius equation, with reaction rate quantified in terms of temperature and activation energy. Activation energy significance becomes appreciable especially in heterogeneous catalysis under the occurrence of compensation effects. Studies point to the fact that variations in surface coverage significantly influence apparent activation energy to produce a dynamic relationship between kinetic factors and catalyst surface dynamics. Therefore, the understanding of activation energy is important for maximizing catalytic processes and creating new catalysts that are more efficient, highlighting its importance in catalysis in kinetics (Farra et al., 2012).

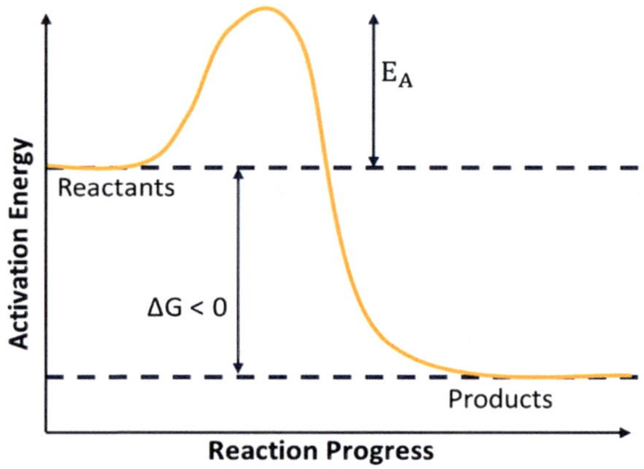

Figure 6.10. *Activation energy.*

Source: https://cdn.kastatic.org/ka-perseus-images/d5ff56983afc-2916e9979c08d81323c331927cfb.png.

Beyond mere theory, the function of activation energy extends even into practical utilization in catalysis. For instance, in electrochemical reactions like water oxidation, charge-trapping intermediates make an appearance, incorporating dynamic mechanisms that are controlled by activation barriers. The interplay between various conditions—such as pH, temperature, and ionic strength—also influences the kinetics of such a reaction by altering the effective activation energy. Recent studies indicate that the degradation of intermediates can reflect two different transition states that are characterized by different activation barriers. These breakthroughs allow scientists to control reaction conditions in a strategic manner to favor specific routes, thus enhancing catalytic efficacy. Hence, knowledge of activation energy and its variation in different conditions is essential to design catalysts that can function under different conditions, particularly in such fields as renewable energy conversion (Aschaffenburg et al., 2019). Overall, the role of activation energy is a fulcrum in the field of catalysis and kinetics. Its influence extends not only to the fundamental understanding of reaction mechanisms but also to the practical design and optimization of catalytic systems. The interdependence between the activation energy and variables such as surface coverage, reaction conditions, and catalyst properties displays a multidimensional landscape to be navigated by scientists in an effort to maximize catalytic processes. By enhancing activation energy manipulation techniques, scientists are able to enhance the efficiency of catalysts, leading to better industrial processes and greener chemical solutions. Activation energy, in addition to providing knowledge for theoretical models, is also of practical significance in catalysis development and thus is a main area of study in ongoing kinetics knowledge (Aschaffenburg et al., 2019; Farra et al., 2012) (Figure 6.10).

6.3. FACTORS AFFECTING CATALYTIC ACTIVITY

The catalytic action of any system could be greatly impacted by the character of the catalyst itself, encompassing its structural features, electronic properties, and coordination sphere. For instance, the reactivity and selectivity of metal-based catalysts can differ immensely according to their oxidation state and ligand surroundings. A fine example of this can be seen with tetranuclear ruthenium complexes that exhibit cooperative catalytic properties in the hydration of nitriles. Addition of electron-withdrawing or donating groups regulates the kinetics of such processes, demonstrating that the electronic composition of substrates is pivotal in catalytic efficiency. Specifically, complexes like $\{[(PCy_3)(CO)RuH]_4(\mu_4\text{-}O)(\mu_3\text{-}OH)(\mu_2\text{-}OH)\}$ enable the catalytic process through a relaxed tetrameric core architecture, allowing catalytic transformations in a significantly different manner from conventional monomeric catalysts, as described by Lindeman et al. (2008). Thermal properties and reaction conditions near the catalyst also play a major role in its activity. Elevated temperatures could accelerate reaction kinetics by adding more energy input for molecular collisions and activation energy barriers. However, high temperatures could lead to deactivation of the catalyst by sintering or thermal degradation. In addition, the presence of solvents, pH, and reaction medium can alter catalytic pathways. For electrocatalytic reactions, which are usually solid-electrolyte gas-phase reactions, ionic and electronic conductivity interaction can dictate the efficiencies, as recently exemplified in studies of mixed conducting materials. Understanding of these thermodynamic principles is essential in order to optimize catalysis, because temperature's effect on catalytic activity generally follows a non-linear trend and must be handled cautiously (Burggraaf et al., 1988). Aside from that, substrate concentration has a deep influence on the reaction rate of catalytic reactions by the principles of the maximum rate of reaction. For most catalytic systems, increasing the substrate concentration causes the reaction rate to increase, typically by Michaelis–Menten kinetics. However, at the limit of saturation, the rate of the reaction can become limited, indicating the accumulation of a saturated catalytic site. This saturation phenomenon indicates the importance of optimizing substrate concentrations for optimal catalytic activity. In addition, cooperative processes, such as the ones observed in benzonitrile hydration with ruthenium complexes, illustrate the complexity of reaction kinetics. In such systems, the periodic alignment of numerous catalytic sites under specific conditions may enhance bulk substrate conversion rates, thereby enriching our understanding on factors governing catalysis and overall kinetic behavior in advanced chemical systems, as shown in the observations reported by Lindeman et al. (2008).

6.3.1. Temperature and Pressure

The relationship of temperature and pressure also plays a fundamental role in the kinetics of catalytic reactions, profoundly influencing the rates of reaction and equilibria involved. As a general rule, rising temperature raises the kinetic energy of molecules, with a resulting tendency to increase the rates of reaction since there will be an increased opportunity for favorable collisions among reactants. Qualitatively, this is explained by the Arrhenius equation, which explains the exponential dependence of rate constants on temperature. In catalytic processes, this is of great importance because most catalysts are also temperature-dependent, with the activation energy often requiring specific thermal environments to produce maximum activity. Hence, it is of utmost importance to understand how modulation of temperature will affect the efficiency of catalysts while designing and implementing industrial catalytic processes where controlled heating can achieve maximum yield and selectivity. In contrast, pressure is a basic parameter that influences catalytic kinetics, particularly in gas-phase reactions. Pressure determines the rates of reaction via numerous effects that tend to lead to changes in the concentration of gaseous reactants, which at high pressures have higher frequencies of collision. In organic compound reactions like 1-octene, as has been found in some studies, there is complex pressure dependence of rates of reaction—showing initial improvement in rate followed by inhibition at high pressure as a result of gas-phase saturation (Delmas et al., 2011). This indicates the necessity of precise pressure control and optimization, particularly for industrial processes of large scale, where ideal conditions must be realized for potentially radical boosts in overall efficiency and profitability of the catalytic process. Besides, so-called compensation effects in heterogeneous catalysis illustrate the exchange-off between surface coverage and temperature in determining the reaction kinetics. Based on experimental observations of catalysts, it is found that apparent activation energy often exhibits linear relationships with variations in surface coverage, which also rely on temperature variations (Farra et al., 2012). This exchange sees a new paradigm for catalysis, showing how slight variations in temperature can have dramatic effects on catalytic efficiency. Therefore, an understanding of the interactions between temperature, pressure, and surface interactions not only aids in the fundamental understanding of catalytic processes but also aids in the improvement of more efficient catalysts and reaction conditions, thus driving innovation in chemical production methods.

6.3.2. Concentration of Reactants

Concentration of reactants is a fundamental chemical kinetic process which, to a great extent, dictates the rate of reaction as well as catalyst performance. As reactant concentrations increase, so do the molecular collisions, which are proportionally commensurate with frequency of successful encounters necessary to perform a reaction. This is a main concept in catalysis, where the design and application of catalysts often entail the tuning of reaction conditions to maximize efficiency. For instance, in zeolite catalytic reactions, the composition and structure of the catalyst can modify the adsorption characteristics of reactants and thereby influence their effective concentration in active sites. The possibility to understand how concentration variables affect the mechanisms and kinetics of catalytic reactions can lead to improved process designs in industry, ultimately ensuring maximal productivity and energy conservation in chemical manufacture. Particularly, the interaction between reactant concentration and catalytic activity can be studied by examining isolated systems, such as hydrocarbon cracking in zeolites. The rates of such conversions are not only a function of the concentrations of the reactants but also of factors such as the rate coefficients and the equilibrium constants for adsorption. Studies have shown that determination of such factors in reality is a matter of more than empirical collection of data under standard conditions; instead, complicated techniques such as configurational-bias Monte Carlo (CBMC) simulations determine how zeolite's structure influences the measured rate coefficient (k_{app}) and its substrates, K_{ads}–H^+ and k_{int} (Bell et al., 2018). Chemists can, using such correlations, enhance catalytic processes using the optimal concentrations to maximize the reaction efficiency. Moreover, the activity of reaction conditions is not only confined in simple systems, but to more complex ones as well, where the nature of intermediates may dictate the courses and speeds of reactions.

For example, in photocatalytic reactions in *n*-$SrTiO_3$, changes in reactant concentrations can lead to the creation of diverse transition state pathways that govern the speeds of product formation. Reaction conditions, i.e., temperature and pH, affect not only the concentration of the reactants but also charge-trapping intermediate reactivity and stability (Aschaffenburg et al., 2019). This delicate interplay indicates that the concentration of the reactants is a vital parameter that controls the kinetic landscape of catalysis. By research and regulation of these concentrations, researchers can optimize catalysts with their intended actions in chemical reactions to further advance both basic research and industrial chemical engineering.

6.3.3. Surface Area of Catalysts

The surface area of catalysts plays an important role in determining their efficiency in catalyzing chemical reactions, particularly heterogeneous catalysis. An increased surface area would usually provide more reaction sites for the reactants to adhere to, thereby speeding up the rate of reaction. This is closely akin to concepts of kinetics, and the quantity of such sites plays an important role in determining the mechanism and pathway of a reaction. Surface area-reactive rate relation can be observed in catalytic reactions, where the greater the surface area of materials, the higher will be their catalytic activity. The surface interactions such as adsorption, desorption, and reaction kinetics have to be comprehended to optimize the performance of catalysts. Thus, surface area maximization-based approaches such as nano-structuring or the utilization of porous materials form the basis for constructing effective industrial catalysts. Moreover, the relationship between surface area and catalytic activity is not merely quantitative but also qualitative and, furthermore, is a function of structural properties of the catalyst. Disordered and dynamic surface phases can significantly affect catalytic activity. Experiments reveal that high catalytic activity in some systems is directly correlated with surface effects involving large compositional fluctuations under reaction conditions. This perspective is well expressed in experiments on catalytic carbon monoxide oxidation on surfaces like $RuO_2(110)$, which have a rich interplay of elementary steps owing to the surface structure of the catalysts. With advanced modeling techniques, such as density-functional theory and kinetic Monte Carlo simulations, researchers can simulate the impact of surface area on reactions and confirm that optimized surface structure can lead to increased turnover frequency with concurrence in experiments (Gross et al., 2005). The surface area also clarifies diffusion phenomena in catalytic reactions. In ionic-electronic mixed-conducting perovskite-type oxides systems, as evidenced by the investigation of $La_{0.6}Sr_{0.4}Co_{0.8}Fe_{0.2}O_3$ membranes, surface activity plays an extremely significant role in oxygen permeation, and hence on methane coupling reactions. The oxygen flux is largely limited by surface processes, which indicates the intricate relationship between surface area and catalytic efficiency. Besides, the studies also indicate that the inclusion of reactants like methane influences selectivity, thus emphasizing the role of surface functionality and structure in determining the pathways of catalytic reactions. This again endorses the idea that control over the surface area is crucial not just for activity enhancement but also for optimization of selectivity of desired products in complex reaction networks (Bouwmeester et al., 1995).

6.4. KINETICS OF CATALYTIC REACTIONS

The kinetics of catalytic reactions play a pivotal role in understanding the specificity and efficiency of chemical processes. Catalysis significantly lowers the activation energy required for reactions, thus enhancing reaction rates without getting used up in the process. Various parameters, such as pressure, temperature, and concentration of reactants, determine these kinetics. For instance, enzymes, which are biological catalysts, have kinetic behavior defined by the Michaelis-Menten model. This model illustrates how substrate concentration influences reaction velocity to achieve maximum rate, thereby yielding information about the efficiency of enzymes. Not only does enzyme kinetics aid in the development of pharmaceuticals, but it also streamlines industrial processes, revealing the critical interplay between catalysis and reaction kinetics in biological as well as artificial processes. Molecular biology advancements in recent times have even better elucidated catalytic dynamics through non-covalent DNA catalytic reactions. These systems utilize ssDNA molecules to catalyze product release through coordination of displacement in multi-stranded complexes. The fault tolerance of these strand displacement-based reactions, as highlighted in research, exhibits them to be resistant to sequence mutation but sensitive to molecular impurities and noise (Aldaye et al., 2009). Their resistance is crucial for synthetic biology applications, where biochemical pathway regulation needs to be accurate. By exploiting the structural anomalies of DNA in catalytic uses, researchers have succeeded in integrating these reactions into more complete synthetic chemical and biological reaction networks, thereby harmonizing fundamental catalysis with current technology. The coarse-graining approach of biocatalysis presents a new viewpoint for reaction kinetics through the coupling of complex mechanisms with larger biochemical networks. This method captures principal stoichiometries and reaction fluxes with thermodynamic consistency, and it is thus particularly beneficial for the analysis of complex reactions (Esposito et al., 2018). Additionally, by describing the relationship between one-way fluxes and reaction forces at a coarse-grained level, researchers can more effectively research complex cell processes. Such a development not only challenges conventional assumptions about flux-force relations in biochemical networks but also contributes to the development of a better understanding of thermodynamic principles for catalysis. The output of such research thus constitutes a central contribution to further theoretical developments and application in catalytic kinetics.

6.4.1. Rate Laws and Equations

In chemical kinetics, rate laws are the core frameworks that describe how the concentration of reactants influences the rate of reaction. They give mathematical equations that describe the rate of reaction as a function of the concentration of the reactant to various powers, known as the reaction order. For any reaction, the rate law can be determined experimentally, and this provides evidence regarding the mechanism of the reaction and nature of the interactions involved. By establishing the influence of change in concentration on the rate, chemists can reason out the underlying processes governing a reaction. Such an investigation is very important in catalysis since catalysts are substances that increase rates of reactions without altering the chemical equilibrium. It makes possible the optimization of more effective catalytic processes, thus enhancing the effectiveness of chemical reactions in practical applications. Mathematical expressions of rate laws at times entail a high-order interaction between reaction mechanisms and thermodynamic principles. Specifically, such a complete understanding of catalysis entails grappling with how expert conditions define the directions of reaction pathways and, finally, rate equations. Such questioning is also supplemented by developments in analytical tools for monitoring reaction dynamics under various conditions. Coarse-graining protocols have recently been utilized to develop thermodynamically consistent theories bridging reaction fluxes and forces to uncover finally how macroscopic phenomena emerge from the microscopic dynamics (Esposito et al., 2018). These findings illustrate that rate laws are more than a description; they are unifying theories including kinetic and thermodynamic aspects of catalysis. Therefore, the study of rate laws improves the knowledge of catalytic proficiency and the optimization of reaction conditions in commercial processes. The inclusion of novel theoretical methods in rate-law studies has opened new avenues for gaining insight into advanced biochemical systems. Researchers now study how variations in concentration, temperature, and pressure affect reaction rates and qualitative properties of the involved catalysts. Utilizing sophisticated computer programs and advanced laboratory techniques, researchers can determine the kinetic parameters that govern the reactions. This computational innovation not only improves our comprehension of classical catalysis but also becomes relevant in the biochemical networks, where enzyme kinetics is of utmost importance. The interplay between reaction forces and fluxes then becomes of special interest in such systems, as it can offer understanding of thermodynamic consistency and the kinetics of biochemical reactions on a larger scale (Pinto et al., 2017). Thus, there is constant feedback between experiment and theory that further refines our understanding of rate laws, making them central to the science of kinetics and catalysis (Figure 6.11).

$$aA + bB \longrightarrow cC + dD$$

$$R = k\,[A]^x\,[B]^y$$

R : Reaction rate

[A], [B], [C], [D] : Concentrations of A, B, C, and D, respectively

a, b, c, d : Stoichiometric coefficients of A, B, C, D, respectively

x, y : Reaction order in A and B, respectively

Figure 6.11. *Rate law.*

Source: https://www.chemistrylearner.com/wp-content/uploads/2023/01/Rate-Law.jpg.

6.4.2. Order of Reaction

Order of reaction is one of the most fundamental terms in chemical kinetics, particularly in catalyzed reactions whose dynamics are determined by a number of factors, including catalyst type and operating conditions. Under normal circumstances, the reaction order is typically determined experimentally and varies from zero to higher integers, with each order defining how the reaction rate depends on reactant concentration. For instance, a first-order reaction exhibits linearity in a relationship between the concentration of a single reactant and reaction rate, while a second-order reaction is indicative of the dependence of reaction rate on two reactant concentrations or on one reactant concentration squared. Chemists apply this knowledge to manage reaction conditions in an efficient manner and maximize the yield of products, thereby establishing the applied relevance of reaction orders in catalytic industrial processes. For heterogeneous catalysis, the order of reaction can be ruled by the character of the catalyst and its characteristics, such as surface area and accessibility of active sites. For example, investigations of transesterification of ethyl acetate with methanol demonstrate how the choice of solid acid catalysts, e.g., gel-type and macroporous resins, critically influences reaction kinetics. This difference in performance is due to the swelling of the resin and the accessibility of the acidic sites, which will directly influence the rate and order of the reaction. Notably, the derived fundamental kinetic model from such results emphasized conformity of active site occupation and associated surface reactions, as a result of which the order of the reaction can often be varied with the specific catalytic system employed, as observed from data correlation attained through experimentation using different types of resins (Clercq et al., 2012). In addition, the presence of compensation phenomena complicates our understanding of catalyzed process reaction orders further. Experimental evidence indicates that surface coverage variations, particularly in heterogeneous catalysis systems, can induce measurable changes in apparent activation energy and pre-exponential factors of the rate of reaction. A detailed exploration of HCl oxidation to Cl_2 on RuO_2 is an exemplar of this relationship, demonstrating how changes in reaction kinetics with variations in surface coverage can balance against one another, therefore influencing the observed order of reaction. These findings not only make traditional notions of reaction kinetics suspect but also underscore the intricate interplay between reaction dynamics and surface chemistry. With advances in catalysis science, a synergistic understanding of reaction order will become critical to the design of efficacious catalytic systems for specific industrial use (Farra et al., 2012) (Figure 6.12).

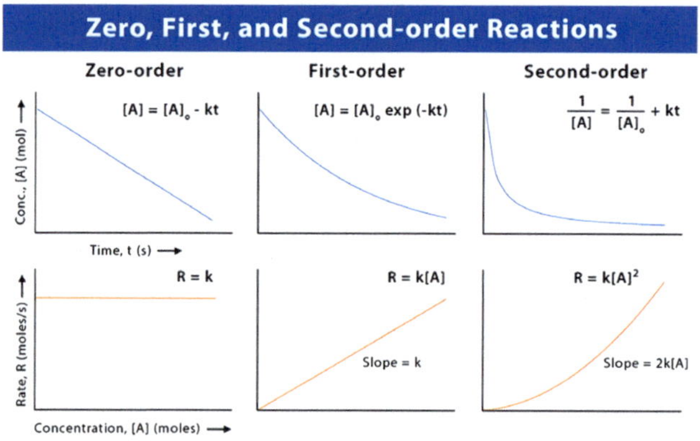

Figure 6.12. *Order of reaction.*

Source: https://www.chemistrylearner.com/wp-content/uploads/2021/05/Zero-First-Second-Order-Reactions.jpg.

6.4.3. Influence of Catalysts on Reaction Rates

The influence of catalysts on reaction rates is a significant consideration in kinetics, where catalysis can significantly enhance the speed of chemical reactions. Catalysts, by reducing the activation energy for other reaction pathways, allow reactions to proceed at a much faster rate than otherwise without them. This enhancement can significantly reduce the period over which products are being created, hence enhancing overall efficiency in chemical processes. For instance, structural properties such as topology and acidity of the catalyst are very important in determining the kinetics of the reaction in zeolite-catalyzed reactions. Sophisticated molecular dynamics techniques elucidate the impact of such factors on the mechanistic pathways, suggesting that variations in temperature and the presence of specific guest molecules can regulate the competition between the different reaction mechanisms, i.e., concerted versus stepwise reactions as highlighted in the case of methanol to hydrocarbons (Bailleul et al., 2016). Aside from altering activation energies, catalyst choice and composition are of the utmost importance to reaction selectivity, which can also affect reaction rates.

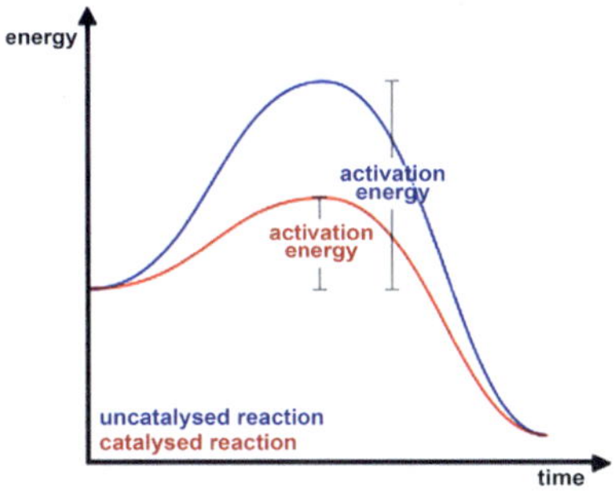

Figure 6.13. *Effect of catalysts.*

Source: https://ch302.cm.utexas.edu/images302/Catalyst_effect.png.

Certain catalysts are specifically engineered to promote the formation of desired products without creating side reactions, thus resulting in increased yields for chemical syntheses. For example, in biphasic hydroformylation of 1-octene with a rhodium catalyst, the system was considerably selective towards linear aldehydes and greatly contributed to the reaction kinetics. The initial rate of conversion was found to be a linear relationship with the catalyst and substrate concentrations but showing more complicated features in respect to gas components, which expresses the importance of reaction conditions on catalytic activity. This perspective points out that understanding the roles of catalysts involves not only an in-depth knowledge of their mechanistic processes but their interactions with reaction components as well (Delmas et al., 2011). Moreover, the ongoing evolution of computational models has revolutionized the understanding of catalysts and their effects on reaction rates. Using first-principle calculations and kinetic modeling, researchers are able to predict how changes in experimental conditions, i.e., pressure or temperature, will affect the overall kinetics. With the use of detailed reaction mechanism studies and modeling, it has been possible to design catalysts for some reactions, thus optimizing conditions for processes to achieve maximum efficiency. This intersection of theoretical knowledge and practical application emphasizes the fundamental role of catalysts in modern chemistry, especially in production where rate of reaction determines productivity. As chemical processes advance, control and understanding of catalysts will remain core in achieving sustainable and efficient reaction conditions, paving the way to new catalytic technology and catalysis-controlled kinetics (Bailleul et al., 2016; Delmas et al., 2011) (Figure 6.13).

6.5. APPLICATIONS OF CATALYSIS

The diverse applications of catalysis are key to accelerating the pace of reaction, rationalizing chemical processing, and ultimately optimizing industrial manufacture. One of the most obvious ways in which catalysis has demonstrated its importance is in the chemical industry, where it enables everything from pharmaceutical production to polymer production. Catalytic technologies enable manufacturers to increase production yield and reduce energy consumption as well as waste. For instance, developments in continuous flow chemistry via catalytic reactions provide more effective processes and safer handling of toxic substances. This is particularly pertinent to the transformation from batch to continuous operations, which introduces the necessity for catalysts capable of operating effectively under various conditions without contravening safety and environmental laws (Elgue et al., 2009). Hence, catalysis not only optimizes economic efficiencies but also addresses sustainability concerns associated with modern chemical production. Another critical application of catalysis is in environmental control technology, or pollution abatement. Catalytic converters in automobiles offer one such application, where precious metal-based catalysts transform toxic exhaust gases, i.e., carbon monoxide and nitrogen oxides, to less harmful emissions. These technologies have contributed significantly to urban air quality improvement, as they reduce the environmental burden of transportation. Moreover, the application of heterogeneous catalysis in the catalytic oxidation of volatile organic compounds (VOCs) has been very successful in industrial processes, where emissions controls call for stringent measures. This innovation illustrates how catalysts can be employed to speed up the kinetics of reactions which otherwise would be too sluggish or wasteful without them, illustrating their pivotal role in linking industrial needs with environmental accountability (Pinto et al., 2017). At a general level, continued innovation and utilization of catalytic technology is crucial in environmental mitigation of important issues.

Aside from these traditional applications, catalysis science is advancing through continuous research towards novel materials and processes. Catalysts are being designed to compensate for specific reaction routes and substrate requirements, thus making hitherto unachievable reactions or enhancing available reactions. The development of silica carbide-reactors, for example, promises to transform batch operation into continuous application in a more secure and sustainable manner (Elgue et al., 2009). This advancement is one facet of an overarching pattern in catalysis in which advanced materials enhance mass and heat transport, ultimately leading to more efficient reactions. Computational methods and machine learning applied to the development of catalysts continue this advancement by allowing rapid catalytic system optimization. With science continuing to push the boundaries of catalysis, it is bound to uncover even more invaluable applications that cooperate in a synergistic fashion to produce industries commercially acceptable and environmentally friendly (Figure 6.14).

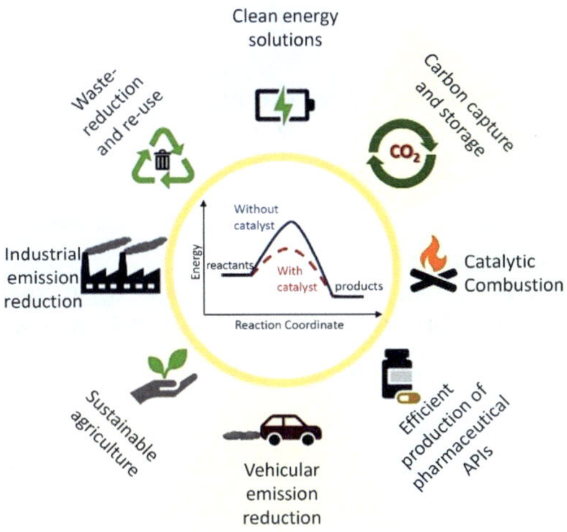

Figure 6.14. *Various applications of catalysis in the modern world.*

Source: https://www.researchgate.net/publication/376577320/figure/fig1/AS:1143 1281244555979@1715944407324/Various-applications-of-catalysis-in-the-modern-world.png.

6.5.1. Industrial Processes

Industrial processes have been remarkably developed through the integration of various catalytic processes making chemical reactions more efficient. One of the best examples is found in the production of fatty acid methyl esters (FAME) from waste cooking oil (WCO). It has key steps such as transesterification, esterification, and hydrolysis that are important to ensure optimal yield and reduce the environmental cost of biodiesel production. Choice of the catalyst matters in such reactions, and heterogeneous catalysts exhibit significant advantages in some designs. For instance, novel process hardware such as cavitational reactors and oscillatory baffled reactors not only speed up the reaction rates but also enable efficient separation of the products, thus embodying a contemporary interest in streamlining industrial operations (Aubin et al., 2013).

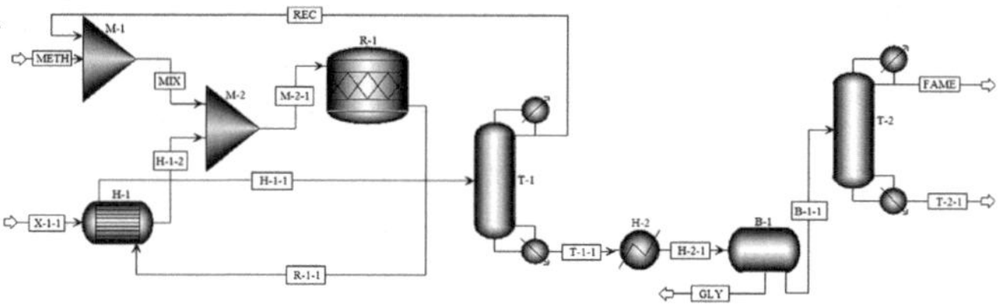

Figure 6.15. *Process flow chart of biodiesel production from pretreated WCO. Legend: M-1, M-2: mixer; H-1: heater; R-1: esterification reactor; T-1: methanol rectification tower; H-2: cooler; B-1: liquid–liquid separator; T-2: biodiesel rectification tower.*

Source: https://www.researchgate.net/profile/Zongyuan-Zhu-2/publication/350733232/figure/fig4/AS:1025464642596865@1621501231267/Process-flow-chart-of-biodiesel-production-from-pretreated-WCO-Legend-M-A-1-M-2-mixer.ppm.

Growing catalytic developments have placed a focus on green chemistry in industrial processing, with the adipic acid being produced more sustainably. Formerly produced using less environmentally friendly methods, newer technologies with sodium tungstate as a catalyst within a microemulsion environment have been shown to be a worthwhile alternative. This novel process includes hydrogen peroxide oxidation, which produces clean adipic acid yields in high amounts and decreases reliance on harmful organic solvents. The exploitation of environmentally friendly solvents like microemulsions improves phase contact and thus enhances reaction efficiency and catalyst and surfactant recovery. This shift not only reflects an emphasis on sustainable method but also illustrates how industrial operations can be adapted to more environmentally conscious means, reducing hazardous emissions over time (Cognet et al., 2012). Incorporating these catalytic processes into industrial application highlights the need for the optimization of conditions in order to achieve workable outcomes. For instance, manipulations of operating conditions such as temperature and reaction time can strongly affect the kinetics of developed and novel reactions in selected environments. Improvement in apparatus and catalyst performance aid in reaching desired product yields efficiently, thus enhancing the feasibility of up-scaling such processes in industries. Moreover, continuous flow systems emphasize the integration of reaction and separation steps further illustrating the ability to optimize overall process effectiveness. Lastly, industry process synergy between catalysis and kinetics conveys the industry's commitment to innovation and sustainability, bringing opportunities of what can be achieved in the future (Aubin et al., 2013; Cognet et al., 2012) (Figure 6.15).

6.5.2. Environmental Applications

In environmental uses, catalysis is a key feature of pollution control and enhanced sustainability. Catalytic transformation of hazardous waste can induce more efficient breakdown and destruction of the pollutants in the environment. For instance, catalysis can promote breakdown of long-lived organic pollutants in water and soil and transformation of such toxic materials into benign chemicals. Through the utilization of varied catalytic materials, from metal nanoparticles to metal oxides, researchers can design targeted solutions for addressing environmental problems. The kinetically dynamic aspect of reaction rates is central to this use, as tracking the speed and efficacy of such catalytic reactions can be used to maximize processes minimizing environmental impacts while promoting cleaner production means. Studies of materials such as titanium dioxide (TiO_2) are on the rise in environmental catalysis, particularly photocatalysis. TiO_2 possesses excellent photocatalytic activity to degrade organic pollutants under UV light exposure. According to DFT calculations, differences in stability of TiO_2 polymorphs, namely rutile and anatase, can influence their catalytic efficiency. Dopant introduction can also stabilize the anatase phase, further enhancing its photocatalytic activity and providing a new pathway for environmental remediation. Through the optimization of the catalytic activity of TiO_2 using doping, researchers are able to tailor its catalytic response, ultimately leading to cleaner forms of remediating contaminated sites and water bodies and thus addressing important environmental issues directly. In addition, ion-exchange resins are presently efficient catalytic agents in certain environmental processes, including transesterification of waste oils to biodiesel. Kinetics of the reactions are determined by the characteristics of the resin used, with gel-type resins performing better than macroporous resins. Swelling properties of these resins, depending on their cross-linking abilities, determine acid site accessibility and, consequently, reaction kinetics. As observed, catalytic performance is strongly dependent on resin type, and material selection is extremely critical in optimizing reaction conditions for the production of fuel from waste feedstocks. Overall, such advancements in catalytic technologies are a crucial step towards sustainable energy solutions and waste management, demonstrating the role of catalysis in addressing significant environmental challenges (Figure 6.16).

Figure 6.16. *Environmental applications.*

Source: https://www.azonano.com/images/Article_Images/ImageForArticle_5881_16371573152371559.jpg.

6.5.3. Pharmaceutical Synthesis

Pharmaceutical synthesis forms the core of drug development and production, where concern is put on efficiency, selectivity, and the environmental impact of the chemical process. At the center of this lies the activity of catalysis, which accelerates chemical reactions, thereby increasing reaction rates and producing better yields at a reduced waste. The kinetic aspect of catalysis is particularly important in the sense that it determines the viability and feasibility of different synthetic routes. By studying the reaction kinetics in pharmaceutical synthesis, researchers can optimize conditions such as temperature and pressure, which translates into a more economic and sustainable production pathway. Thus, advances in catalytic methods not only decrease the cost of production but also enhance the therapeutic activity of the resulting pharmaceutical compounds, making catalysis a crucial technology in the field of medicinal chemistry today. Enzyme processes are at the fore in drug synthesis in catalysis due to their selectivity and mild conditions of reaction. Enzymes are biocatalysts that have the ability to perform targeted reactions that could be inefficient or challenging with traditional chemical catalysts. For instance, the use of ene reductases in synthesis pathways offers a biocatalytic approach that allows the asymmetric hydrogenation of activated alkenes to be used directly in the synthesis of valuable chiral product. Use of Ppo-Er1, a newly found enzyme from *Paenibacillus polymyxa*, is an illustration of the potential of enzymes to facilitate high conversion levels while being substrate specific under various operating conditions. This expansion of the scope of biocatalytic tools not only supplements other procedures but also with the increased pressure for more sustainable synthesis procedures within the pharmaceutical industry, reducing reliance on harsh reagents and conditions (Aregger et al., 2020). Also, since information regarding the kinetics of enzymatic reactions is crucial for the optimization of these biocatalytic processes. For example, the enantioselectivity of Ppo-Er1 was confirmed to be high and exhibited remarkable specificity constants, indicating the potential of the enzyme as a robust competitor catalyst for drug synthesis. The evidence of Michaelis–Menten kinetics allows us to understand efficiency and interaction of enzymes with the substrates, enabling optimal fine-tuning of reaction conditions for effectiveness. All this data ultimately culminates in producing better drug candidates with less side effect. The employment of such biocatalysts in the manufacture of medicines is congruent with the industry's need to increase productivity while not compromising on the standards of sustainability and conservation of the environment, hence demonstrating how catalysis shapes future trends in drug discovery (Pinto et al., 2017) (Figure 6.17).

Figure 6.17. *Pharmaceutical synthesis.*

Source: https://www.arborpharmchem.com/wp-content/uploads/2024/12/Pharmaceutical-Synthesis-Solutions-5-Proven-Strategies-for-Safer-Drugs-1.jpg.

6.6. CHALLENGES IN CATALYSIS

In catalysis science, practitioners are faced with an array of challenges that hamper the evolution of efficient catalytic systems. One of the biggest challenges is the inherent complexity of catalytic processes, comprising various substrates, intermediates, and reaction routes. Each catalyst must be specially crafted to operate efficiently under specific conditions, but the unstable nature of reaction dynamics is likely to generate suboptimal results. Moreover, traditional catalysts are also likely to be low in selectivity and stability, making it even harder to develop sustainable catalytic processes. As researchers strive to push the boundaries of catalytic technology, understanding such problems is needed in order to make breakthroughs that could result in highly effective industrial applications. A different problem in catalysis issues involves materials that are crucial to developing catalysts. The new developments also highlighted the potential of novel materials, such as nanostructured electrochemical interfaces, to revolutionize catalytic efficiency. However, their synthesis and characterization also present challenges of their own. Composition and structural changes have the potential to introduce variability in performance and make reproducing results more difficult. Although high-resolution microscopy techniques have a lot of promise for the study of such interfaces, one has to comprehend them deeply in terms of the fundamental aspects of nanomaterials and how they are integrated into useful applications. The paradigm shift in the materials science discipline requires scientists to consider not only catalytic performance but also how material design plays a role in catalytic performance and potential. Electrochemical catalysis is a pertinent example of challenges of modern catalytic design, particularly of lowering carbon dioxide. The design of effective catalysts for lowering carbon dioxide is presently based on measures that must navigate significant obstacles, including the knowledge gap in reaction mechanism and efficient transfer of reactants and products. Techniques such as vapor-phase CO_2 reactant addition have emerged as emerging technologies that enhance reaction rate and quality of local reaction conditions. However, translating such outcomes into commercially scalable and profitable technologies is plagued by issues. The overcoming of such challenges will require an interdisciplinarity that is based on breakthroughs across all sciences for the creation of solid technologies and thus make it possible to seek out sustainable chemical processes, as exemplified in the ongoing endeavor at uniting theoretical vision with real application in the area of catalysis (Bentley et al., 2019; Hahn et al., 2019).

6.6.1. Catalyst Deactivation

Catalyst deactivation is a significant issue in catalysis because it directly influences the efficiency and lifetime of the catalytic process. It occurs when active sites on a catalyst get blocked or poisoned, reducing the ability to catalyze reactions. This inactivation can be caused by a variety of mechanisms, including adsorption of impurities, sintering of the catalyst particles, or structural changes that affect accessibility of active sites. Understanding the specific mechanisms responsible for inactivation is important in order to optimize catalytic systems and maximize general reaction kinetics. For instance, impurity chemisorption can cause catastrophic losses of catalytic activity, necessitating the development of measures to minimize these effects and maximize catalyst lifetime (Froment et al., 2009). During studies on some reactions, the effects of deactivation of the catalyst will be more catastrophic. For instance, hydrodechlorination of dichloromethane and trichloromethane, the catalytic performance can drastically differ based on the reaction mixture. Under dichloromethane alone, the catalyst remained steady-state active for several hours; with the introduction of trichloromethane, there was extensive deactivation within a few hours. This highlights the stability of reaction environments where different reactants not only crosstalk with each other but also with the catalyst itself, leading to either enhanced performance or rapid deactivation. These types of observations call for detailed studies of the operating conditions in order to achieve optimum catalyst formulations and deliver steady performance (Kovalchuk et al., 2012). Moreover, deactivation kinetics of the catalyst are not merely an afterthought to the primary catalytic reaction but are related inherently to the dynamics of the system as a whole. The mode of transformation of an active into a deactivated state can be characteristic depending on various variables, such as chemical composition of reactants and physical properties of the catalyst. For example, models of unimolecular and bimolecular reactions in the context of poison catalyst illustrate how catalytic activity can undergo radical transitions to end up causing power-law decay in reactant concentration. This shows how deactivation of a catalyst can leads to distorting reaction kinetics very much in the long run, ultimately impacting yield and selectivity in industrial applications. A better understanding of catalyst deactivation is hence needed to create more robust and efficient catalytic systems (Froment et al., 2009).

6.6.2. Selectivity Issues

Selectivity issues are of prime importance in catalytic reactions because they have a direct impact on chemical reaction yield and efficiency. Intersection of the principles of kinetics and thermodynamics normally determines in which direction a reaction will go, and hence it is necessary to deeply understand both these concepts. Neglecting consideration may witness the formation of undesirable by-products, leading to less yield and more consumption of resources. Experimentation has found that selectivity can be enhanced by changing the conditions of the reaction, such as pressure and temperature, or by employing some catalysts that favor certain pathways. With the application of Campbells degree of rate control, recently extended in studies, it is possible to systematize the evaluation of selectivity in complex reaction networks, allowing chemists to optimize their new process designs, particularly in heterogeneous catalysis where multiple pathways tend to be present in such systems (Andreasen et al., 2011). Selectivity can also be a critical criterion in making the catalytic reaction industrially viable. For example, in the new chemistry of silicone, the discovery of catalysts capable of converting raw material into effective product with minimum byproducts is imperative. Bluestar Silicones, which is at the forefront in its field, has come to appreciate the need for state-of-the-art technology that ensures safety as much as efficiency in chemical processing. Their work and development project aims to move from batch to continuous processes, demonstrating the pragmatic value of selectivity in practical applications. Through the design of new heat exchanger reactors that are able to be operated under perfect conditions, problems related to corrosion, mass transfer, and general catalytic efficiency can be addressed directly. This contrast between selectivity and practicability emphasizes the worth of selectivity problems in catalysis, as better processes not only increase productivity but also reduce environmental effects (Elgue et al., 2009). In these perspectives, selectivity problems remain an important area of concern for further research in the area of catalysis and kinetics. As chemists probe into increasingly complex reaction mechanisms, the guidelines for selectivity must be refined in order to drive innovation. The union of advanced theoretical constructs, e.g., the degree of kinetic and thermodynamic control, enables finer scrutiny of reaction conditions and catalyst selection. But the successful transference into practical applications is prompted by a cycle of reciprocity, i.e., experimentation and adjustment. Both fundamental knowledge and technological innovation will be utilized in overcoming selectivity challenges, such that catalytic reactions are not only efficient but also sustainable in the long term for industrial applications. While reaction systems become increasingly complex, so should the approaches being employed to enhance selectivity in catalytic systems (Andreasen et al., 2011; Elgue et al., 2009).

6.6.3. Economic Factors in Catalyst Development

Catalyst development is increasingly being dictated by economic factors that dictate their viability and application in various sectors. Cost is a significant factor, particularly in mass production operations where the cost of catalyst material can play a significant role in total profitability. For instance, expensive noble metals used in catalytic applications can potentially drive-up production costs, and substitute materials need to be studied that provide comparable performance with better affordability. In addition, economic rewards provided by governments and institutions for sustainable technologies, particularly renewable energy, propel catalyst innovations. The focus on catalyst development that is productive, selective, and durable, noted in various studies, is not merely scientific but also a manifestation of industry demands for cost-cutting and energy-reducing efficiency in energy conversion and chemical synthesis (Luna et al., 2019). The integration of economic considerations in catalyst synthesis is also viewed in terms of environmental sustainability. As industries are increasingly under pressure to adopt greener practices, the need for energy-efficient and low-waste catalysts that facilitate environmentally friendly processes becomes the priority. Low-energy-consuming catalysts and low-waste catalysts can significantly reduce operational expenses and environmental imprints. For instance, efforts towards designing catalysts to enable recycling of carbon dioxide into valuable products reflect a double benefit: stimulating technological advancement as well as economic and environmental issues. Improved catalytic processes incrementally can lower energy-intensive industries' costs considerably, thereby enhancing their competitiveness. For this purpose, the economic viability of uncomplicated catalytic reactions heretofore supported by costly processes is increasingly recognized as a primary driver in modern catalysis (Luna et al., 2019). Second, improvements in catalyst technology, such as photocatalysis, demonstrate the intersection between economic factors and catalyst effectiveness. The finding of approaches like controlled periodic illumination demonstrates a forward-looking response to economic needs via greater catalytic reaction efficiency at lower expenses. This method, originally speculated upon for use in boosting semiconductor photocatalytic reactions, demonstrates the potential of revolutionary technologies to decrease resource use while generating a maximum return. Financial feasibility is critical in the estimate of whether such breakthroughs will scale by ultimately deciding upon the validity of industry-wide application. So, it becomes imperative for engineers and researchers to understand the economic climate so as to contribute to the field of catalysis so that the burgeoning technologies are not only scientifically possible but economically viable as well.

SUMMARY

In reviewing the findings of this chapter on catalysis in kinetics, one can see that the intricate relationship between substrate binding and catalytic efficacy is the golden key to regulation of reaction rates. The exploration of mechanisms such as acid-base catalysis not only outlines the chemical principles at stake but also their applicability in biological processes. For instance, the study of *Plasmodium falciparum* glyoxalase II (GloII) demonstrates the role of individual amino acids in interaction dynamics with an effect on catalytic turnover. The significance of pH and ionic strength, as depicted in the activity of the enzyme, documents the delicate balance governing enzymatic process. These comments provide the impetus for introducing the need for a critical appreciation of the manner in which molecular changes can lead to variations in catalytic efficiency and enzyme function across various different biological environments (Elia et al., 2009). Finally, the applicability of heterogeneous catalysis introduces an additional sophistication of importance in the study of kinetics. The formation of a model that considers disorder in catalytic surfaces illustrates the means through which adsorption rate variation can bring about non-trivial kinetic behaviors. This contrasts with homogeneous surface models, which suggest more straightforward kinetics without steady-state behavior. These results suggest that the characterization and modeling of disorder in catalysis would lead researchers to discover new routes to efficiency and selectivity in catalytic reactions. This viewpoint opens a wider window for research into heterogeneous systems in the future, extending the limits of our conceptual understanding of catalytic processes outside mainstream paradigms (Ben-Avraham et al., 1995). From these insights, researchers are better able to inform catalyst design and use across a wide range of disciplines from environmental science to medicine. Ultimately, the conclusion drawn from this thorough review of catalysis in kinetics emphasizes that synergistic coordination is crucial to make catalytic procedures less mystifying. The intersection of biochemical knowledge from work like GloII and theoretical explanations of heterogeneous catalysis represents the range of strategies employed for enhancing rates and efficiencies of reactions. The implications of these findings are far-reaching beyond the academic community because they are of enormous potential use in real-world applications like drug discovery and industrial catalysis. Since scientists continue to probe the boundaries of catalysis, a greater understanding of the guiding principles governing these reactions will surely lead the way to innovative solutions to key issues, reinstating the central role of catalysis in propelling scientific innovation and technological progress. The dynamic investigation of catalysis highlights the intricate relationships between reactants and other entities in chemical reactions. The focus is on the activation energy that entails the amount of energy required for reactants to change into products. Catalysts play a crucial role in lowering the level of energy, thereby enhancing the rate of reaction. Familiarity with these dynamics is essential, and foremost, it is necessary to comprehend the nuances of responses occurring in real systems. Such study is of particular relevance within the field of enzyme catalysis, where substrate-protein interactions can dramatically affect the net efficacy of the catalytic process, as noted in recent research detailing the role of entropic contributions to substrate assembly and disassembly mechanisms (Jiménez-Ángeles et al., 2010). This implies that catalytic performance often depends on both kinetic and thermodynamic factors. Compensation effects in heterogeneous catalysis also warrant mention since they define inherent correlations that can determine reaction kinetics. Specifically, the effects result from a linear correlation between apparent activation energy and logarithm of the pre-exponential factor. Such interactions suggest that surface coverage oscillations can significantly impact reaction rates, as seen in experimental evidence of HCl oxidation over RuO_2 catalysts (Farra et al., 2012). This interaction between activation energy and entropic considerations on surface coverage determines the complex balance that dictates catalytic efficiency. Hence, knowledge of these compensation effects reveals a new insight in the interpretation of kinetic behaviors in catalysis and how varying external conditions can affect reaction pathways and eventually catalyze innovations in catalytic design. Lastly, the role of solvent interactions and how they contribute to catalytic processes cannot be overstated. As has been explained through Monte Carlo simulations of substrate-lock mechanisms in interactions, the environment of a solvent has a profound effect on the way substrates assemble and react under catalytic conditions (Jiménez-Ángeles et al., 2010). This points to the fact that even without attractive forces, entropic effects can still provide avenues for self-assembly, highlighting the role of solvent in catalytic action. By syntheses of these disparate factors—activation energy, compensation effects, and solvent interactions—researchers can build a rich picture of catalysis kinetics. This multi-faceted approach not only enriches theoretical constructs but also holds out promise for application in the design of more efficient catalysts for many chemical processes.

REFERENCES

1. Aldaye, C., Zhang, D. Y., Winfree, E., Feldkamp, U., Flamm, C., Frezza, B., et al. (2009). Robustness and modularity properties of a non-covalent DNA catalytic reaction. https://core.ac.uk/download/4885209.pdf (accessed on 10th September 2025).
2. Aleem, A., Mukhtar, A., Alvarez, G., Deschamps, J., & Holman, T., et al. (2019). Probing the electrostatic and steric requirements for substrate binding in human platelet-type 12-lipoxygenase. https://core.ac.uk/download/305124116.pdf (accessed on 10th September 2025).
3. Andreasen, A., & Stegelmann, C. (2011). Finding the key transition states and intermediates controlling net reaction rates and selectivity. https://core.ac.uk/download/pdf/290114.pdf (accessed on 10th September 2025).
4. Andreasson, J. O. L., Block, S. M., Greenleaf, W. J., Savinov, A., et al. (2020). Comprehensive sequence-to-function mapping of cofactor-dependent RNA catalysis in the *glmS* ribozyme. https://core.ac.uk/download/323062787.pdf (accessed on 10th September 2025).
5. Aregger, D., Buller, R., & Peters, C. (2020). Characterization of the novel ene reductase Ppo-Er1 from *Paenibacillus polymyxa*. https://core.ac.uk/download/289121447.pdf (accessed on 10th September 2025).
6. Aschaffenburg, D. J., Chen, X., & Cuk, T. (2019). Selecting between two transition states by which water oxidation intermediates on an oxide surface decay. http://arxiv.org/abs/1903.01675 (accessed on 10th September 2025).
7. Assadi, M. H. N., Hanaor, D. A. H., Li, S., & Sorrell, C., et al. (2012). *Ab initio* study of phase stability in doped TiO_2. http://arxiv.org/abs/1210.7555 (accessed on 10th September 2025).
8. Aubin, J., Mazubert, A., & Poux, M. (2013). Intensified processes for FAME production from waste cooking oil: A technological review. https://core.ac.uk/download/18540047.pdf (accessed on 10th September 2025).
9. Bailleul, S., De Wispelaere, K., & Van Speybroeck, V. (2016). Towards molecular control of elementary reactions in zeolite catalysis by advanced molecular simulations mimicking operating conditions. https://core.ac.uk/download/55855326.pdf (accessed on 10th September 2025).
10. Bančáková, A. (2014). Effects of detergents on activity, thermostability and aggregation of immobilized lipases. https://core.ac.uk/download/30296208.pdf (accessed on 10th September 2025).
11. Bazant, M. Z., & Fraggedakis, D. (2020). Tuning the stability of electrochemical interfaces by electron transfer reactions. http://arxiv.org/abs/2003.03057 (accessed on 10th September 2025).

12. Bell, A. T., Head-Gordon, M., Janda, A., Lin, L., et al. (2018). Understanding Brønsted-acid catalyzed monomolecular reactions of alkanes in zeolite pores by combining insights from experiment and theory. https://core.ac.uk/download/158345582.pdf (accessed on 10th September 2025).
13. Ben-Avraham, D., Considine, D., Clément, E., Forgacs, G., Grinstein, G., Yablonskii, G., Campbell, I. M., et al. (1995). Heterogeneous catalysis on a disordered surface. http://arxiv.org/abs/cond-mat/9506122 (accessed on 10th September 2025).
14. Bentley, C. L., Kang, M., & Unwin, P. R. (2019). Nanoscale surface structure–activity in electrochemistry and electrocatalysis. https://core.ac.uk/download/195265565.pdf (accessed on 10th September 2025).
15. Bergel, A., Erable, B., & Féron, D. (2012). Microbial catalysis of the oxygen reduction reaction for microbial fuel cells: A review. https://core.ac.uk/download/12043853.pdf (accessed on 10th September 2025).
16. Bouwmeester, H. J. M., Elshof, J. E. ten, & Verweij, H. (1995). Oxidative coupling of methane in a mixed-conducting perovskite membrane reactor. https://core.ac.uk/download/pdf/11452991.pdf (accessed on 10th September 2025).
17. Branco Pinto, D., Constales, D., Marin, G., Yablonsky, G., et al. (2017). Kinetic intersections as a source of information on rate coefficients. https://core.ac.uk/download/91288153.pdf (accessed on 10th September 2025).
18. Burggraaf, A. J., Gellings, P. J., & Koopmans, H. J. A. (1988). Electrocatalytic phenomena in gas phase reactions in solid electrolyte electrochemical cells. https://core.ac.uk/download/pdf/11473689.pdf (accessed on 10th September 2025).
19. Cognet, P., Lesage, G., Poux, M., Quesada-Peñate, M., et al. (2012). Green process for adipic acid synthesis: Oxidation by hydrogen peroxide in water micromulsions using benzalkonium chloride C12-14 surfactant. https://core.ac.uk/download/18540237.pdf (accessed on 10th September 2025).
20. De Clercq, J., Thybaut, J., & Van de Steene, E. (2012). Experimental kinetic study of transesterification of ethyl acetate with methanol catalyzed by gel and macroporous acidic ion exchange resins. https://core.ac.uk/download/55729062.pdf (accessed on 10th September 2025).
21. De Luna, P., Dinh, C. T., Kim, D., Li, Y., et al. (2019). Designing materials for electrochemical carbon dioxide recycling. https://core.ac.uk/download/287622524.pdf (accessed on 10th September 2025).
22. Delmas, H., Deshpande, R. M., Julcour-Lebigue, C., Kelkar, S., et al. (2011). Kinetics of hydroformylation of 1-octene in ionic liquid–organic

biphasic media using rhodium sulfoxantphos catalyst. https://core.ac.uk/download/12043525.pdf (accessed on 10th September 2025).

23. Elgue, S., Ferrato, M., Gourdon, C., Jolivet, A., et al. (2009). Silicon carbide equipment for process intensification of silicon reactions. https://core.ac.uk/download/12041266.pdf (accessed on 10th September 2025).

24. Elia, A. C., Hamilton, D. S., Penninckx, M. J., Ridderström, M., & Thornalley, P. J. (2009). *Plasmodium falciparum* glyoxalase II: Theorell-Chance product inhibition patterns, rate-limiting substrate binding via Arg(257)/Lys(260), and unmasking of acid-base catalysis. https://core.ac.uk/download/19418380.pdf (accessed on 10th September 2025).

25. Esposito, M., Rao, R., & Wachtel, A. (2018). Thermodynamically consistent coarse graining of biocatalysts beyond Michaelis–Menten. https://core.ac.uk/download/154761372.pdf (accessed on 10th September 2025).

26. Farra, R., Hevia, M., Knop-Gericke, A., López, N., et al. (2012). *In situ* surface coverage analysis of RuO_2-catalysed HCl oxidation reveals the entropic origin of compensation in heterogeneous catalysis. https://core.ac.uk/download/210619064.pdf (accessed on 10th September 2025).

27. Froment, G. F., Aarão Reis, F. D. A., Rice, S. A., & Mattos, T. G. (2009). Phase transitions and crossovers in reaction–diffusion models with catalyst deactivation. http://arxiv.org/abs/0906.3150 (accessed on 10th September 2025).

28. Gross, A., McQuarrie, D. A., Dumesic, J. A., Reuter, K., Reuter, K., Reuter, K., Reuter, K., et al. (2005). First-principles kinetic Monte Carlo simulations for heterogeneous catalysis, applied to the CO oxidation at RuO_2(110). http://arxiv.org/abs/cond-mat/0510234 (accessed on 10th September 2025).

29. Hahn, C., Higgins, D., Jaramillo, T. F., Weber, A. Z., et al. (2019). Gas-diffusion electrodes for carbon-dioxide reduction: A new paradigm. https://core.ac.uk/download/211386580.pdf (accessed on 10th September 2025).

30. Jiménez-Ángeles, F., Odriozola, G., Lozada-Cassou, M., Michaelis, L., & Odriozola, G. (2010). Entropy driven key-lock assembly. http://arxiv.org/abs/1010.3000 (accessed on 10th September 2025).

31. Kotanjac, Z. S., Niederer, J. P. M., & Versteeg, G. F. (2007). Influence of the pressure on the product distribution in the oxidative dehydrogenation of propane over a Ga_2O_3/MoO_3 catalyst. https://core.ac.uk/download/pdf/11474268.pdf (accessed on 10th September 2025).

32. Kovalchuk, V. I., McDonald, M. A., Rhodes, W. D., Ковальчук, V. I., et al. (2012). Dechlorinative oligomerization of multiply chlorinated

methanes catalyzed by activated carbon supported Pt-Co. https://core.ac.uk/download/38634314.pdf (accessed on 10th September 2025).

33. Krakauer, D., Smith, E., Krishnamurthy, S., & Fontana, W. (2007). The stochastic behavior of a molecular switching circuit with feedback. https://core.ac.uk/download/pdf/7633734.pdf (accessed on 10th September 2025).

34. Lawton, L. A., Prabhu, R., Robertson, P. K. J., Tokode, O., et al. (2015). Controlled periodic illumination in semiconductor photocatalysis. https://core.ac.uk/download/33591761.pdf (accessed on 10th September 2025).

35. Lindeman, S. V., Yi, C. S., & Zeczycki, T. N. (2008). Kinetic, spectroscopic, and X-ray crystallographic evidence for the cooperative mechanism of the hydration of nitriles catalyzed by a tetranuclear ruthenium-μ-oxo-μ-hydroxo complex. https://core.ac.uk/download/213085466.pdf (accessed on 10th September 2025).

36. N/A. (2025). *ECUT (Energy Conversion and Utilization Technologies Program). Biocatalysis Project.* https://core.ac.uk/download/pdf/42839622.pdf (accessed on 10th September 2025).

37. Scholey, J. M. (2013). Compare and contrast the reaction coordinate diagrams for chemical reactions and cytoskeletal force generators. https://core.ac.uk/download/323071498.pdf (accessed on 10th September 2025).

CHAPTER 7

FAST REACTIONS

Contents

Introduction 249

7.1. The Basics of Chemical Kinetics 253

7.2. Types of Fast Reactions 258

7.3. The Role of Temperature 263

7.4. The Influence of Concentration 268

7.5. Techniques for Studying Fast Reactions 273

7.6. Applications of Fast Reaction Kinetics 278

Summary 282

References 285

INTRODUCTION

Kinetics, particularly in relation to fast reactions, plays a critical role in the study of biochemical processes that are attained at phenomenal speeds. Rapid reactions are generally responsible for various biological processes, such as enzyme-catalyzed processes and metabolism, wherein selectivity and efficiency of reaction are crucial. It is important for drug development, synthetic biology, and metabolic engineering applications to accurately model and predict the behavior of such rapid kinetics. Investigating the dynamics involved, researchers can see how these rapid reactions alter the overall network behavior, which further impacts larger biological systems. On top of that, the profound knowledge of fast reactions enables scientists to develop new techniques for controlling biochemical pathways, which becomes increasingly important on the path toward the development of targeted drugs and productivity improvement in biological processes. To appropriately study fast reactions, sophisticated mathematical modeling techniques are absolutely necessary. Classical approaches will not work in the context of biochemical networks with a variety of types of kinetics, like Michaelis-Menten kinetics for enzymatic activity. Projection techniques have already demonstrated potential for investigating subnetwork behavior and memory effects in simplified models but can be applied much more broadly based on more sophisticated frameworks. For instance, by constructing larger networks that explicitly describe enzymes and their respective complexes, researchers can achieve more accurate descriptions of kinetics involved. This approach gives rise to enhanced computational efficiency and predictive power when investigating the fast kinetics inherent in biochemical networks (Bishop et al., 2016). All these are pointers to a revolutionary jump in the discipline yielding more reliable instruments for further research. In addition, the model reduction problem in biochemical reaction networks is an uppermost consideration when contemplating fast kinetics. Owing to the intricacy of these systems, especially where they involve polynomial or rational reaction rates, model simplification without loss of essential information is a top priority. The tropical equilibration problem offers the platform through which reduction is possible since it allows one to determine leading dynamics in highly complex systems. By employing this method, researchers can iterate their models to fit them exactly to the underlying dynamics of fast subsystems without breaking the conservation laws in the network. More importantly, this method not only streamlines analytical methods but also enhances the stability of simulations that investigate fast reaction kinetics, thereby increasing the possibility of research in computational systems biology (Grigoriev et al., 2015). Their combination is a significant leap forward in kinetic research that is paving the way towards new opportunities in biological research and development.

Definition of Kinetics

Kinetics is the science of how chemical reactions take place and the mechanisms involved. It is core to the understanding of numerous scientific disciplines, including chemistry, biology, and materials science. Kinetics not only looks at how quickly a reaction happens, but also at what controls those rates, such as concentration, temperature, and the presence of catalysts. In the case of rapid reactions, the subject of kinetics is more pertinent, as the quick changes may bring involved dynamics that require sensitive scrutiny and modeling. The knowledge gained from kinetic study is critical to know in applications from industrial chemical processes to biological ones, where rapid reactions are a prime consideration of cellular processes and metabolic pathways. One of the core areas of kinetics is identifying reaction mechanisms, or detailed descriptions of the individual steps that constitute a chemical reaction. Mechanisms can be revealed by experiment or theory and enable chemists to construct predictive models of behavior under different conditions.

Figure 7.1. *The kinetics of chemical reactions.*

Source: https://i0.wp.com/freescience.info/wp-content/uploads/2024/12/reaction-dynamics-the-pathways-of-chemical-reacti-56406-adt1_2.jpg?w=1200&ssl=1.

Specifically, Tangential Stretching Rate (TSR) is a fundamental analytical tool in the investigation of explosive reactions and ignition phenomena as it relates the most significant energy scales to specific states of the chemical mixture (Ciottoli et al., 2015). By quantifying participation indices of different modes of reactions, it is possible to ascertain which reactions are dominant in the explosive or relaxation phase, thus allowing for increased knowledge on reaction dynamics under rapid conditions. In addition, the connection of kinetics and systems biology highlights the complexity involved in biochemical reactions, particularly in the context of a living cell. Kinetics in this context are likely to deal with large networks of biochemical reactions that have various time scales and equilibrations between sets of variables. This level of complexity means novel approaches such as tropical analysis can be employed, which enables one to reduce models by identifying dominant reaction paths and scaling kinetic models down thereto (Grigoriev et al., 2012). With the incorporation of these advanced modeling techniques, scientists are able to give insights into the rates at which reactions in living systems take place, in a sense arriving at the root principles of cellular organization and function. This integrated understanding of kinetics is vital in fields ranging from drug discovery to synthetic biology, where fast biochemical reactions are the basis of new therapies and engineered systems (Figure 7.1).

Importance of Studying Fast Reactions

The study of fast reactions is at the core of the field of kinetics since they are responsible for determining the rates and mechanisms of the majority of chemical conversions. Fast reactions, which generally occur on a timescale from milliseconds to microseconds, necessitate sophisticated analysis techniques to adequately characterize their dynamics. Understanding these rapid processes not only adds to our knowledge of fundamental chemical principles but also aids in developing more efficient chemical processes. For instance, in biochemical systems, rapid reaction kinetics is extremely crucial in assisting us to understand how enzymes catalyze reactions at phenomenal speeds, which has an influence on metabolic pathways and signaling mechanisms. The implications of research on these rapid processes extend widely to other areas of science, influencing drug discovery and environmental chemistry, among others, and therefore making them integral to the advance of modern science. Rapid reactions also form the core of advanced areas of research, such as stochastic modeling of biological systems. New work has highlighted the requirement of stochastic processes in cellular dynamics to account for the behavior of small numbers of molecular species at high rates of reaction. For example, studies show that stochastic modeling is important to account for noise in biological pathways, where small changes can lead to large changes in cellular responses (Burrage et al., 2005). Application of these models in simulating genetic regulatory networks offers an opportunity to explore complex interactions that occur during swift reactions. Consequently, the study of swift reactions becomes integral in an attempt to dissect the complexity of biological processes, as well as supplementing computational methods employed in simulating biological systems. Furthermore, the fast responses in the realm of material science and corrosion phenomena are valuable in which the longevity and stability of materials rely on fast kinetic processes. A stochastic cellular automaton model has been employed to study corrosion with fast electrochemical half-reactions, with a focus on interactions between diffusion and reaction rates (Larbi et al., 2017). This model shows how local conditions cause differing corrosion rates, with rapid reactions having a dramatic effect on material properties and integrity. Understanding these rapid processes is important in creating successful corrosion-resistant materials and technologies. This makes this study relevant to both basic scientific information and applications for enhancing material performance in a broad range of environments.

7.1. THE BASICS OF CHEMICAL KINETICS

Chemical kinetics is at the heart of rates of reaction study, especially in fast reactions where time itself is a crucial factor. Chemical kinetics provides the recipe for determining how quickly reactants become products, a situation that is vital in scientific processes from industrial synthesis to environmental science. A reaction's rate can be influenced by numerous parameters such as temperature, concentration, and catalysts. For example, an increase in the temperature typically accelerates kinetic energy and therefore accelerates reaction rate. For fast reactions, small changes in such conditions can significantly influence the kinetics, supporting the need for tight control and monitoring in experimental apparatus. In addition, full appreciation of such principles allows chemists to predict reaction behaviors and design optimized conditions for desired outcomes. This predictive capability is of great value in fields such as pharmacology and materials science in which reaction rates have the potential to have a significant impact on development processes. The contribution of sophisticated instrumentation and techniques to chemical kinetic investigation cannot be overstated, particularly for fast reactions. It has been established through experiments that sophisticated numerical techniques are required to accurately describe and predict complex physical processes associated with chemical conversions. For instance, as was considered in recent combustion system studies, the necessity to develop new ideas describing fundamental aerothermodynamic phenomena is urgent. The investigation was motivated by the stringent design requirements in aerospace applications, which necessitate precise prediction of flame stabilization, combustor aerodynamics, and multiphase reaction dynamics (Mularz et al., 2025). By utilizing computation combined with experimental data, researchers can ascertain better configurations and optimize reaction efficiency in various industrial processes, ultimately facilitating more sustainable methods of chemical production and energy conversion. Besides computational advances, catalyst selection and reaction conditions are also crucial in accelerating reaction rates. Recent studies on the transesterification of waste cooking oil to fatty acid methyl esters have highlighted effective methods of promoting chemical reactions with novel process equipment. The use of processes such as microwave reactors, cavitational reactors, and reactive distillation has been proven to enhance the efficiency of transesterification and esterification reactions significantly (Aubin et al., 2013). Besides that, integration of reaction and separation processes within continuous flow systems is also becoming a successful method, particularly in achieving maximum yields from complex feedstocks. The application of catalysts, e.g., heterogeneous ones possessing good performance characteristics, also contributes to the enhanced kinetics of such reactions. This highlighted the interdisciplinary nature of chemical kinetics, where chemical science and engineering knowledge combine to facilitate easy and effective chemical conversions (Figure 7.2).

Figure 7.2. *Basic concepts in chemical reactions.*

Source: https://i0.wp.com/freescience.info/wp-content/uploads/2024/12/reaction-dynamics-the-pathways-of-chemical-reacti-56406-adt1_1.jpg?w=1200&ssl=1.

7.1.1. Rate of Reaction

Knowledge of the reaction rate is important in kinetics, particularly in maximizing efficiency in chemical reactions. Reaction rate will dictate the rate at which the reactants are converted to products and can be controlled by factors like concentration, temperature, catalysts, and physical forms of the reactants. For instance, increasing the concentration of reactants typically leads to a higher probability of collisions, thus accelerating the reaction rate. This principle is foundational in both laboratory settings and industrial applications, where optimizing reaction conditions is paramount for achieving desired yields and economic viability. The ability to manipulate reaction rates has the potential to deliver revolutionary processes in material science and pharmacy, highlighting the need to investigate kinetic parameters within reaction mechanisms. Recently, the study of self-replicating systems illustrates how reaction rates are central to the development of new chemical constructs. In general, an understanding of dispersion in reaction timescales can provide insight into the emergence of new structures from elementary particles. It has been proven in experiments that spontaneous self-replication is mainly controlled by the inherent distribution of reaction rates and bound-state energies among reactants, indicating that physical conditions that are favorable can be developed to facilitate such rapid reactions. With these criteria, experimental chemists are better positioned to create the systems for self-replication, achieving high yields and efficiency in the synthesis of complex molecules. This coupling of self-replication and kinetics makes clear the broader implications of reaction rate knowledge to dream up new chemical technologies (England et al., 2018). Furthermore, the significance of reaction rates also extends beyond typical chemical reactions into complex phenomena such as photochemical processes in planetary atmospheres. For example, research on the photochemistry encountered on exoplanets, like 51 Eridani b, describes how reactions initiated by ultraviolet light can lead to the accumulation of various atmospheric compounds. The reaction rate of sulfur compounds depending on various temperatures describes the haze formation processes and the likelihood for observable occurrences in diverse environmental conditions. By quantifying the transformation and condensation rates of elemental sulfur, researchers can make conclusions regarding the atmospheric chemistry of 51 Eridani b and other planetary atmospheres across a broad range of temperatures. These investigations demonstrate the pivotal role that the understanding of reaction rates has in deciphering complex systems for governing the behavior of celestial bodies (Marley et al., 2016) (Figure 7.3).

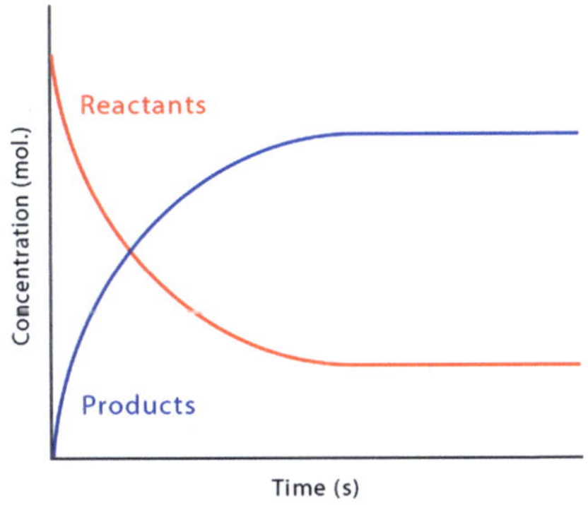

Figure 7.3. *Understanding the rate of reaction.*

Source: https://www.chemistrylearner.com/wp-content/uploads/2022/11/Reaction-Rate.jpg.

7.1.2. Factors Affecting Reaction Rates

The rate of a chemical reaction is influenced by a multitude of factors, each having a precise mechanism through which it affects the rate of reaction. Among these determinants is concentration; an increase in reactant concentration increases the rate of collision between reactant molecules. An increase in collision rate usually increases the rate of reaction as there are more molecules to react and productize. In addition, temperature is also a vital parameter because greater temperatures impart greater kinetic energy to molecules, resulting in firmer collisions that overcome the activation energy barrier to enable the reaction. The proportionality between temperature and reaction rate is in accordance with the Arrhenius equation, which illustrates the mathematical relationship of such parameters. Recognition of these primary factors offers the foundation for understanding the dynamics of the reaction rates of various chemical reactions. The second critical factor in predicting reaction rates is the existence of catalysts, which are substances that speed up reactions without being consumed in the reaction. Catalysts perform this function by providing an alternative reaction route with reduced activation energy, thereby increasing the rate of effective collisions towards product formation. This process is especially manifest in enzymatic reactions, whereby biological catalysts reduce energy demands, permitting life-giving processes to occur at room temperature. The scope of catalytic processes extends far beyond laboratory settings to industry where efficiency and speed are maximized. For example, the creation of catalytic converters in vehicles illustrates how manipulation and understanding of reaction rates can provide valuable environmental dividends through the minimization of noxious emissions. In this context, catalysts are a classic parameter in reaction condition optimization and speeding up reaction rates. In addition, pressure plays a significant role in changing reaction rates, particularly for gaseous reactants. Pressure, by virtue of its increase, actually increases the density of gas molecules, resulting in increased collisions, thus enhancing reaction rates. This phenomenon is most evident in gas-phase reactions in which pressure alteration modulates the dynamics of molecular interaction. The correlation between pressure and reaction rate is further highlighted using the theory of the Tangential Stretching Rate (TSR), which takes into consideration the local composition and conditions under a reaction. The TSR describes how reaction mechanisms function at various pressures, especially for systems with complex kinetics. Through knowledge of such parameters, including those set in terms of the TSR, scientists are able to gain a full picture of the rates of reaction across a broad spectrum of chemical environments, ultimately resulting in more effective kinetic models and methodologies (Alberts et al., 2007; Ciottoli et al., 2015) (Figure 7.4).

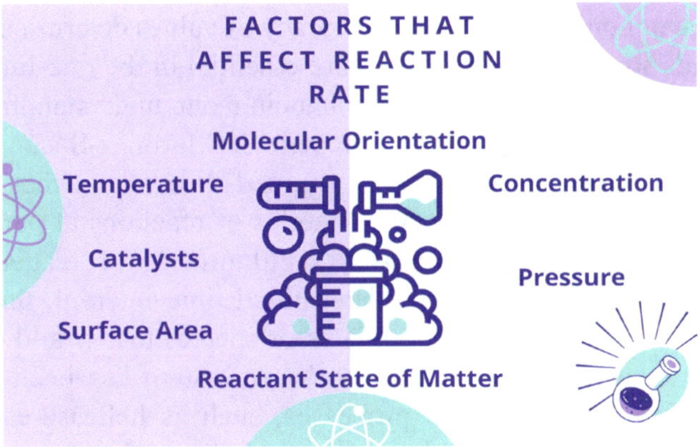

Figure 7.4. *Factors that affect reaction rate.*

Source: https://sciencenotes.org/wp-content/uploads/2020/10/factors-that-affect-reaction-rate-1024x683.png.

7.1.3. Measurement techniques in kinetics

The measurement techniques applied in kinetics are critical to the understanding of the rapid reaction behavior, providing reaction mechanisms and rate constant data. Cyclic voltammetry (CV) is one of the most widely used techniques and offers a versatile tool to investigate electron transfer (ET) kinetics at the nanoscale. With the application of nanoscale electrodes such as gold nanowires, researchers can easily study the electrochemical reaction kinetics. This technique includes the use of microcapillary systems where a redox-active mediator undergoes interaction with the nanoscale electrodes, and rates of heterogeneous ET can be measured. Surprisingly, experiments have revealed that in such instances, intrinsic rate constants can reach astounding magnitudes, such as k_0 up to ca. 1 cm s^{-1}, indicating the success of these techniques in investigating the high-speed dynamics of electrochemical reactions (Alvarez-Ruiz et al., 2011). Another strong measurement technique in kinetics is stopped-flow spectrometry, which is particularly well adapted to the exploration of solution reaction kinetics under rapid conditions. Using this technique, reactants may be mixed under conditions of control and the resulting nearly instantaneous subsequent monitoring of the course of the reaction by spectroscopic means. The rapid mixing brings about a quantifiable change in fluorescence or absorbance such that rate constants for single steps within a reaction mechanism can be measured. The PcrA helicase ATPase cycle, for example, can be characterized by fluorescence-based stopped-flow kinetics to resolve single rate constants.

This technique not only enhances the accuracy of values determined kinetically but also reveals structural changes that are essential in the rate-limiting phases of enzyme reactions and, therefore, enhancing our understanding about the biochemical processes that control fast reactions (Martinez-Senac et al., 2009). Quenched-flow techniques supplement stopped-flow approaches in the sense that they allow scientists to probe the kinetics of reactions at particular times after mixing. This technique involves rapid mixing of reagents followed by stopping the reaction after a predetermined time interval, thus making it possible to observe transient intermediates or species that would otherwise be too fast to be detected. Quenched-flow determination has been of great use in examining intricate biological processes, such as helicase-catalyzed ATP hydrolysis, in which the kinetic data not only indicate the steps occurring but also the impact of substrates on rates of reaction. The correct measurements obtained through quenched-flow lend a firmer grasp of the behavior of fast reactions under physiological conditions and therefore render kinetic models for predicting behavior in more extensive systems more realistic (Martinez-Senac et al., 2009).

7.2. TYPES OF FAST REACTIONS

Sudden responses are of crucial significance in most chemical reactions, particularly in kinetics, where they are usually divided into numerous categories. Among the foremost categories are bimolecular reactions, wherein two reactant molecules collide to form products. They are crucial in living organisms, where molecular interaction quickly determines the destiny of significant cellular processes. The bimolecular rate constants tend to be high because the likelihood of molecule encounters is more in proportion to an increase in their speed. The clarification of such fast reactions provides insights into mechanisms in intricate systems, e.g., metabolic pathways and signal cascades, where promptness in responses is vital to cellular function and homeostasis. Again, a significant class is the class of unimolecular reactions, where one reactant molecule is transformed into products without being compelled to strike another molecule. They are commonly encountered within the dynamics of gases and characterized by their rapid rate due to the built-in instability within the reactant, which can rearrange its bonds or decompose into the products within a nanosecond. The energy profile of a unimolecular reaction typically has several transition states that regulate the conversion rate of reactants into products. Besides, the study of unimolecular reactions integrates understanding from statistical mechanics to gain better insight into reaction pathway probability factors under fast kinetic regimes. Spatial dynamics introduce another level of complexity in fast reactions, particularly in heterogeneous chemical reaction networks. When analyzing systems where molecular movement is rapid in comparison to reaction rates, it becomes apparent that spatial redistribution of molecules significantly influences reaction outcomes. Research indicates that in single-scale reaction dynamics, fast molecular movement averages the effects across varied species concentrations, leading to predictable reactant interactions. In contrast to this, the situation is more intricate in multi-scale systems, where the differential rates of molecule types drastically influence the expected dynamics. Describing real biological processes, which may incorporate conserved quantities and complex space-time issues in rapid reaction kinetics, illustrates the intricacy of such fast processes (Burrage et al., 2004; Pfaffelhuber et al., 2015).

7.2.1. Elementary Reactions

Elementary reactions are used to describe simple processes in chemical kinetics that determine the simplest forms of reactions that occur on the molecular level. Elementary reactions typically consist of straightforward contact between reactants in order to form products without going through any intermediate steps, thereby providing an exact description of the mechanism of the reaction. One of the significant characteristics of elementary reactions is that their rate laws are explained according to their stoichiometry, making it easy to make predictions about the behavior of the reaction. From these simple reactions, chemists can infer bigger and more complex reaction mechanisms. This elementary understanding is particularly critical in developing more complex models with several pathways and interactions leading to a deeper understanding of reaction dynamics in industrial and biochemical processes. The elementary reactions play a larger role than simple kinetics in real-world applications across many fields, such as materials science and biology. In biochemical networks, for example, elementary reactions are part of process dynamics modeling in cell signaling and metabolic pathways. Difficulty with analysis given by these intricate networks—e.g., combinatorial explosion and parametric uncertainty—often prompts the need for model simplification in order to be able to determine key dependencies and interactions. After application of the dominance principle among model components as newly developed in literature, the role played by reactions in affecting network behavior can be organized in order, hence enabling more effective simplifications of intricate systems into manageable models. This scheduling enables investigation of dynamic behavior without loss of accuracy and enhancing our capability to predict and control biological systems in actual application (Gorban et al., 2012). Further, the application of elementary reactions to nonlinear multicomponent diffusion reveals a further scope of their significance in advancing scientific information. By drawing concepts from chemical kinetics, scientists can develop systematic plans to construct nonlinear diffusion equations that regulate multicomponent interactions. These must comply with thermodynamic limitations, which necessitate stringent formulation of kinetic laws confined to elementary processes. The ability to visualize such responses with cell-jump formalism not only aids in the comprehension of transport processes but also aids in the formulation of useful tools for numerical simulations in complex situations. Following the development of this branch of science, the information gained in the research of elementary reactions is continually directing several scientific fields in return, thereby serving as their fundamental importance in advancing our knowledge of fast reactions and their intricate mechanisms within kinetics (Gorban et al., 2011) (Figure 7.5).

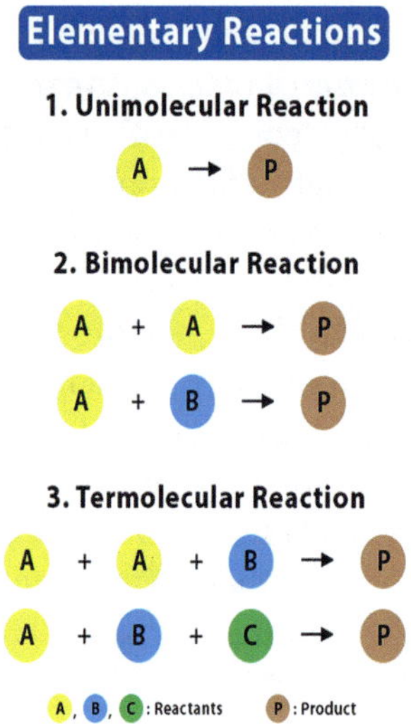

Figure 7.5. *Elementary reactions.*

Source: https://www.chemistrylearner.com/wp-content/uploads/2021/09/Elementary-Reaction.jpg.

7.2.2. Chain Reactions

Chain reactions are of central significance in chemical kinetics, particularly in fast reactions, where the speed and efficiency of reaction pathways can fundamentally determine the overall system behavior. Chain reactions exhibit a cascade process, in which the products of a step in one reaction serve as reactants for the succeeding step, creating a self-perpetuating sequence of reactions. The mechanism of chain reactions typically depends on the presence of radicals, highly reactive molecules with the potential to cause further reactions with other molecules. This renders such a reaction scheme hard to interpret with a critical mind and the likelihood of rates varying under varying conditions, which can be hugely influenced by spatial issues and availability of reactants. Significantly, chain reactions underlie numerous applications, ranging from industrial processes to biochemical cycles, demonstrating their significance in the comprehension of fast kinetics. Chain reaction dynamics can be dramatically affected by the spatial mobility of molecules in a reaction network, where fast spatial mobility may induce reaction averaging over an inhomogeneous system. In those systems described by single-scale reaction dynamics, mobility effects converge to produce an averaged reaction behavior. However, in the case of multiple scales, the rate of molecular motion compared to the effective reaction dynamics significantly decides the resultant reaction pathway. This is particularly pertinent in conservation law and interaction constraint dynamics between reactants, potentially creating intriguing emergent behavior in the kinetic formalism. As suggested by more recent research, the interaction between spatial dynamics and reaction rates offers strong leverage for the optimization of conditions for target reactivity, and hence for the practical implementation of such systems in chemical and engineering applications (Pfaffelhuber et al., 2015). With the shifting of the understanding of chain reactions, particularly within the scope of ignition chemistry, scientists have introduced parameters such as the Tangential Stretching Rate (TSR) for the differentiation of the energetic scales in a reactive system. The parameter is employed for the identification of reaction pathways by quantifying individual mode contributions in the local composition and mixture state. The TSR, being additive in nature, provides an overall view of how different reaction channels contribute to both explosive and relaxation regimes of a chain reaction. The TSR, when combined with the Computational Singular Perturbation (CSP) method, provides easy analysis of kinetic mechanisms, especially helpful in complex cases of multi-step reactions. This theoretical background not only assists with diagnostics but also enhances the predictive power, which can transform approaches in chemical kinetics, particularly in treating intricate chain reaction systems (Ciottoli et al., 2015).

7.2.3. Catalytic Reactions

Catalytic reactions are very important within chemical kinetics, particularly in increasing the rate at which reactions occur. Facilitators, such as catalysts, provide an alternative path through which reactions occur, hence lowering the activation energy required for the reaction to occur. The characteristic allows catalytic reactions to proceed at significantly faster rates compared to non-catalytic reactions. For instance, in heterogeneously catalyzed reactions, solid catalysts are brought into contact with gaseous or liquid reactants at the surface to enable efficient conversion of reactants. These effects are significant in industrial applications, where reaction rates are optimized to maximize productivity and reduce energy consumption. The foundational concepts underlying catalytic behavior, especially high-temperature behavior, need to be discerned in order to optimize processes such as the catalytic partial oxidation of methane, a focus of current research in materials science (Tiggelaar et al., 2004). Nucleic acid reactions are also examples of new applications of catalysis in contemporary chemistry.

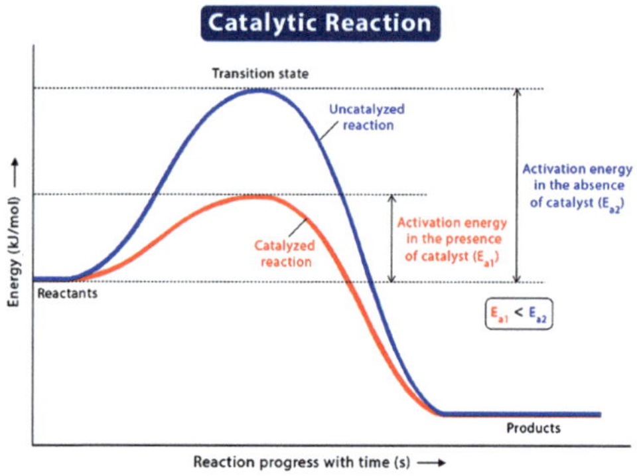

Figure 7.6. *Catalytic reactions.*

Source: https://www.chemistrylearner.com/wp-content/uploads/2021/11/Catalytic-Reaction.jpg.

Current research has illustrated the capacity of DNA as a catalyst through mechanisms like toehold-mediated strand displacement. This is a method of intentional design of DNA sequences involved in swift and reversible reactions that are enabled by the strategic positioning of short single-stranded pieces known as toeholds. Through such mechanisms, scientists have gained an understanding of the kinetics of these biological reactions, modeling them as multi-step processes that are able to predict reaction behavior with good accuracy depending on variations in DNA sequences. The successful application of such catalytic techniques in the construction of nanodevices points to the concept that catalysis is not just confined to traditional chemistry but also migrates into the domain of biochemistry and molecular engineering (Winfree et al., 2009). The design and application of catalysts dictate accelerated kinetics in most areas. Not only do catalysts accelerate the rates of reaction but also contribute to the sustainability of chemical processes as they can make reactions take place under milder conditions. This is particularly relevant in green chemistry practice development, where energy consumption needs to be minimized and waste also. By the employment of catalysts, scientists are able to come up with better routes in the synthesis of complex molecules while restricting the impact of chemical manufacture on the environment. The development of new catalytic reactions, such as silicon-technology-based microreactors for gas-phase reactions, is an area of emerging intensified chemical kinetics with high precision and control over reaction parameters maintained (Tiggelaar et al., 2004). As more work on catalysis is done, the relevance for basic science as well as industrial applications continues to be profoundly transformative (Figure 7.6).

7.3. THE ROLE OF TEMPERATURE

Temperature also plays a crucial role in regulating the speed of chemical reactions, particularly kinetics. As temperature increases, kinetic energy of molecules is boosted, and reactants will collide more frequently and collide with higher energy. It is crucial because reactant molecules have to collide with sufficient energy to overcome the activation barrier for something to occur. Consequently, even a small rise in temperature can produce an enormous acceleration in reaction rates. The relationship between reaction rate and temperature is quantitatively accounted for by the Arrhenius equation, which illustrates how higher temperature tends to lower the activation energy that molecules going through a reaction face. Thus, one must understand the temperature-dependent properties of reactions for uses ranging from industrial processes to environmental chemistry. The effect of temperature on kinetics of reaction is intersected by other environmental conditions that could decide the nature of the products that are produced. For instance, under photochemistry in planetary atmospheres, the change in temperature is a significant factor governing what paths of reaction are favored. More specifically, under photochemistry of sulfur compounds, the effective temperature is an important deciding factor. Experiments have revealed that the passage through different stages of sulfur, i.e., vapor to solid, is directly associated with temperature boundaries. The condensation critical temperature of sulfur occurs between 280 and 320 K, for instance, and extensive modeling suggests that this may be observed in a variety of atmospheric conditions on a range of celestial bodies. Hence, the role of temperature is not only confined to kinetics; it also dictates chemical diversity among atmospheres, including off-Earth atmospheres in systems like warm Jupiter and brown dwarfs (Marley et al., 2016). Secondarily, the interaction of temperature with reaction mechanism could reorganize the chemical landscape of objects in space, affecting our understanding of extraterrestrial atmospheres. For example, in the upper layers of brown dwarf atmospheres, there is an interesting interplay between effective temperature and molecular abundances. 500 K to 1100 K temperatures have a remarkable effect on equilibrium positions of molecules like methane (CH_4) and carbon monoxide (CO). Beyond these temperatures of material alteration, a trend against CH_4 toward CO is observed due to mechanisms like atmospheric mixing and surface gravity. These differences highlight the delicate kinetic-thermodynamic balance that governs the chemistry of the atmosphere, which informs our understanding of future life-supporting worlds and conditions that may prevail there (Marley et al., 2014). Correspondingly, understanding temperatures role in the kinetics of chemical reactions widens our understanding from the lab setting into the realm of astrophysics (Figure 7.7).

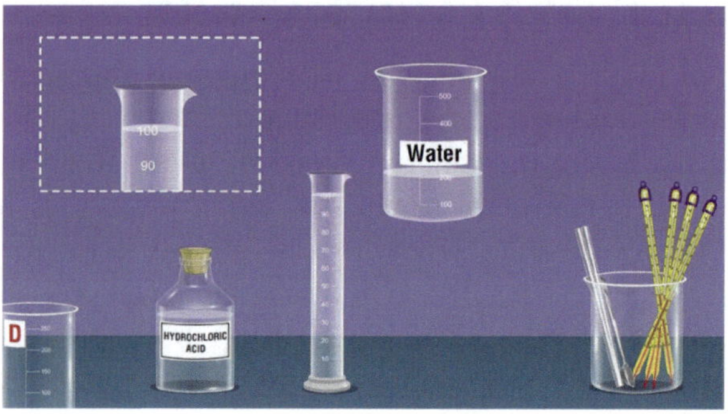

Figure 7.7. *Measure the change in temperature during chemical reactions.*

Source: https://i.ytimg.com/vi/6G4955krGl4/maxresdefault.jpg.

7.3.1. Effect of temperature on reaction rates

Temperature affects reaction rates heavily and is one of the important parameters of chemical kinetics. As temperature increases, so does the energy of molecules and hence their speed. The increased energy allows reactants to collide more strongly and with greater energy, which is necessary in order to overcome the barrier of activation energy. Hence, according to Arrhenius equation, rate constant of the reaction is higher at higher temperatures, leading to high rate of reaction. Temperature influence is mostly observed in various chemical reactions, in which minute differences can lead to drastic changes in reaction kinetics, showing the intricate relationship between thermal energy and molecular interactions. Within specific reactions, temperature dependence of the rate can be revealing of key mechanistic information. For instance, research on the photochemical reaction of sulfur demonstrates the effect of temperature on elemental sulfur formation from H_2S. Condensation-vapor transition of sulfur is close to the effective temperature of approximately 700 K, which is a key level at which reaction mechanisms may change substantially. Such temperature-induced transition points hold profound meaning in understanding chemical behavior across a series of conditions and assist further general comprehension of fast reactions in various kinetic environments, including planetary atmospheres and industrial processes (Marley et al., 2016). Reaction kinetics of more complex systems, including those that involve nitrogen bases, also exhibit the exponential influences of temperature on reaction rate. The selectivity of the piperidine conversion over isolated catalysts is observed to vary markedly with temperature and pressure, exhibiting a delicate interplay between kinetics and thermodynamics. At high temperatures and at high pressures of hydrogen, fast alkyl transfer reactions dominate, with considerable contrast in comparison to conditions at lower temperatures. This phenomenon highlights that reaction mechanisms, activation energies, and product distribution are all related to temperature variations, ultimately emphasizing its important place in chemical reaction kinetics (Mars et al., 1973) (Figure 7.8).

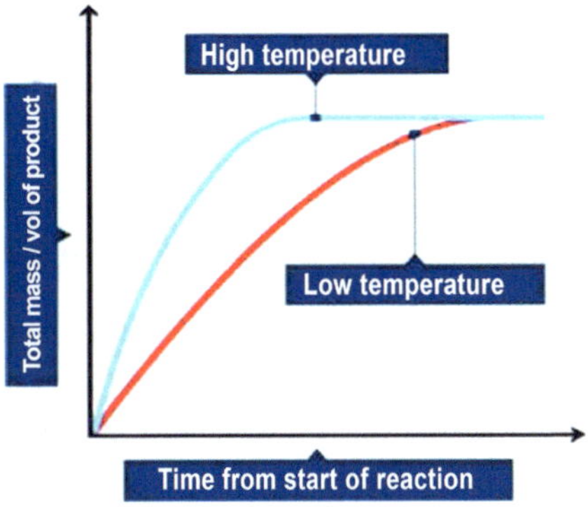

Figure 7.8. *Influence of temperature on reaction rates.*

Source: https://qsstudy.com/wp-content/uploads/2017/08/Influence-of-Temperature-on-Reaction-Rates.jpg.

7.3.2. Arrhenius Equation

The Arrhenius equation is one of the fundamental cornerstones of chemical kinetic research, namely concerning the speed of fast reactions. The equation explains quantitatively how temperatures influence rates of reaction and can provide information concerning the energy barriers which must be overcome for a reaction to proceed. Mathematically, the Arrhenius equation is expressed as $k = Ae^{-E_a/(RT)}$, where k is the rate constant, A is the pre-exponential factor, E_a is the activation energy, R is the universal gas constant, and T is the absolute temperature. On the basis of this equation, the temperature dependence of reaction rates can be determined, showing that even small increases in temperature will cause a tremendous increase in the reaction rate. This observation becomes key in the context of quick reactions where controlling temperature can optimize reaction efficiency and product yield. Outside theoretical models, the Arrhenius equation's usability also extends to numerous pragmatic applications, from industrial processes to ecological research. Atmospheric chemistry, for instance, uses the Arrhenius equation to model the kinetics of molecular reactions in different ranges of temperatures.

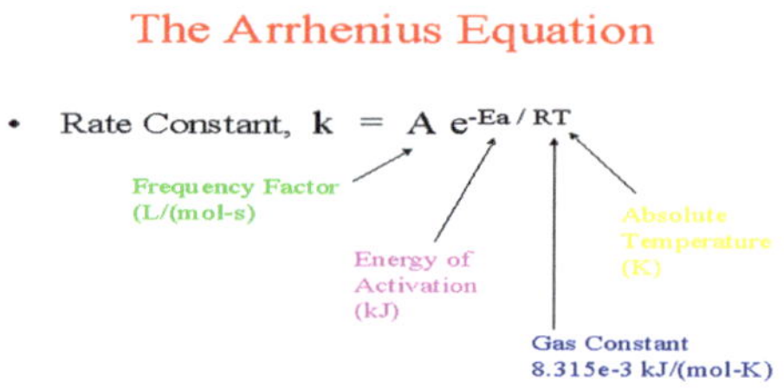

Figure 7.9. *The Arrhenius equation.*

Source: https://people.hws.edu/craig/chem120/kinetics/img017.GIF.

One study of the disequilibrium of light molecules in atmospheres of solar-composition brown dwarfs used a kinetics-based model to predict behaviors at a range of temperatures. Findings established that the conditions offering equal levels of activation energy were significant in accounting for the transition between substances such as methane (CH_4) and carbon monoxide (CO) within these regimes as a direct point of reference to principles found within the Arrhenius equation (Marley et al., 2014). The equation thus not only enhances our theoretical understanding but also offers plausible channels through which real-world prediction of chemicals can be accomplished. Secondly, determination of activation energy through the use of the Arrhenius equation is essential in the calculation of the efficiency of various chemical reactions, such as those that are exothermic and rapid. For example, an investigation on vegetable oil transesterification revealed that measurements of activation energy, such as the determined 51.0 kJ/mol, provided critical information on the kinetics of reactions (Dubreuil et al., 2011). This paper demonstrates how the Arrhenius equation is responsible for the development of rapid analytical methods for monitoring reaction dynamics in real time. By using techniques such as Near InfraRed spectroscopy coupled with computational kinetics, researchers can track fast reaction dynamics and enhance the knowledge of reaction mechanisms. Consequently, integration of the Arrhenius equation into such experimental systems highlights its relevance to both theoretical and practical realms of chemical kinetics (Figure 7.9).

7.3.3. Temperature Dependence of Activation Energy

Temperature dependence of activation energy is relevant when considering the kinetics of fast reactions, particularly those involving molecular transformation. As temperature increases, kinetic energy between molecules also increases, which is likely to surpass the energy barrier that accompanies activation. This phenomenon can be quantitatively described through the Arrhenius equation, which predicts the possibility of formulating reaction rates as a function of temperature and activation energy. The higher temperatures reduce the activation energy required for a reaction and therefore enhance the rate of reaction. This encounter not only illustrates the dynamic nature of reaction, but also marks the central role played by thermal energy in facilitating molecular collisions and subsequent reaction, with far-reaching application in realms ranging from organic synthesis to biochemistry. Recent studies emphasize the multifaceted role played by temperature in dictating faster reaction kinetics through control over the activation energy. Specifically, advanced techniques such as ultrafast T-jump have been employed in studying processes such as ligand exchange in metal complexes. In aqua cobalt (II) complex reactions, for instance, time scales in reactions suggest a delicate temperature dependence where the stepwise mechanism is very much different. Though under structural retention reaction rates can accelerate, these rates are bottle-necked by structure changes under various thermal conditions. Bimolecular kinetics of the reactions in the above reactions describe complex interactions driven by changes in enthalpic and entropic parameters, showing how temperature can have a strong influence on the energy landscape of high-speed processes, highlighting the salient features of chemical and biologic processes (Zewail et al., 2008). Furthermore, the study of mixed-conducting perovskite oxides determines temperature as a parameter in the activation energy of electrochemical systems, such as solid oxide fuel cells (SOFC). In a study of $SrTi_{1-x}Fe_xO_{3-}\delta$, the oxygen surface exchange rate was found to be poorly correlated with standard measures of electronic and ionic conductivity at certain temperatures. This would indicate that temperature generates not only conductivity variations but also affects the relative concentration of minority electronic species, which play crucial roles in reaction dynamics. Furthermore, the surface enrichment of strontium alters the constant of oxygen exchange rate, for which significant differences are observed with varying temperatures. Such observations indicate the significant interaction between temperature, activation energy, and reaction kinetics, particularly in complex systems where both structural as well as electronic contributions dominate fast reactions (Jung et al., 2011) (Figure 7.10).

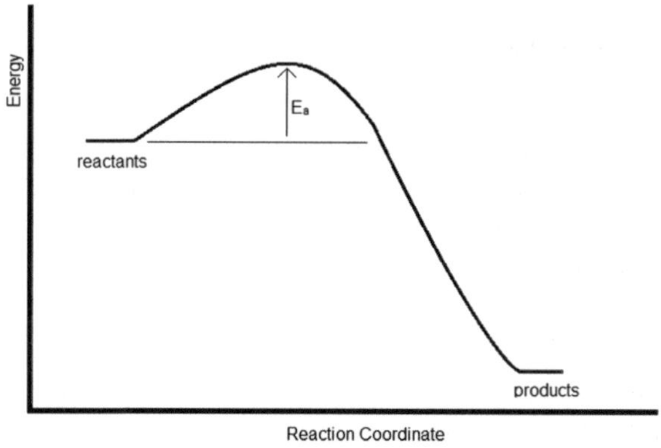

Figure 7.10. *Reaction profile for an elementary step over an activated barrier of height E_a.*

Source: https://chem.libretexts.org/@api/deki/files/113377/Arr.png?revision=1&size=bestfit&width=592&height=404.

7.4. THE INFLUENCE OF CONCENTRATION

The concentration effect is a highly significant aspect of reaction kinetics because it makes rate values in chemical reactions. Higher reactant concentrations tend to favor greater collision frequency among molecules with higher chances of successful encounters to drive reactions forward.

The case of gas-liquid reactions is such that the absorption of carbon dioxide into solution provides this specific focus with the consequence that higher concentrations significantly determine the rate of absorption. Studies indicate that such processes involve intricate numerical calculations to adequately model the dynamics involved, the shortcoming of the simplest models that ignore the complexity of concentration effects.

The gas absorption into sterically hindered amine solutions analysis demonstrates that previously set rates are sufficient to describe observed behavior and improved insight in concentration effect is essential in proper modeling of reaction kinetics in various cases (Bosch et al., 1989). In addition, the solid-state polymerization (SSP) of nylon 6 is an insightful example of understanding concentration effects in another state of matter. The SSP kinetics indicate that the process does not proceed via normal melt kinetics but rather reaction rates are significantly dominated by autocatalyzing acid chain end group diffusion. The studies reveal that the molecular weight of nylon 6 at the beginning of the process has a significant influence on the observed reaction rates during SSP. Specifically, higher initial concentrations of reactive groups enable quicker reaction, showing that concentration distribution is a decisive factor for overall kinetics. This shows how fundamental it is to consider concentration not only in liquid reaction but also in solid-state reaction, where molecular interaction differs substantially (Amirtharaj et al., 1982). The use of concentration extends not only to accelerating the rate of reaction but also to deciding on the mechanism and route of reaction. In the majority of cases, changing the concentration of one of the reactants can lead to changes in the dominant reaction pathways so that new products are produced. This is more evident in complex reaction networks where intermediates may be produced at different concentrations based on substrate availability. As shown in SSP investigations, kinetics possess other regions of operation governed by concentration dynamics and induced changes in temperature. It is by understanding the change brought about by concentration in kinetics and mechanisms that chemists are able to design better reactions and condition best for desired outputs. It is therefore important to identify the nature of concentration as multi-dimensional in theoretical and practical implications in the field of chemical kinetics (Amirtharaj et al., 1982; Bosch et al., 1989) (Figure 7.11).

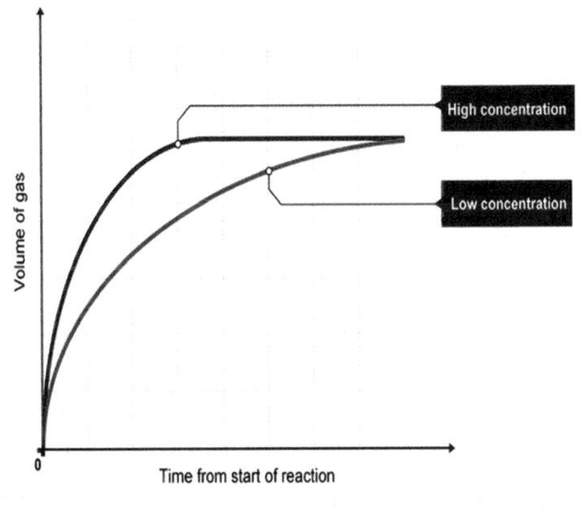

Figure 7.11. *Effect of concentration.*

Source: https://bam.files.bbci.co.uk/bam/live/content/zh6f7nb/large.

7.4.1. Reaction Order and Concentration

Appreciation of reaction order against concentration is crucial for describing the kinetics of fast reactions. Reaction order is the way in which the rate of a chemical reaction changes with the concentration of reactants. In a first-order reaction, rate is linearly dependent upon one reactants concentration, and it implies that the changes in concentration will affect the reaction rate linearly. Conversely, second-order reactions may be dependent either to the second of one reactant or on concentrations of two reactants. The ability to predict reaction rate from concentration not only provides valuable insights into reaction mechanisms but also enables efficient chemical process design. For instance, control of concentration can lead to spectacular increases in reaction rate, essential in industries relying on rapid conversion steps. Reaction network dynamics would introduce complexities that could mask the connection between reaction order and concentration. An extension of standard models, such as those involving Michaelis-Menten kinetics, might reveal subnetwork dynamics. These reaction networks consist of unary and binary reactions and enzymatic dynamics, which provide insights to help in understanding collective concentration effects on reaction order. Significantly, extension of projection approaches to such intricate networks can provide equations that capture memory effects inherent to biochemical systems. This is a crucial comprehension as it facilitates our ability to make predictions when deciding how conditions in reactions affect overall kinetics in systems where concentrations change dynamically, with good alignment in terms of governing principles of fast reactions in kinetics (Bishop et al., 2016). One of the significant uses of the investigation of the concentration and reaction order is in the synthesis processes of biodiesel, where the reaction does not go to completion much. Kinetic studies of the reaction of biodiesel have identified mass-transfer limitations as significant determinants of the methyl ester end-point concentration. By employing mass-transfer based models for the examination of drop size and concentration relation, it is evident that intensified mixing intensity results in desirable rates of conversion. It not only demonstrates how concentration influences reaction rate but also the requirement for the physical conditions to be optimized in a way that results can be achieved. Therefore, understanding the order of reaction and its dependence on concentration unlocks avenues for enhancing methods in industries that depend on fast, efficient, and high-yield chemical reactions (Kendall et al., 2009).

7.4.2. Rate Laws and Their Derivation

Derivation of rate laws forms the core of reaction kinetics since it sets the stage for connecting reactant concentration to product formation rate. The relationship can be expressed in numerical terms; the rate law of a reaction usually takes the form of a mathematical equation that incorporates the order of the reaction per reactant. The reaction order is found experimentally and describes the degree to which the rate depends on the concentration of each of the reactants. Generally symbolized by the exponent in the rate equation, the coefficients can differ from one reaction to another, indicating different kinds of kinetic behavior. In quick responses, the precise identification of these orders becomes important as small variations in concentration can lead to significant deviations in reaction rates, particularly in complex biochemical networks where different pathways may be playing a role simultaneously. The evolution of rate laws is directed by the reaction mechanisms that determine how fast the reactions proceed. For instance, in complex reaction networks, identification of reaction intermediates and fast cycles is key to constructing good rate equations. Techniques such as the method of tropical equilibration allow a more immediate interpretation of rate laws under these circumstances, allowing model reductions that allow abstraction of necessary information from large biochemical systems. This methodology enables the solution of non-linear equations for reaction kinetics and the alleviation of computational loads. The application of these advanced methodologies becomes imperative when handling systems in which non-linearities and rapid equilibria complicate the derivation of conventional rate laws, thereby extending our knowledge of dynamic phenomena in chemical and biochemical systems (Grigoriev et al., 2015). Furthermore, the kinetic features of what happens in fast reactions tend to recognize the thermodynamic constraints imposed by conditions of equilibrium in biochemical reaction cascades. For instance, the complexation of heat shock proteins such as Hsp70 with substrates demonstrates how fast reactions are practically non-equilibrium thermodynamic processes. In the research, the ultra-affinity of the proteins is regulated by ATP hydrolysis, a process which occurs under conditions where conventional equilibrium assumptions are not held. This not only highlights the intricacy involved with the attainment of accurate rate laws in fast processes but also indicates the significance of cofactors in facilitating reactions in the beyond scope of equilibrium. Therefore, awareness of the thermodynamic constraints on affinities helps one understand the all-important fluctuations in reaction rates, particularly in biological processes in which reactions occur at high velocity and regulation remains of utmost significance (Hartich et al., 2017).

7.4.3. Impact of Concentration on Fast Reactions

The influence of concentration on fast reactions is a key characteristic when studying reaction rates, particularly in determining how the variation of concentration influences rate of reaction. For the majority of chemical reactions, their rate of formation of products from the reactants is directly proportionate to their concentration, a correspondence stipulated by the law of mass action. This principle holds that when the concentration of one or more reactants is raised, the rate of collision and the subsequent conversion to products also rises, thus accelerating reaction rates. In such systems where reactants are mixed in definite ratios, higher concentration can even create conditions leading to reaction kinetics non-linearities, most specifically in the case of bimolecular reactions where the presence of multiple reactants can interfere with one another's path towards product. Understanding concentrations role is therefore essential in predicting and controlling reaction behavior in laboratory and industrial applications. Further, concentration influences not only the overall rate of reaction but also specific mechanisms that play a role in deciding the transition from reactants to products. Concentration in fast reactions, such as those that occur in atmospheric and biochemical systems, can result in huge variances in the rate laws employed to describe them. For instance, elevated reactant concentrations can lead to conditions described by the Damköhler number, a dimensionless quantity that compares the reaction rate to transport process rates like diffusion.

Concentration has been shown to create rapid increases in the rate of reaction, resulting in strong stretching conditions that alter mixing patterns and impose localized hotspots of reactivity. Thus, comprehending the relationship between concentration and such reaction mechanisms is crucial in explaining intricate behavior behind rapid chemical transformations under different conditions, as explained in reaction interface dynamics-literature (Bandopadhyay et al., 2016). Apart from this, discussing specific instances of rapid reactions also shows how concentration plays a great role in the kinetics of chemical reactions. For instance, the reaction process of hydroperoxy radicals with acetone illustrates how concentration changes can have a profoundly dramatic effect on reaction rates but even on equilibrium constants for reactions involved. Experiments conducted in controlled environments have shown that at given temperatures and concentrations, the reaction will become very rapid to approach an equilibrium very quickly, establish the cause-and-effect relationship between concentration and reaction kinetics.

As suggested, the prompt decrease in hydroperoxy concentration upon reaction with acetone suggests that high concentrations of acetone favor rapid initial reactions, while subsequent processes might differ based on concentration levels in the course of the reaction. Concentration is thus a root parameter dictating the kinetics of rapid reactions and their broader environmental ramifications (Atta et al., 2011).

7.5. TECHNIQUES FOR STUDYING FAST REACTIONS

Awareness of rapid reactions in kinetics requires a vast array of methods through which scientists are able to identify and study rapid processes accurately. One of such methods is the use of rapid mixing methods, enabling the rapid addition of reactants under controlled conditions by scientists, facilitating scientists to observe reaction kinetics in real-time. These methods, such as stopped-flow and continuous-flow systems, provide critical data about reaction mechanisms by capturing transient species created in the early phases of a reaction. By employing sophisticated detection techniques, like UV-Vis spectroscopy or fluorescence, researchers are able to monitor concentration changes during course of reactions, thereby gathering critical data to input into kinetic models. In addition, the use of fast cameras and advanced computational techniques broadens the capability to image these transient processes, which is a major advancement in experimental kinetics research. Another significant area of exploring fast reaction studies is understanding the dynamics of enzyme-catalyzed reactions, where characteristic timescales are necessary to determine global reaction behavior. Through the application of scaling analysis, scientists can identify several regimes arising depending on the given timescales of the reaction components. For instance, in a Michaelis–Menten system, the analysis recognizes significant transitions from early fast transients to steady-state kinetics with specific focus on the significance of optimizing timescale selection in experimental design. By examining more complex reaction mechanisms, such as sequential enzyme-catalyzed systems, scientists are able to gain deeper insights into how various timescales interact and lead to a wide range of dynamic regimes. Such an analysis not only elucidates the individual dynamics of fast reactions but also hints at the broader application of such techniques in a range of nonlinear dynamical systems, as in the research into enzyme kinetics (Eilertsen et al., 2019). Additionally, the challenge posed by biochemical networks to the study of fast reactions underscores the utility of model reduction techniques to simplify complex systems. These systems are often multi-scaling with the elements reacting on a range of timescales, so it is more challenging to analyze them. By employing concepts such as dominance between model elements, researchers are able to simplify big models without losing crucial dynamical behavior. This hierarchical organization allows a tractable analysis to be preserved with key features of the original system. Techniques like quasi-steady state and quasi-equilibrium approximations are techniques of practical simplification in models, especially in linear and nonlinear networks. These techniques are not only useful in kinetics but also useful in other fields like computational biology, where information regarding dynamic interaction between complex networks has great relevance to systems biology (Gorban et al., 2012) (Figure 7.12).

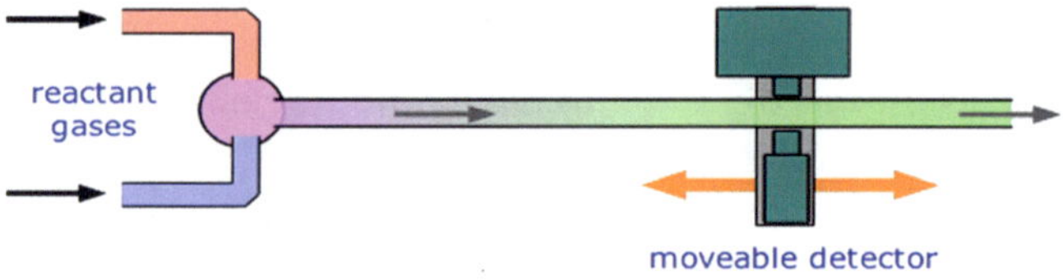

Figure 7.12. *A continuous flow fast kinetic system.*

Source: https://chem.libretexts.org/@api/deki/files/17495/flowexpt.png?revision=1&size=bestfit&width=482&height=147.

7.5.1. Stopped-Flow Spectroscopy

Stopped-flow spectroscopy is a vital technique for the examination of rapid kinetic processes, and it permits scientists to trace the dynamics of biochemical reactions that occur on a timescale of milliseconds or even microseconds. Stopped-flow spectroscopy involves a rapid mixing of the reactants in a specified environment and an immediate measurement of the absorption or fluorescence characteristics of products. By applying this technique, scientists can examine complex mechanisms of reaction pathways, making conclusions about reaction intermediates and transient species that would otherwise be invisible to them. The temporal resolution of stopped-flow spectroscopy is so great that it can investigate processes such as enzyme catalytic activity, conformational change of molecules, and ligand-binding dynamics, thereby improving our understanding of basic biochemical processes. The application of stopped-flow spectroscopy has significantly enhanced our knowledge of numerous varied biochemical transitions through the account of time-dependent kinetics. For instance, protein research has demonstrated how rapidly conformational change may be achieved on ligand binding. The implementation of microfluidics with the use of stopped-flow techniques has enabled more precise mixing as well as transcending diffusional constraints that are typically present with manual mixing procedures. With this advanced equipment, researchers have been able to investigate the kinetics and the calcium ion-stimulated conformational change in calmodulin with distinct rates of transition that can differ quite dramatically between homologous domains of the protein, as revealed by experiments with characteristic time constants of approximately 490 μs and 20 ms for different domains (Andersson et al., 2008).

Furthermore, stopped-flow spectroscopy has greatly aided the study of metal complex formation in biological systems through the demonstration of intricate mechanisms governing cellular signaling pathways. For example, dinitrosyl iron complex (DNICs) formation is evidence of the efficacy of the method in studying rapid reaction events involving nitric oxide and ferrous ions. The kinetic investigation of these reactions not only underlines the two-stage reaction kinetics characteristic of thiols but also better explains how such complexes are formed at physiological conditions. The knowledge gleaned from stopped-flow experiments adds importantly to our biological understanding, showing how fast reaction may result in crucial regulatory processes within cells, especially in the case of inflammation and oxidative stress (Augusto et al., 2019). Through these developments, stopped-flow spectroscopy continues to be a cornerstone in the field of chemical kinetics (Figure 7.13).

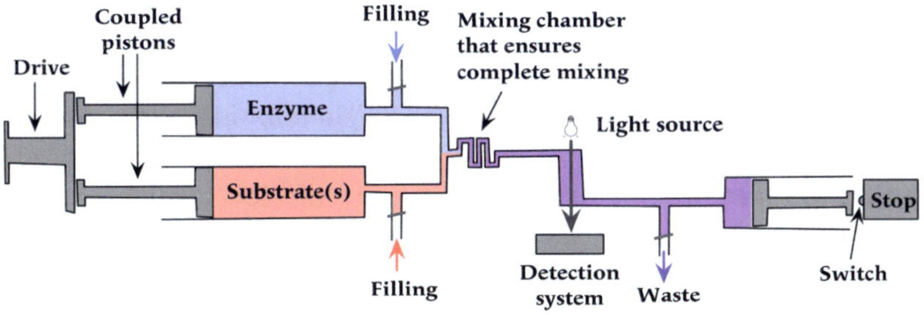

Figure 7.13. *Diagram of a stopped-flow instrument.*

Source: https://upload.wikimedia.org/wikipedia/commons/7/73/Diagram_of_stopped_flow_equipment.png.

7.5.2. Flash Photolysis

Flash photolysis is a powerful experimental technique employed to study fast chemical reactions and kinetics. Flash photolysis entails the photolytic generation of active intermediates with high speed through a short pulse of ultraviolet or visible light. This enables scientists to form transient species in an instant and observe their reaction in real time. The ability to produce and subsequently investigate these intermediates allows scientists to determine valuable kinetic parameters and reaction mechanisms for fast reactions. Flash photolysis has been crucial in illuminating an assortment of systems, including atmospheric chemistry, where it assists in the understanding of fundamental reactions that control ozone depletion and pollutant formation. Flash photolysis has become increasingly significant in atmospheric chemistry due to its capability to analyze complex reaction processes that are accountable for such effects as ozone depletion. For instance, reactions involving bromine oxide (BrO) and chlorine oxide (ClO) give significant insights regarding stratospheric chemistry. There are various pathways, such as the reactions provided by BrO + ClO to produce halogens and other oxygen species, which illustrate the simplicity of such systems. The presence of metastable intermediates in such reactions calls for an exhaustive study via flash photolysis in conjunction with techniques like mass spectrometry. These methods enable researchers to interrogate the reaction kinetics as well as the spectroscopic features of the intermediates so that their implication in atmospheric chemistry and resulting influence on ozone concentration can be better understood, as elucidated by Friedl et al. (2025). Furthermore, flash photolysis finds application in other fields aside from atmospheric science, particularly in organic chemistry where it aids research on the dynamics of photochemical processes. Through flash photolysis, the intricacies of electron transfer processes and biradical or cyclic product formation following intramolecular processes are unraveled. For instance, research on N-(selenomethyl)alkyl-phthalimides demonstrates the ability of flash photolysis to separate competitive processes following electron transfer. By ultrafast probing of ultra-short-lived transients and their ultra-short lifetimes and populations, researchers are provided with insight into the effect of molecular structure on the ultra-fast outcome of reactions, like cyclization and the consequent product formation. This assessment puts into perspective the broader impact of flash photolysis in uncovering reaction mechanisms and kinetics within a broad range of chemical topographies (Argüello et al., 2015) (Figure 7.14).

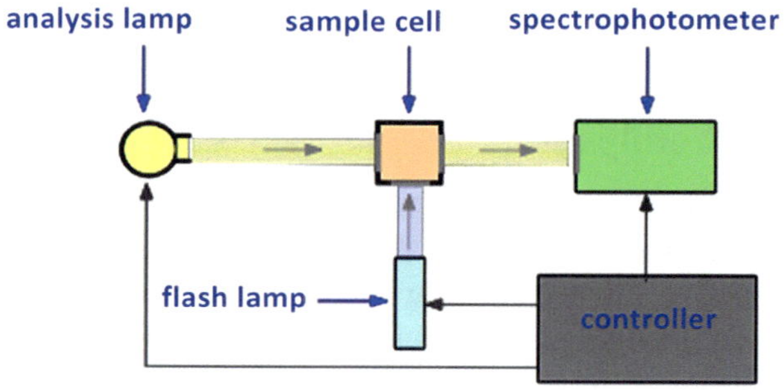

Figure 7.14. *A flash-photolysis relaxation experiment.*

Source: https://chem.libretexts.org/@api/deki/files/17501/flashphoto.png?revision=1&size=bestfit&width=389&height=196.

7.5.3. Rapid Mixing Methods

Rapid mixing methods are a necessary part of the study of fast reactions in kinetics, especially where the time scales of the process are very short. Traditional methods, such as stop-flow and quench-flow, have limitations on the minimum observation time that they can achieve; most operate on the millisecond scale. New techniques based on miniaturized equipment are being developed to overcome this limitation. Another major innovation is the use of an etched silicon chip with "T" mixers coupled by a short channel. The structure not only reduces mixing chamber volume but also reduces the distance between the chambers, substantially reducing dead times to as low as 110 microseconds. Such fast-mixing capabilities are crucial in the study of reactions occurring on time scales below a millisecond, thereby advancing understanding of complex kinetic processes like protein folding and enzyme reactions in biochemical systems (Bökenkamp et al., 1998). In biochemistry, fast mixing techniques have been crucial in the study of proteins' conformational transitions. One such application is a microfluidic mixer for the study of calmodulins kinetics under the process of calcium binding. By facilitating homogeneous, fast mixing, this device is able to, for the first time, realize efficient decoupling of hydrodynamic focusing from diffusive mixing, allowing researchers to access tens of microseconds time scales. Utilizing this setup in conjunction with advanced imaging techniques, like multiphoton microscopy, researchers are able to monitor directly fast protein transitions, like in calmodulin. This approach has provided information on widely differing kinetic rates for the various parts of the protein, such that conformational motions can also range greatly in their timescales, with time constants from hundreds of microseconds to a few milliseconds. Such quantitative kinetic characterization is important to an understanding of biomolecular interaction dynamics and biological function (Andersson et al., 2008). The integration of rapid mixing methods into kinetic studies not only enables us to observe more swiftly progressing reactions but also promotes the validity of data acquisition under real-time settings. With the reduction of the possibility for artifacts generated by slower mixing protocols, researchers can provide more authentic results that faithfully reflect reaction kinetics. In applications beyond simple biochemical reactions, such as in nanomaterial synthesis or catalysis, these methods facilitate access to the transient state and intermediates otherwise inaccessible. Ongoing development towards increasingly sophisticated microfluidic systems is highly promising, as such devices continue to push the frontiers of what may be achieved in kinetic research. Generally, rapid mixing methods are principal tools in the unfolding of rapid reaction complexities, ushering in new discoveries in the sciences.

7.6. APPLICATIONS OF FAST REACTION KINETICS

The study of rapid reaction kinetics is at the core of various scientific and industrial technologies, particularly biochemical networks. The networks play a vital role in modeling the dynamic phenomena of cellular processes like signaling, metabolism, and regulation of gene expression. However, analyzing the kinetics of large biochemical models would prove to be challenging due to parametric uncertainty and combinatorial explosion of potential interactions. Through novel approaches, such as the concept of model element dominance, researchers are able to provide a hierarchical organization of these elements based on their influence on network behavior. The simplification through this allows for the examination of complicated systems in a more conceivable fashion, thereby providing the key to enhanced computational biology and, ultimately, better cellular function understanding (Gorban et al., 2012). Industry has applications of fast reaction kinetics equally as significant. For instance, companies like Bluestar Silicones are engineering the shift from continuous processes to batch processes for chemical production. This shift calls for the development of new technologies offering high productivity as well as addressing safety and environmental problems involved in chemical reactions.

In order to achieve this transition, researchers have come up with innovative types of reactors, such as a heat exchanger reactor made from silicon carbide plates that exhibit improved heat and mass transfer performance. Such prototypes prove their feasibility through various industrial reactions, and they result in more effective and ecologically sound system designs, ultimately making the importance of kinetics in advancing chemical engineering and process designs more sensitive (Elgue et al., 2009). In addition, fast reaction kinetics also plays a vital role in optimizing laboratory and industrial processes. By applying principles learned from kinetically studied systems, chemists can be able to tune reaction conditions for an individual synthetic path, maximizing yield and avoiding formation of unwanted byproducts.

Fast determination of reaction time and evaluation of the consequences of varying parameters can play a significant role in advancement in fields such as pharmaceuticals, where early production of effective drugs is vital. Besides, multiscale-based modeling techniques are capable of simplifying intricate reaction networks, thus leaving doors open for researchers to adapt their technique to specific applications.

This implication of rapid reaction kinetics in realistic scenarios points to its significance not only in theoretical chemistry but also to addressing actual issues in the chemical and biological industries (Gorban et al., 2012).

7.6.1. Industrial Processes

In industrial processing, the shift from batch to continuous flow is becoming increasingly recognized as a route to increased productivity and efficiency. Pressure for sustainable processing, in particular in chemical production where environmental concern and safety are paramount, drives the trend. New technologies, such as those now being developed for Bluestar Silicones, follow the trend. The synthesis of a new heat exchanger reactor constructed using silicon carbide plates is illustrative of the potential of advanced materials to deal with corrosion resistance and improve heat and mass transfer capability. Not only do these advancements simplify the ease of process operation changes, but they also highlight the need for further research into such technology. Through the advancement of industrial reactor development with consideration for safety and resource consumption, industries can significantly minimize their impact on the environment while maximizing production in various chemical reactions, which is the focus of the kinetics of fast reactions. Process intensification is also a significant aspect of ongoing industrial processes, particularly in biofuel generation from waste feedstocks like waste cooking oil (WCO). Employing state-of-the-art equipment such as microwave reactors and reactive distillation is a significant innovation in optimizing the transesterification and esterification reactions required to produce fatty acid methyl esters (FAME). Such advancements enable not just enhanced reaction rates, but better separation capacity too, thereby reducing reaction times in general and enhancing yield. The integration of heterogenous catalysts and continuous flow systems in essence shows the use of principles of kinetics in an industrial setting to make processes more efficient. Such developments demonstrate the need to integrate reaction and separation phases in the end, ultimately leading to a more efficient and economically viable approach towards the production of biofuel, which is crucial for meeting burgeoning energy needs responsibly. Finally, process intensification technologies are a testament to the dynamic nature of industrial chemistry, revolutionizing the conventional ways of doing things at its core. With increasing numbers of industries adopting continuous processes, stable systems are needed that integrate several stages of operation. Approaches such as cavitation and oscillatory baffling have been proven to be viable means of enhancing mixing and reaction efficiency at the cost of product quality. This innovation not only enhances the kinetics of fast reactions through improved mass transfer but also provides a way to supplement overall goals of sustainability in industrial processes. The versatility of application of these new systems to a variety of chemical processes, combined with ongoing research into efficiency improvement, reflects the two-way relationship between industrial chemistry and chemical kinetics. Hence, embracing these innovations is key to businesses that aim to remain competitive while accommodating regulatory and environmental demands, ensuring future growth and compliance in a changing landscape.

7.6.2. Environmental Chemistry

It is important to comprehend the intricacies of environmental chemistry in relation to the kinetics of rapid reactions. The rapid chemical conversions are fundamental to ecological equilibrium and regulation of biogeochemical cycles. For instance, the degradation of contaminants, such as nitrogen oxides, is initiated by highly dynamic reactions in various environmental compartments, such as the atmosphere and aquatic systems. These are regulated by numerous factors, such as temperature, pH, and light, which can dramatically speed up or decelerate the reaction rates. By growing research, kinetic studies reveal these processes more definitively, and what is learned from them can continue to shape environmental policy and remediation processes. The intricacy of such quick reactions places a premium on ongoing research, particularly as the impact of anthropogenic contribution to chemical reactions grows increasingly central to halting worldwide environmental issues. The environmental chemistry use of nitrogen species is only one example of why learning their kinetics and stability is so important, particularly with respect to early Earth conditions. Current studies indicate that dissolved nitrogen oxide anions (NO_x^-), for example, nitrate (NO_3^-) and nitrite (NO_2^-), may have played significant roles in prebiotic chemistry, including the evolution of life. Specifically, these nitrogen species have been shown to have roles in required processes like RNA oligomerization and protometabolism generation. However, studies determine that the levels of NO_x^- predicted in early Earth oceans were significantly lower than initially calculated, primarily due to their unstable nature in reducing environments. UV photolysis and reaction with reduced iron (Fe^{2+}) processes can greatly minimize these levels, contradict earlier assumptions and point to the need for more insight into the kinetic pathways of these essential species (Babbin et al., 2019). Improvement of our understanding in environmental chemistry about fast reactions is of the utmost importance for various reasons, including the fight against climate change and pollution. Fast reactions have the tendency of regulating the transport and fate of various chemical species within the ecosystem and impacting air and water quality. Kinetic studies allow scientists to predict the development of such reactions in the future under changing environmental conditions, particularly as regards anthropogenic factors. The effects on global warming and the well-being of ecosystems are important. For instance, accurate simulation of reactions involving nitrogen can lead to better management of nutrient loading in bodies of water, thereby avoiding eutrophication and its effects. Therefore, it is necessary that future research targets the kinetics of key environmental reactions, ultimately linking laboratory findings to broader ecological contexts and dictating means by which key environmental issues are addressed (Pinto et al., 2017) (Figure 7.15).

Figure 7.15. *Understanding the environmental chemistry.*

Source: https://blogger.googleusercontent.com/img/b/R29vZ2xl/AVvXsEhIhJPV_5Ut9CPuY4K4Bfplss3xx3ciXqEuUhCVZssoOhOTPhJSrf-c4e6wp6uoFHvQ5Tkxg-gaJQGUj58e4SD4eLLZYLBrkh8WCqvEqF51dHIxW9lz05g4Kzo7hzrPZ2qDdB-2wWpJwzjXt3oB8R6N-6Y1k2uDMM952gMkTzF3lBOSALLvng7ieNdg4odto/s767/Environmental%20Chemistry%20-%20Short%20Handwritten%20Notes%20%5BPDF%5D%20%F0%9F%93%9A.jpeg.

7.6.3. Biochemical Reactions

The intricacy of biochemical reactions demands a sound understanding of kinetics involved, particularly in fast reactions where time scales are an absolute necessity. Fast reactions may have enzyme-catalyzed reactions that are characterized by their potential to process substrates swiftly into products. In these instances, traditional models such as the Michaelis-Menten equation provide the basics of reaction rates under saturated conditions. Yet, as biological systems become increasingly complex, the need to incorporate beyond the elementary models becomes obvious. Novel approaches, particularly subnetwork dynamical ones, allow for the introduction of memory effects essential for an adequate description of biochemical reaction temporal evolution. The findings indicate that such complex networks can significantly model how quickly enzymatic reactions influence greater metabolic pathways and cellular functions, broadening the application of kinetic models in biochemistry (Bishop et al., 2016). Large biochemical networks also present the challenges that necessitate new means of reducing models, which can facilitate synergy among fast and slow reactions. The dominance hypothesis between model elements highlights the hierarchical nature of biochemical interactions, where there are reactions that contribute more to the overall network activity. The hierarchical nature facilitates the simplification of complex models without compromising key kinetic behaviors. According to multi-scaleness principles, individual reaction elements may be segregated and evaluated independently using time constants, including faster reactions without compromising contextual consistency. This strategy not only assists in circumventing computational constraints but also makes the prediction of cell behavior under various physiological conditions more accurate. Hence, advances in model reduction techniques are crucial to our enhanced understanding of fast reactions in dynamic biochemical networks (Gorban et al., 2012). Understanding biochemical reactions within the context of fast kinetics also determines the importance of enzyme dynamics to metabolic control. Maximum enzymic activity is often crucial for biological homeostasis, where the immediate regulation of metabolic fluxes is necessary to adapt to environmental changes. Besides acting as catalysts, enzymes are also regulatory proteins that control the rate of biochemical reactions through various pathways. Thus, the kinetics of rapid reactions involves not only quantitative requirements of reaction rates but also qualitative requirements of enzyme interactions and regulatory processes. By shedding light on these processes' dynamics, scientists can develop comprehensive biochemical models that direct therapeutic interventions in metabolic diseases and other diseases. Hence, a qualitative understanding of fast biochemical reactions is vital in theoretical and applied biochemistry, highlighting the interdependence of kinetic studies and biological function (Bishop et al., 2016).

SUMMARY

In an overview of discussion regarding rapid responses in kinetics, the importance of an examination of implications of mechanisms in various catalytic reactions must be taken into consideration. Rich awareness of the reaction dynamics, such as an examination of piperidine conversion over $CoO–MoO_3–Al_2O_3$ catalysts, highlights the contribution of factors such as temperature, pressure, and reaction time. The ring-opening in two stages followed by alkyl transfer has been recognized as the mechanism and illustrates how kinetic parameters can influence the formation of products. The observation that the reaction is first-order in piperidine as well as hydrogen but –1 order with regard to overall nitrogen base pressure emphasizes the complexity involved in reaction routes. The dependence on equilibrium constants and activation energies also confirms the significance of scrutinizing catalyst performance under various conditions carefully, highlighting the intricate interaction of kinetics for fast reactions (Mars et al., 1973). Moreover, biochemical networks provide knowledge about the extended applications of kinetic principles beyond traditional chemical reactions, describing how such formats can reproduce intricate biological processes. As highlighted by cellular dynamics research, key problems such as parametric uncertainty and network complexity can be surmounted using advanced techniques like model reduction. By applying dominance principles between system elements, researchers are able to simplify complex networks while retaining essential dynamic characteristics. Not only does this validate kinetic theories' applicability in biological systems but also serves as an example of innovative problem-solving strategies in computational biology. Therefore, the understanding of such dynamics offers researchers the means to treat complex biochemical interactions efficiently, demonstrating the integrative role of kinetics across disciplines (Gorban et al., 2012). Generally, the investigation of fast reactions in kinetics covers a wide array of catalytic and biological reactions, demonstrating the inherent nature of reaction dynamics in different scenarios. Through the integration of knowledge of catalytic reactions and advances in biochemical networks, one is able to imagine the fundamental principles governing chemical and biological conversions. Through continued refinement of methods of probing kinetic behavior, such integration affords improved prediction and modeling of reaction outcomes. The exchange between theoretical models and experimental data will surely enhance our knowledge of kinetic systems and thus enable innovations in fundamental science and technological applications. Lastly, further debating rapid reactions in kinetics not only advances scientific knowledge but also inspires innovations in the future (Gorban et al., 2012; Mars et al., 1973).

Research on rapid reactions in kinetics reveals a variety of crucial insights into the mechanisms and dynamics of such reactions. The reaction between nitric oxide (NO) and ferrous ions and a series of thiols, such as glutathione and cysteine, is one such observation. The formation of dinitrosyl iron complexes (DNICs) exhibits a two-step reaction mechanism with rapid pre-equilibria followed by an autoreduction step that is rate-determining. This critical analysis is informed by stopped-flow kinetics and density functional theory research, which demonstrates the complexity of DNIC assembly in biological systems. In the end, these studies suggest that interactions between ferrous ions and thiols facilitate the rapid assembly of DNIC under inflammatory conditions, with the important roles of these complexes demonstrated in physiological as well as pathological processes (Augusto et al., 2019). In addition to the biochemical interactions highlighted in DNIC structures, recent advancements in modeling biochemical reaction networks significantly enhance our understanding of kinetics. Projection methods allow researchers to study dynamics of subnetworks, addressing memory effects of previous reactions appropriately. The approach has been limited to unary and binary reactions in the past, but recent developments include Michaelis-Menten kinetics to model enzymatic reactions explicitly. By constructing a broad network that engages enzymes as well as their complexes, such a structure ultimately gives a closed-form description of subnetwork dynamics. These advances emphasize the accuracy and effectiveness of these methods in describing biological systems, particularly when it comes to complicated enzymatic reactions that are undergoing high kinetics (Bishop et al., 2016). The consequences of rapid responses in kinetics are not restricted to fundamental science, but they have echoes in numerous applications in medicinal chemistry and biological engineering. The rates of most of the most important processes are generally dictated by rapid reactions, which in turn influence drug metabolism and design of therapeutic agents. This information allows drug formulation to be optimized by pharmaceutical scientists and drug efficacy maximized. Moreover, insights from kinetics research can direct the biosensor and other diagnostic instrumentation design that monitors biochemical processes in real time. With the incorporation of DNIC formation knowledge and advanced modeling of reaction networks, researchers can further enhance our understanding of fast reactions and hence contribute to numerous scientific and medical endeavors.

REFERENCES

1. Alberts, B., Phillips, A., Blossey, R., Bundschuh, R., Cherry, J. L., De Jong, H., Fall, C. P., et al. (2007). Compositionality, stochasticity and cooperativity in dynamic models of gene regulation. http://arxiv.org/abs/q-bio/0702016 (accessed on 10th September 2025).
2. Alvarez-Ruiz, B., Amatore, C., Baranski, A. S., Bard, A. J., Beriet, C., Campbell, S. A., Chen, S., et al. (2011). Electrochemistry at nanoscale electrodes: Individual single-walled carbon nanotubes (SWNTs) and SWNT-templated metal nanowires. https://core.ac.uk/download/9561586.pdf (accessed on 10th September 2025).
3. Amirtharaj, J., Gaymans, R. J., & Kamp, H. (1982). Nylon 6 polymerization in the solid state. https://core.ac.uk/download/pdf/11474314.pdf (accessed on 10th September 2025).
4. Andersson, A., Baird, B., Bayley, H., Drum, B., Rhoades, E., Eaton, W. A., et al. (2008). Conformational changes of calmodulin upon Ca^{2+} binding studied with a microfluidic mixer. https://core.ac.uk/download/4879681.pdf (accessed on 10th September 2025).
5. Argüello, J. E., Heredia, A. A., Oksdath Mansilla, G., Peñeñory, A. B., et al. (2015). Photochemistry of *N*-(selenoalkyl)-phthalimides: Formation of *N,Se*-heterocyclic systems. https://core.ac.uk/download/159300430.pdf (accessed on 10th September 2025).
6. Aubin, J., Mazubert, A., & Poux, M. (2013). Intensified processes for FAME production from waste cooking oil: A technological review. https://core.ac.uk/download/18540047.pdf (accessed on 10th September 2025).
7. Augusto, O., Ford, P. C., Iretskii, A. V., Truzzi, D., et al. (2019). Dynamics of dinitrosyl iron complex (DNIC) formation with low molecular weight thiols. https://core.ac.uk/download/286028126.pdf (accessed on 10th September 2025).
8. Babbin, A. R., Ranjan, S., Rimmer, P. B., Sasselov, D., et al. (2019). Nitrogen oxide concentrations in natural waters on early Earth. http://arxiv.org/abs/1902.07740 (accessed on 10th September 2025).
9. Bandopadhyay, A., Le Borgne, T., Dentz, M., Méheust, Y., et al. (2016). Enhanced reaction kinetics and reactive mixing scale dynamics in mixing fronts under shear flow for arbitrary Damköhler numbers. http://arxiv.org/abs/1604.08092 (accessed on 10th September 2025).
10. Ben Cheikh Larbi, F., Chaussé, A., De Lamare, J., Di Caprio, D., et al. (2017). Contribution of cellular automata to the understanding of corrosion phenomena. http://arxiv.org/abs/1710.01545 (accessed on 10th September 2025).

11. Bishop, C. M., Haldane, J. B. S., Henri, V., Rubin, K. J., Michaelis, L., Murray, J., Sollich, P., et al. (2016). Michaelis–Menten dynamics in protein subnetworks. https://core.ac.uk/download/45319691.pdf (accessed on 10th September 2025).

12. Bökenkamp, D., Desai, A., Marzluff, E. M., Mayo, S. L., et al. (1998). Microfabricated silicon mixers for submillisecond quench-flow analysis. https://core.ac.uk/download/4889488.pdf (accessed on 10th September 2025).

13. Bosch, H., van Swaaij, W. P. M., & Versteeg, G. F. (1989). Gas–liquid mass transfer with parallel reversible reactions—I. Absorption of CO_2 into solutions of sterically hindered amines. https://core.ac.uk/download/pdf/11473934.pdf (accessed on 10th September 2025).

14. Branco Pinto, D., Constales, D., Marin, G., Yablonsky, G., et al. (2017). Kinetic intersections as a source of information on rate coefficients. https://core.ac.uk/download/91288153.pdf (accessed on 10th September 2025).

15. Burrage, K., & Tian, T. (2004). Effective simulation techniques for biological systems. https://core.ac.uk/download/81387.pdf (accessed on 10th September 2025).

16. Burrage, K., & Tian, T. (2005). Parallel implementation of stochastic simulation for large-scale cellular processes. https://core.ac.uk/download/81556.pdf (accessed on 10th September 2025).

17. Ciottoli, P. P., Grenga, T., Martelli, E., Paolucci, S., et al. (2015). Dynamical system analysis of ignition phenomena using the tangential stretching rate concept. https://core.ac.uk/download/54521520.pdf (accessed on 10th September 2025).

18. Davis-Van Atta, C., Grieman, F. J., Noell, A. C., Okumura, M., et al. (2011). Determination of equilibrium constants for the reaction between acetone and HO_2 using infrared kinetic spectroscopy. https://core.ac.uk/download/4891946.pdf (accessed on 10th September 2025).

19. Dubreuil, B., Li, Y., Prat, L. E., Richard, F., et al. (2011). On-line monitoring of the transesterification reaction between triglycerides and ethanol using near infrared spectroscopy combined with gas chromatography. https://core.ac.uk/download/12042713.pdf (accessed on 10th September 2025).

20. Eilertsen, J., Schnell, S., & Stroberg, W. (2019). Characteristic, completion or matching timescales? An analysis of temporary boundaries in enzyme kinetics. http://arxiv.org/abs/1809.03961 (accessed on 10th September 2025).

21. Elgue, S., Ferrato, M., Gourdon, C., Jolivet, A., et al. (2009). Silicon carbide equipments for process intensification of silicon reactions. https://core.ac.uk/download/12041266.pdf (accessed on 10th September 2025).

22. England, J. L., & Sarkar, S. (2018). Design of conditions for emergence of self-replicators. http://arxiv.org/abs/1709.09191 (accessed on 10th September 2025).
23. Friedl, R. R., & Sander, S. P. (2025). Studies of ClO and BrO reactions important in the polar stratosphere: Kinetics and mechanism of the ClO + BrO and ClO + ClO reactions. https://core.ac.uk/download/pdf/42830246.pdf (accessed on 10th September 2025).
24. Gorban, A. N., Barenblatt, G., Bertini, L., Blesgen, T., Boudart, M., Briggs, T., et al. (2011). Quasichemical models of multicomponent nonlinear diffusion. http://arxiv.org/abs/1012.2908 (accessed on 10th September 2025).
25. Gorban, A. N., Noel, V., Radulescu, O., Zinovyev, A., et al. (2012). Reduction of dynamical biochemical reaction networks in computational biology. https://core.ac.uk/download/pdf/82880909.pdf (accessed on 10th September 2025).
26. Grigoriev, D., Noel, V., Radulescu, O., Vakulenko, S., et al. (2012). Tropicalization and tropical equilibration of chemical reactions. https://core.ac.uk/download/223016261.pdf (accessed on 10th September 2025).
27. Grigoriev, D., Radulescu, O., & Vakulenko, S. (2015). Model reduction of biochemical reaction networks by tropical analysis methods. http://arxiv.org/abs/1503.01414 (accessed on 10th September 2025).
28. Hartich, D., Nguyen, B., De Los Rios, P., Seifert, U., et al. (2017). Thermodynamic bounds on the ultra- and infra-affinity of Hsp70 for its substrates. https://core.ac.uk/download/148029434.pdf (accessed on 10th September 2025).
29. Jung, W., & Tuller, H. L. (2011). Investigation of cathode kinetics in SOFC: Model thin film $SrTi_{1-x}Fe_xO_{(3-\delta)}$ mixed conducting oxides. https://core.ac.uk/download/4894528.pdf (accessed on 10th September 2025).
30. Kendall, K., & Slinn, M. (2009). Developing the reaction kinetics for a biodiesel reactor. https://core.ac.uk/download/pdf/1631264.pdf (accessed on 10th September 2025).
31. Marley, M. S., & Zahnle, K. J. (2014). Methane, carbon monoxide, and ammonia in brown dwarfs and self-luminous giant planets. https://core.ac.uk/download/pdf/189597261.pdf (accessed on 10th September 2025).
32. Marley, M. S., Morley, C. V., Moses, J. I., Zahnle, K., et al. (2016). Photolytic hazes in the atmosphere of 51 Eri b. https://core.ac.uk/download/pdf/76423622.pdf (accessed on 10th September 2025).

33. Mars, P., Neyens, W. J., & Sonnemans, J. (1973). The mechanism of pyridine hydrogenolysis on molybdenum-containing catalysts: IV. The conversion of piperidine. https://core.ac.uk/download/pdf/11471850.pdf (accessed on 10th September 2025).
34. Martinez-Senac, M. M., Slatter, A. F., Toseland, C. P., Webb, M., et al. (2009). The ATPase cycle of PcrA helicase and its coupling to translocation on DNA. https://core.ac.uk/download/30707869.pdf (accessed on 10th September 2025).
35. Mularz, E. J. (2025). Future fundamental combustion research for aero propulsion systems. https://core.ac.uk/download/pdf/42845245.pdf (accessed on 10th September 2025).
36. Pfaffelhuber, P., & Popovic, L. (2015). Scaling limits of spatial compartment models for chemical reaction networks. http://arxiv.org/abs/1302.0774 (accessed on 10th September 2025).
37. Tiggelaar, R. M. (2004). Silicon-technology based microreactors for high-temperature heterogeneous partial oxidation reactions. https://core.ac.uk/download/pdf/11458733.pdf (accessed on 10th September 2025).
38. Winfree, E., & Zhang, D. Y. (2009). Control of DNA strand displacement kinetics using toehold exchange. https://core.ac.uk/download/4883495.pdf (accessed on 10th September 2025).
39. Zewail, H., Brink, C., Wan, C., Ma, H., Ma, Y., Osinsky, A., Richens, D., et al. (2008). Dynamics of ligand substitution in labile cobalt complexes resolved by ultrafast T-jump. https://core.ac.uk/download/4879033.pdf (accessed on 10th September 2025).

CHAPTER 8

REACTION DYNAMICS

Contents

Introduction 289
8.1. Fundamental Principles of Reaction Dynamics 294
8.2. Types of Chemical Reactions 298
8.3. Theoretical Frameworks 303
8.4. Kinetics and Thermodynamics 308
8.5. Experimental Techniques in Reaction Dynamics 313
8.6. Applications of Reaction Dynamics 317
Summary 322
References 325

INTRODUCTION

Reaction dynamics is a wide and intricate sequence of processes that characterize the ways particles interact in many systems. First, reaction dynamics are characterized as equations describing a sequence of processes representing the time evolution of chemical concentrations as they react via diverse reaction channels. These processes occur in intricate networks involving discrete elements that dynamically interact. These reaction networks are ubiquitous in most fields such as biochemistry, pharmacokinetics, and social sciences. Information on such interactions enables the study of crucial phenomena such as synthesis and degradation in biological processes or reaching consensus in social networks and thus represents a multidisciplinary framework key to scientific progress and engineering applications (Abou-Kandil et al., 2012). Significantly, mathematical modeling of such reaction dynamics most commonly employs master equations that capture the probabilistic nature of processes at issue. Nonlinearity in many reactions makes the challenge in analysis and computation of solutions to models more difficult.

Figure 8.1. *Understanding reaction dynamics: Exploring the pathways of chemical reactions.*

Source: https://i0.wp.com/freescience.info/wp-content/uploads/2024/12/reaction-dynamics-the-pathways-of-chemical-reacti-56406-featured_1_0.jpg?fit=1000%2C666&ssl=1.

A sophisticated understanding of Markovian dynamics emerges as a central component in this respect. With careful application of numerical and computational methods, researchers can approximate numerically solutions to master equations of population dynamics, thereby uncovering complex systems' patterns of behavior. More recently, there has even been the suggested application of potential energy landscapes to analyze stationary behaviors of Markovian reaction networks, further refining the theoretical frameworks of the dynamics of such complex interactions (Abou-Kandil et al., 2012). Also, observations based on chemical reaction network theory help us understand stability and equilibria in such systems greatly. By structuring reaction dynamics in terms of consensus dynamics and making use of concepts such as the complex-balanced assumption, scientists are able to extract key results that shed light on the thermodynamic behavior of these networks. Critical matrices like the balanced Laplacian matrix crop up in this description and enable straightforward derivation of core inequalities employed in defining equilibrium states. Moreover, integration of such basic concepts as Kirchhoff's Matrix Tree theorem provides a solid foundation for reaction networks involving various inflows and outflows to analyze steady-state stability. Using these analytical approaches, the fundamental principles of reaction dynamics can be investigated systematically and offer vital information about the behavior of the competing processes in natural and constructed systems (Figure 8.1).

Definition of Reaction Dynamics

Reaction dynamics is the phenomenon of studying the rates and mechanisms of chemical reactions and the transformations that the reactants undergo to yield products. Reaction dynamics essentially tries to comprehend how a variety of parameters such as temperature, pressure, and concentration influence the rate and routes of chemical reactions. This field lies at the heart of understanding not only simple chemical reactions but also applications thereof in other fields, ranging from drugs to materials science. An example is reaction network building, which is instrumental in unraveling complex biochemical processes sustaining life. Understanding reaction dynamics further optimizes industrial processes so that reactions can be better engineered, thus reducing costs and environmental impacts. Therefore, a critical examination of the structure and outcomes of reaction dynamics is of extreme significance to both theoretical research and applications. In studying reaction dynamics, researchers would seek to explore reversible reactions, as well as irreversible reactions, and how they impact system behavior. Reactions are defined through rate laws, which express how reaction rates are dependent on reactant concentrations. Knowledge of steady-state distributions is of immense value concerning the dynamics of equilibrium systems and along non-equilibrium processes. Non-weakly reversible chemical reaction networks (CRN) are of special interest. These networks possess unique dynamic features that differ significantly from weakly reversible networks, broadening the scope of chemical kinetics. By studying product-form steady-state distributions, especially in biological systems and robots, one can gain insight into complex interactions and emergent behavior within the networks, echoing the correspondence between single parts and system dynamics as a whole (Hoessly et al., 2019). From the foundational principles of reaction dynamics, one can better understand open chemical networks, particularly metabolic processes that regulate cell functions. The thermodynamic explanation of these networks is grounded in mass-action kinetics and conservation laws but goes beyond them when outside forces such as constant substrate concentrations come into play. This extrapolation takes the system out of equilibrium and results in non-linear behaviors with irreversible cycles. The analysis of affinities in these networks uncovers complicated relationships that regulate material and energy flow. By deciphering the connections between simple affinities and broken conservation laws, scientists are well-positioned to better recognize the efficiency and limitations that govern metabolic networks and therefore provide the entrance to forthcoming theoretical and applied chemistry advancements (Cercignani et al., 2014).

Importance in Chemistry and Biology

In chemistry and biology alike, chemical reaction dynamics underlie the understanding of essential life processes. Reaction dynamics entail examination of how chemical reactions occur and how the rate, mechanism, and pathway of a reaction are regulated. This information not only explains rudimentary biochemical processes such as metabolism and enzyme catalysis but fuels pharmaceutical discovery and biotechnological application as well. For instance, integrating high throughput metabolomic data has facilitated that scientists can build kinetic models describing such behavior at a genomic scale. In this regard, separation of the Jacobian matrix into chemical, kinetic, and thermodynamic information—is significant as described by Jamshidi et al. (2008)—for building a complete picture of the biological systems and their reaction dynamics, thereby marking its significance in chemistry as well as biology. Besides, the boundaries of traditional reaction dynamics have begun to get cloudy with the emergence of quantum biology, which explores the interplay between quantum mechanics and biological systems. Radical-ion-pair reactions, specifically, illustrate the manner in which quantum information principles can elucidate biological mechanisms, such as bird navigation by magnetic fields or photosynthetic electron transport. Recent studies have proven that conventional theories may be fallibly portraying these processes and, therefore, may violate the Ozawa bound, a theoretical limit on quantum systems' entropy. Instead, newer theories that involve the inclusion of quantum measures of coherence adhere to such boundaries and enhance understanding of how quantum phenomena function within biological processes, as noted by Kominis et al. (2017). This paradigm revolution not only improves our theoretical comprehension of reaction dynamics but also opens up new research avenues in chemistry as well as biology. The implications of reaction dynamics reach far beyond theoretical study; they also fundamentally impact practical applications in medicine and environmental science. For example, technological advances in kinetic modeling allow the development of specific drug therapies with improved efficacy and lower side effects since it is possible to predict biochemical network behavior toward various compounds. Moreover, insights from thermodynamic characteristics of biological processes direct the development of green biotechnologies, such as biofuels and biodegradable plastics. As the merging of disparate data sets develops further, by means such as described by Jamshidi et al. (2008), the field will increasingly apply these findings to guarantee innovations in harmony with the world's sustainability goals. A profound grasp of reaction dynamics within chemistry and biology thus becomes important to addressing many of today's most difficult problems society faces.

Overview of Key Concepts

It is also important to understand the principles behind chemical reactions in reaction dynamics. Some of the key areas of study here are the study of reaction rates, mechanism, and the role of reaction intermediates. Reaction rate, which is how fast the reactants are transformed into products, is a function of many parameters like temperature, concentration, and catalyst availability. Reaction mechanism explains the sequential process through which reactants are transformed into products, with particular focus on the intermediate species present during this reaction. By analyzing these elements, reaction chemists can identify how variations in concentration or conditions in the environment of reactants affect reaction outcomes. Thus, an in-depth understanding of these blocks not only enhances the ability to manipulate chemical reactions but also sets the stage for even more intricate analysis and modeling assignments in reaction dynamics. Recent advances in computational chemistry have been of immense help to the cause of reaction dynamics by enabling scientists to model and examine complex reaction networks better. Programs like KiNetX is a milestone in this area, which is a meta-algorithm that can perform automated kinetic analysis of chemical reaction networks from electronic structure calculations. One main advantage of KiNetX is that it can systematically scan reaction pathways without ease of handling model-inherent errors inherent in electronic structure computation. For instance, by restricting the search for suitable elementary reaction steps to kinetically relevant species, KiNetX reduces computational complexity and enhances the transparency of the resulting kinetic models. Such developments illuminate the intricate relations between reaction networks such that chemists can untangle the underlying mechanisms and thereby gain a clearer insight into the involved dynamics. Secondly, the addition of uncertainty analysis to kinetic models is another major development in reaction dynamics. Notably, KiNetX is a high-level methodology to transmit free-energy uncertainties throughout its analytical domain. This is facilitating a robust investigation of how variability in individual rate constants could affect aggregate reaction dynamics and ultimately progress toward improved product distribution and reaction pathway prediction. By the application of programs like AutoNetGen, which produces chemistry-like reaction networks with diversity, scientists can test the significance of uncertainty propagation across a broad set of scenarios. These studies give data important for reaction behavior under a wide array of conditions and provide a statistical basis for the refinement of theoretical models. All together, these advances underscore the increasing sophistication of reaction dynamics as a science grounded on both theoretical principle and practical applicability.

8.1. FUNDAMENTAL PRINCIPLES OF REACTION DYNAMICS

Some of the fundamental principles of reaction dynamics are the governing concepts that regulate the nature and direction of chemical reactions. One of the main components of the subject is an understanding of reaction kinetics that focuses on how quickly a reaction happens. The rate of a reaction is influenced by various factors like temperature, concentration, and the presence of catalysts. This delicate interplay governs how molecules react and transform, thus affecting the yield and effectiveness of chemical reactions. Mathematical modeling is also essential in simulating reaction behavior, particularly in intricate biochemical systems where traditional methodologies might fail to stand up to the challenge. Through a deep exploration of these principles, scientists can uncover the delicate mechanisms of reaction dynamics, ultimately leading to insightful application in biotechnology and materials science. The structural and thermodynamic characteristics inherent to reaction systems strongly influence feedback into our understanding of reaction dynamics. One of the greatest interest areas is the construction of genome-scale kinetic models based on intensive integration of biochemical data. This offers an explicit avenue towards the decomposition of complex networks, as seen in the decomposition of the Jacobian matrix. Here, the reduction into chemical, kinetic, and thermodynamic information allows scientists to chart and understand the multi-dimensional behavior of biological systems. Secondly, the recognition of structural analogies existing between stoichiometric matrices and gradient matrices allows for the construction of the model. By discussing these basic building blocks, scientists can create a dynamic model that includes the reality of reaction mechanism and facilitates the construction of predictive tools required for experimental design and verification (Jamshidi et al., 2008). The second dimension provided to the discussion of reaction dynamics is the contribution of thermodynamic principles to biochemical systems. Though often overlooked, the significance of Gibbs free energy and chemical potential are critical to the validity of such models. Systemic approaches, such as the bond graph technique, have contributed significantly towards the development of models that obey thermodynamic laws to justify accurate representation of energy transformation in such networks. This modular and scalable approach not only facilitates model reduction with preserved essential properties, but it also allows for direct incorporation of stoichiometric information into simulations. Emphasizing thermodynamic consistency in biochemical cycles demonstrates the convergence among energy and reaction dynamics considerations, resulting in holistic models that can predict behavior in complex biological systems and furthering our comprehension of metabolic processes (Crampin et al., 2014).

8.1.1. Reaction Rates and Their Measurement

Measurement of reaction rates is a fundamental component for understanding chemical dynamics. Reaction rates describe the rate at which reactants are converted into products and are essential in predicting the operation of chemical systems. Central to these measurements is the concept of the reaction rate coefficient, which quantifies the rate of a reaction under a set of circumstances. There are various techniques available for measuring reaction rates, including spectroscopic techniques, by which reactant and product concentrations may be followed in real time. Advanced techniques like time-resolved frequency comb spectroscopy have significantly enhanced the ability to examine complex reactions, and data on transient species of crucial importance for accurate reaction dynamics calculation, specifically for atmospheric hydroxyl radical and carbon monoxide reactions, as is evident in HOCO formation and isomerization dynamics studies (Bjork et al., 2017). In addition to the inherent quantitative aspects, mechanistic knowledge of reaction pathways also contributes to our knowledge of reaction rate. The behavior of intermediates during a reaction has a tendency to define the pathway and eventually the rate of product formation. This dynamic interaction among intermediates can be regulated by pressure, temperature, and physical state of the reactant. For instance, investigations of charge-trapping intermediates in catalyzed reactions indicate that reaction pathways exist in lieu of the primary reaction pathway when the conditions change, further confirming that environments affect reaction rates. These are also essential to the synthesis of successful catalysts and optimization of reaction conditions, as demonstrated in water oxidation reactions on semiconductor surfaces where activation barriers and kinetics were found to be pH and ionic strength-dependent experimental parameters (Aschaffenburg et al., 2019). The technical applications of reaction rate determination also transfer to environmental science, materials chemistry, and biochemistry. Chemists can predict the lifespan of pollutants in the atmosphere, estimate the catalytic converter effectiveness, and define the best conditions for industrial processes based on reaction rates knowledge. This is particularly relevant in energy storage and conversion, where catalytic processes are essential to optimize performance. As research advances in these fields, merging the latest kinetic measurement with theory will make a deeper understanding of reaction dynamics increasingly possible and lead to innovations in scientific study as well as industrial processes. By utilizing advanced techniques and methods, researchers can make more complicated reaction mechanisms clearer, eventually leading to more efficient and more sustainable chemical processes (Figure 8.2).

$$\text{RoR} = \frac{\text{reactant used}}{\text{time}}$$

$$\text{RoR} = \frac{\text{product formed}}{\text{time}}$$

Figure 8.2. *Formulas for rate of reaction.*

Source: https://www.elevise.co.uk/uploads/9/8/0/2/98021560/published/screen-shot-2020-12-31-at-15-14-01_1.png.

8.1.2. Factors Affecting Reaction Rates

Understanding what governs rates of reaction is valuable in the extension of chemical kinetics. Temperature is one of the primary driving forces, as it has a significant influence on molecular motion. Raising the temperature enables reactant molecules to gain kinetic energy, therefore making more effective and more frequent collisions. This phenomenon, known as the Arrhenius effect, illustrates how increased thermal energy makes it more probable to overcome the activation energy barrier required in the initiation of a reaction. Additionally, temperature changes can affect the rates of biochemical reactions, as with most enzyme-catalyzed reactions that usually have an optimal range of temperatures regarding maximum activity. Thus, the dynamics of temperature and reaction rates yield the standard examples of critical dynamics of chemical and biological systems that affect everything from industrial processes to metabolic pathways. Concentration of reactants is another important factor that affects the rate of reactions. A rise in concentration of reactants increases the likelihood of molecular collisions, thereby increasing the chances of productive interactions. This interaction is encapsulated by the collision theory, which describes reaction rates to be directly proportional to the rate of collisions with adequate energy. A sample in gas reactions is that if the volume of a reaction mixture is decreased, concentration and thus the rate of reaction are enhanced. This phenomenon does not occur just in gas-phase reactions but also in solutions, where molarity plays the central role. Understanding how concentration influences kinetics will have real-world applications in everything from pharmaceutical formulations to environmental chemistry, where the concentration of pollutants dictates reaction rates and degradation processes. Last but not least, catalysts help regulate the rate of reaction by lowering activation energy such that reactions happen at a higher speed without being consumed in the process. Catalysts operate by providing an alternate reaction pathway, which typically involves the formation of lower energy transition states. In complex biogeochemical systems, as modeled by HPx, the role of catalysts is even more important because they can regulate the rates of organic matter decomposition and nutrient cycling in soils. An example of some enzyme-catalyzed activities in soil microbes is how biocatalysts can greatly enhance metabolic rates, their effects directly translating to ecosystem well-being and function. Since catalytic processes are understood, scientists are better able to understand and regulate reaction dynamics for engineering and environmental applications, as seen through the technological advancements with regards to platforms such as HPx (Jacques et al., 2018) and types of computational modeling software (Alberts et al., 2007) (Figure 8.3).

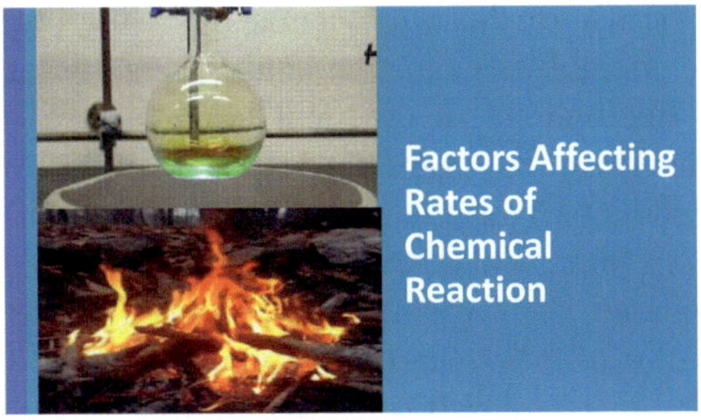

Figure 8.3. *Factors affecting rates of chemical reaction.*

Source: https://cdn.slidesharecdn.com/ss_thumbnails/factors-affecting-rates-of-chemical-reaction1-230528053821-3c327ed2-thumbnail.jpg?width=640&height=640&fit=bounds.

8.1.3. The Role of Catalysts

In the complex world of chemical reactions, the catalysts are the key go-betweens that increase the rate of reaction without ever undergoing any long-term alteration themselves. Their underlying function lies in the lowering of the activation energy barrier, the energy barrier that the reactants have to navigate in order to be turned into products. This minimization enables reactions to proceed faster and in an optimized state, a demand which is especially important in most industrial applications where resource and time optimality is critical. Catalysts also have the capability of selectively controlling reaction pathways and thus enable desired compound synthesis and lower by-products. Such selectivity is very useful in fields such as pharmaceutical production, where key reactions must take place in order to enable good drug synthesis. Therefore, an understanding of the fundamental catalyst dynamics enriches our overall knowledge of reaction kinetics and their use. The mechanisms by which catalysts operate often involve subtle molecular-scale interactions. Scientists apply advanced techniques, e.g., atomistic molecular dynamics simulations, to examine such interactions at a more detailed level. As illustrated in recent experiments, such simulations allow for a full understanding of functional dynamics in enzymes, wherein conformational fluctuations contribute significantly to catalysis. For instance, the dynamics may change based on varying molecular environments, influencing substrate recognition and reaction efficiency. These results have added depth to the knowledge of catalytic reactions, demonstrating how different protonation forms of substrates may affect catalytic effectiveness, illustrated in work on tryptophan synthase (TRPS) and other vital enzymes (Chang et al., 2016). Hence, the use of these molecular modeling tools is paramount to explaining the fundamental mechanisms behind catalytic action as well as to maximizing their effectiveness in practical implementation. In addition, the catalyst has more than just acceleration; many control the direction of reaction to more favorable products. The concept of allosteric regulation, where binding of one molecule at one site influences the activity at another, demonstrates how certain catalysts modulate their own activity as a result of external factors. This is most vivid in studies of β-ketoacyl ACP synthase I (KasA), where quantum mechanics/molecular mechanics (QM/MM) approaches have discovered the complex dynamics of the catalytic residues in varying conditions (Chang et al., 2016). Such work is indicative of the novel methodologies being used to better understand catalytic dynamics. Finally, the constant quest for catalysts not only enhances our basic scientific knowledge but also enhances our capability to design reactions with improved efficiency and selectivity, both applicable to industrial processes as well as scientific research.

8.2. TYPES OF CHEMICAL REACTIONS

The study of chemical reactions shows that they are broadly categorized into a number of types based on their mechanisms and effects. Synthesis reactions involve the combination of two or more reactants to form a product. It is fundamental in inorganic and organic chemistry and provides the basis of a number of biological processes. For example, in photosynthesis, carbon dioxide and water are converted to glucose and oxygen. Synthesis reactions explain the mechanism through which complex molecules are synthesized from simple substances, an important field of research in biochemistry and materials science. Besides, these reactions can be regulated by the temperature and pressure conditions, which contribute to the dynamicity of chemical reactions and emphasize the interrelatedness of reaction types in various scientific studies. Alternatively, decomposition reactions are another significant class, in which one compound is broken down into two or more simple products. It is a very crucial type of reaction in industrial applications as well as in nature. Decomposition reactions could be spontaneous, such as the decomposition of hydrogen peroxide to oxygen and water, or could require external energy, such as heat, light, or electricity. The dynamics of reactions have a tremendous role in deciphering mechanisms like nutrient cycling in the environment, where complex organic compounds are decomposed into simple molecules that could be reused by biological organisms.

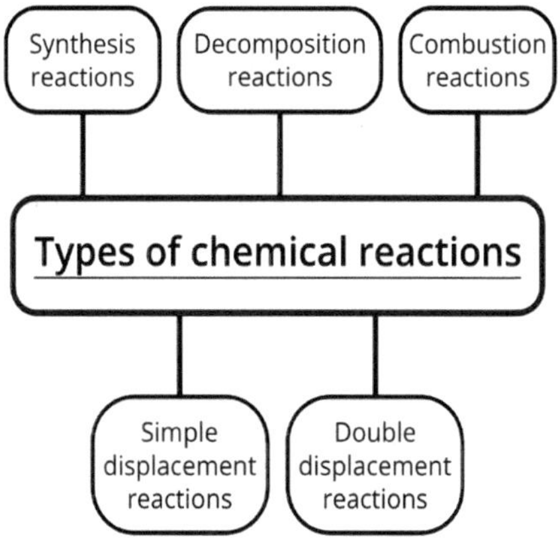

Figure 8.4. *Types of chemical reactions.*

Source: https://psiberg.com/wp-content/uploads/2021/11/Featured-image-6.svg.

The confrontation of decomposition with other reaction types, such as synthesis and combustion, illustrates that chemical transformations are cyclical in nature and how they influence environmental processes along with industrial efficiency. Finally, combustion reactions are a chemical reaction process where a fuel is rapidly oxidized, usually by the release of energy in the form of heat and light. These answers are crucial in a wide range of applications, including power generation in power plants and internal combustion engines. Combustion dynamics is complex, with the interaction of reactants and products, as well as how reaction conditions like temperature and pressure influence them. Current advances in the science of combustion processes have made tremendous strides towards higher efficiency and lower emissions, pointing to the relevance of chemical reaction types in addressing contemporary environmental issues. Moreover, understanding gained from research on stochastic processes in chemical reactions has significant ramifications in industries like pharmaceuticals, where reaction kinetics can have significant implications on drug development and effectiveness (Burrage et al., 2004; Furusawa et al., 2000) (Figure 8.4).

8.2.1. Synthesis Reactions

Synthesis reactions, the key component of reaction dynamics, are reactions wherein smaller reactants couple to form more elaborate products. Synthesis reactions are key components of organic and inorganic chemistry and are the cause of synthesis for many compounds required for life and technology. Synthesis reactions in organic synthesis are generally driven by the need to synthesize complex molecules that will serve as drugs, agrochemicals, or materials. Synthesis reaction kinetics may vary widely based on the energy profile of the reactants, presence of catalysts, and reaction conditions. Understanding these factors facilitates better prediction of product synthesis and improvement of reaction pathways and thereby better outcomes in laboratory as well as industrial applications. Synthesis of polyynes, particularly triacetylene, through bimolecular gas-phase reactions is a classic example of a synthesis reaction.

The reaction between the ethynyl radical (CCH) and diacetylene (HCCCCH) is one such reaction that is fast with no entrance barrier, showing how efficient and spontaneous some synthesis reactions can be. This complex mode of synthesis through which such a reaction takes place shows the relationship between molecular structure and reaction dynamics, as the reaction goes through a process of indirect scattering where intermediate complexes are established during a single act of collision. The relevance of the polyynes lies beyond mere molecular fascination; their promise as starting materials for even more advanced molecules underlines the worth of the knowledge of synthesis reactions in enabling chemical exploration and application (Mebel et al., 2009). The study of synthesis reactions also reveals intriguing implications to planetary chemistry, particularly in exceptional environments like Titan, the moon of Saturn. Polyynes formed in Titans atmosphere can be potentially utilized to protect against harmful ultraviolet radiation, the way ozone does on earth. Understanding the processes that lead to the formation of such chemicals informs us about prebiotic chemistry and the potential for life elsewhere in the universe. While as researchers uncover these processes using experimental, theoretical, and modeling investigations, they expose the underlying reaction dynamics that govern complex organic molecule formation in extraterrestrial atmospheres. Essentially, studying synthesis reactions not only serves to elucidate chemical principles but also propels our comprehension of planetary-scale chemical evolution (Pinto et al., 2017) (Figure 8.5).

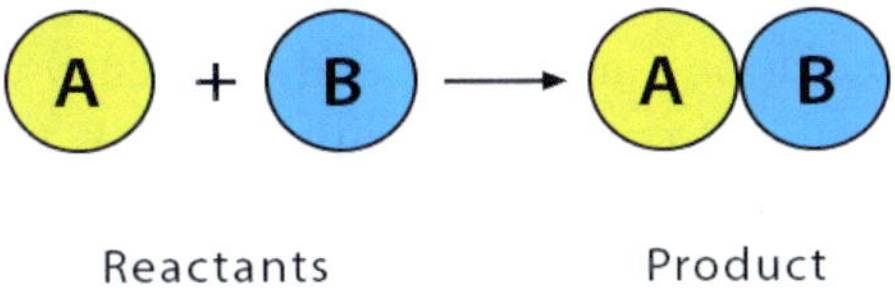

Figure 8.5. *Synthesis reaction.*

Source: https://www.chemistrylearner.com/wp-content/uploads/2020/08/Synthesis-combination-reaction.jpg.

8.2.2. Decomposition Reactions

Decomposition reactions are a key aspect of reaction dynamics in the theory of chemical transformations according to complexity and diversity of chemical transformations. Decomposition reactions are reactions in which a single compound breaks down to produce two or more simpler products that can range from elemental species to smaller molecular species. Perhaps the most important aspect of decomposition reactions is the way in which external factors such as pressure and temperature can have a profound effect on reaction mechanisms and pathways. For instance, in the case of hot liquid nitromethane, the decomposition mechanism varies under different compressions; low densities favor unimolecular C–N bond rupture, while higher densities favor reaction towards complex fragment formation like CH_3NO through H-transfer and N–O bond rupture, demonstrating the dependency of decomposition mechanisms on different environmental conditions (Goddard et al., 2011). The investigation of the thermal decomposition of energetic materials like nitromethane reveals intricate reaction dynamics that contribute to the knowledge of energetic materials. Recent work with molecular dynamics simulations employing the ReaxFF reactive force field describes the reaction sequence of bond breaking and forming at elevated temperatures. At 3000 K, for example, the initial step is an intermolecular proton transfer that results in the production of CH_3NOOH and CH_2NO_2. Conversely, at lower temperatures, the initial reaction typically involves isomerization combined with the cleavage of the C–N bond to yield methyl nitrate. Such information accounts for the kinetic behavior and accompanying thermodynamic alterations upon decomposition, as the product H_2O is the site at which exothermic chemistry commences (Goddard et al., 2011). Decomposition reaction kinetics are far more complex than the mere creation of products; they comprise a broad spectrum of kinetic and energetic phenomena caused by molecular interaction and environmental considerations. Through macroscopic understanding of these kinetics, scientists are able to predict the fate of reactions and tailor conditions for desired products. For example, pressure may cause dramatic variations in the rate and efficiency of decomposition, as illustrated by the transition from low-density channels to high-density routes that alter product distribution and energy release patterns. Such data are not only important for theoretical purposes but also for any practical utilization in, for example, materials science and explosives technology. Thus, knowledge of decomposition reactions provides a foundation support in the comprehension of the complexity of reaction dynamics of chemical systems (Goddard et al., 2011) (Figure 8.6).

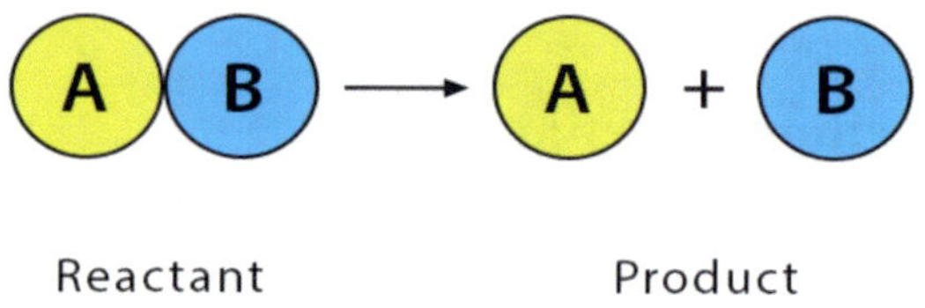

Figure 8.6. *Decomposition reaction.*

Source: https://www.chemistrylearner.com/wp-content/uploads/2020/08/Decomposition-Reaction.jpg.

8.2.3. Redox Reactions

Redox reactions, or reduction-oxidation reactions, are essential in the majority of chemical and biological processes. Redox reactions entail electron transfer between two compounds, altering their oxidation states in a basic sense. In a redox reaction, one species loses electrons, and its oxidation state is raised, and the other gains electrons and reduces its oxidation state. This electron transfer is essential for the energy production in cellular respiration and photosynthesis, proving the significance of redox reactions in sustaining life. In addition, understanding these reactions widens the scope of innovation for batteries and fuel cells, where controlled electron flow is essential in an effort to store and transform energy effectively. The mechanisms of redox reactions can be complex, depending on several variables like concentration, temperature, and presence of catalysts. Other than these conditions, electron affinity and molecular geometry of the reactants are also essential because they determine how easily electrons can be transferred. For instance, quantum chemical studies of molecular geometries and energies provide the basis for understanding how specific configurations predispose a molecule to certain redox activity, affecting rates and directions of the reaction. These investigations enhance our understanding of reaction mechanisms with the caveat that a careful evaluation of global parameters must be made in order to most accurately discern the dynamics at play in these reactions. This kind of analysis is able to uncover molecules' tendency to participate in redox processes under various conditions, a factor integral to predicting reactivity in complex systems (Reiher et al., 2016). In biological systems, integration of big data sets allows the modeling of the complicated networks of redox reactions that regulate metabolic pathways. Building kinetic models from high throughput metabolomic and fluxomic data allows scientists to investigate these reactions dynamically and deepens our knowledge of cellular function. The mathematical formalism for the investigation of such models points to key features, including the structural relation between stoichiometric matrices and their consequences for electron transfer dynamics. With the comprehension of such intricacies in such systems, researchers are now capable of confronting new challenges in computational examination that ultimately lead to an increased understanding of network phenomena. The addition of kinetic modeling within the examination of redox reactions represents a significant advance in the discipline, particularly with regard to the understanding of how such reactions orchestrate biological processes at the genomic scale (Jamshidi et al., 2008) (Figure 8.7).

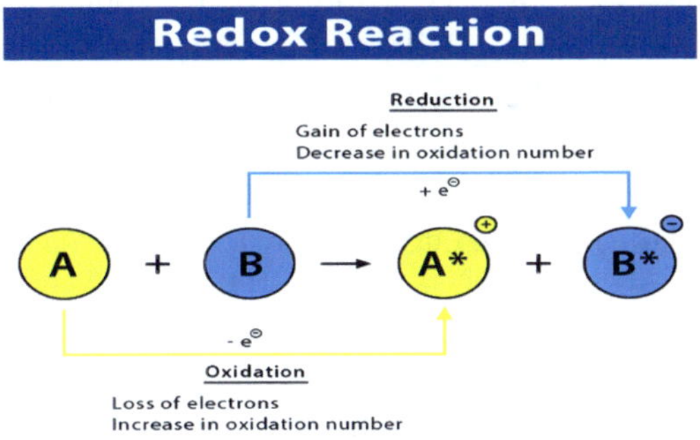

Figure 8.7. *Redox reaction.*

Source: https://www.chemistrylearner.com/wp-content/uploads/2020/09/Redox-Reaction.jpg.

8.3. THEORETICAL FRAMEWORKS

Study of theoretical models in reaction dynamics is needed to analyze the underlying mechanisms governing chemical reactions. Theoretical models provide a quantitative approach to studying the variability of reaction rate factors, product yield, and the efficiency of the catalyst. The models form the foundation of experimental and computational research, enabling researchers to predict the outcome of reactions through known theory. For instance, collision dynamics concepts and transition state theory give information regarding how molecular interactions dictate reaction courses. The combination of these theoretical models allows chemists not just to determine the mechanism of reactions, but also to construct predictive models that are required in industrial processes and hence fill the gap between application and basic research. One of the most striking research emphases among theoretical models is catalytic processes, particularly for zeolite catalysts. The characteristic structure of acidic zeolites is a key player in defining their catalytic role. While researchers venture into specifics, understanding the connection between the structure of the zeolites and Brønsted proton activity becomes the focal point. Theoretical models achieve this by providing one with the tools to account for the role played by acidity and pore structure in altering reaction kinetics. Alkane dehydrogenation and cracking, for example, illustrate how the equilibrium constants for adsorption can be significantly influenced by zeolite architecture. Furthermore, configurational-bias Monte Carlo (CBMC) simulations embedded in these theoretical models enhance their precision by enabling the estimation of such key parameters as K_{ads}-H^+ without being founded on measurements of ambient temperature only (Bell et al., 2018). In reaction dynamics, the framework encompasses the reaction mechanisms and the importance of kinetics and thermodynamics. It is crucial in catalytic reaction optimization to know the parameters affecting measured rate coefficients (k_{app}). The interaction between K_{ads}-H^+ and the intrinsic rate coefficient (k_{int}) indicates the complexity of reaction mechanisms in zeolite frameworks. Through revisiting and enhancing theoretical approaches, scientists can more accurately predict how structural changes in zeolites affect catalytic performance in industrial applications. This multi-disciplinary approach, marrying concepts from thermodynamics and kinetics with cutting-edge simulation techniques, has avenues for innovation in the catalyst field and the maximization of chemical conversions. This innovation is of critical significance, particularly when industries pursue environmentally friendly and efficient processes (Glowacki et al., 2013).

8.3.1. Collision Theory

Subsidiary to Collision Theory is the understanding of how particles interact when they collide, the fundamental concept that underlies Reaction Dynamics. According to this theory, chemical reactions will be obtained only by collisions between reactant particles possessing sufficient energy and proper orientation.

Collision dynamics can seriously affect reaction rates, as in the case of nuclear interactions where reaction barriers in energy must be overcome. For instance, in Coulomb barrier heavy ion collisions, unique mechanisms dominate the process, including fusion and nucleon transfer. The competition of these mechanisms illustrates the intricacies involved in particle interactions as well as their capability to impact reaction channels, which underscores the importance of collision dynamics in reaction channel prediction. In addition, advanced theoretical models like the Time-dependent Hartree-Fock (TDHF) approach provide informative understandings of the complexity involved in collision events, particularly in nuclear physics. This methodology allows researchers to explore circumstances such as competition between nucleon transfer and fusion reactions in intermediate mass systems, such as the $^{16}O + ^{208}Pb$ reaction. The illustrated preferability of proton transfer over fusion signals the counterintuitive character of these collisions, with reduced Coulomb repulsion enhancing interaction chances. Moreover, correlation pairings play a significant role in these processes, bringing out the subtlety of collision theory beyond naive schemes, as summarized by Avez et al. (2012). However, the intricacies of exact modeling of collision dynamics extend beyond simple particle interaction. Latest studies have raised significant challenges in mapping cluster dynamics, fragment characteristics in damped collisions, and sub-barrier fusion by quantum tunneling. With continued advancement of the theory of nuclear reactions, the demand for advanced models that can effectively address these challenges becomes increasingly pertinent. Approaches to integrating microscopic quantum theories into explaining heavy-ion collisions are important for enhancing predictive capacity, as considered in recent studies. Addressing these complexities not only enhances the theory structure, but it also consolidates the experimental analyses that have been conducted in allied disciplines, as discussed by Simenel et al. (2014).

8.3.2. Transition State Theory

In chemical kinetics, Transition State Theory (TST) is the core theory that describes how chemical reactions progress through transient states. The theory establishes the transition state as a peak on the energy landscape, which is the position of highest potential energy reactants must traverse in order for a reaction to take place. One of the most significant implications of TST is that the rate of reaction is inherently connected to the transition states' stability. The greater the stability of the transition state, the lower the reaction energy barrier and thus the faster the rate of reaction. This stability is defined by the molecular interactions for proceeding towards the transition state, and the nature of involved reactants, setting the stage for more study of reaction dynamics and kinetics. Empirical work has validated the applicability of TST, particularly via experimentation on involved reactions such as the one with hydroxyl radicals and carbon monoxide. Observations here reveal the reactivity of cis- and trans-HOCO intermediates and their short lifetime on reaction. That the past history of these intermediates can be followed in real time using advanced techniques such as time-resolved frequency comb spectroscopy is what makes possible a measurable window into branching and isomerization kinetics. These nuanced examinations show how various external conditions such as pressure and temperature play a vital role in identifying the energetics of the transition state. This is a testament to the fact that TST, in addition to shedding light on the intrinsic nature of reaction dynamics, also has implications for applied uses in chemistry in the atmosphere and combustion processes, as illustrated by Bjork et al. (2017). Moreover, the significance of dynamic fluctuations of enzymatic activity favors TST to further expand the understanding of reaction dynamics in biological systems. Recent advances in single-molecule enzymology have been able to show how conformational dynamics of enzymes may lead to fluctuations in catalytic competence. These fluctuations are referred to as dynamic disorder and show that the transition states of enzyme reactions are not rigid but subject to substantial variation over time. These findings underscore the importance of TST vis-à-vis true biochemical phenomena, in which the rate of oscillations can determine substrate concentration crucially. These findings have implications that reinforce the underlying philosophy of TST, which suggest that this wisdom can be used to make new predictions about the action of enzymes in complex cellular situations, which displays the unifying potential of TST in the overall scenario of reaction dynamics as provided by Gupta et al. (2017) (Figure 8.8).

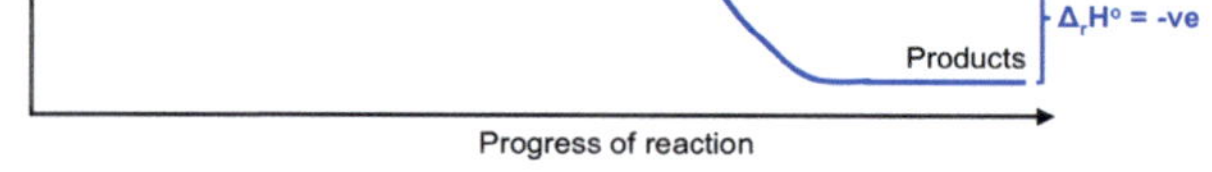

Figure 8.8. *Transition state theory.*

Source: https://monomole.com/wp-content/uploads/transition-state-theory-768x394.png.

8.3.3. Reaction Mechanisms

Understanding reaction mechanisms underlies reaction dynamics because it informs us regarding pathways and steps that govern chemical conversions. At the molecular level, any given reaction can be explained in terms of elementary steps describing how reactants are converted to products. Mechanisms are inclined to specify the transient intermediates formed along the way, which can play a big role in the reaction kinetics and thermodynamics. For instance, in zeolite-catalyzed reactions, the acid nature and topology of the catalyst play a pivotal role in determining how methanol is transformed into hydrocarbons. Through the use of advanced ab initio molecular dynamics tools, one is able to better mimic reaction conditions and study competition between other mechanistic channels, e.g., concerted versus stepwise methylation reactions as described by Bailleul et al. (2016).

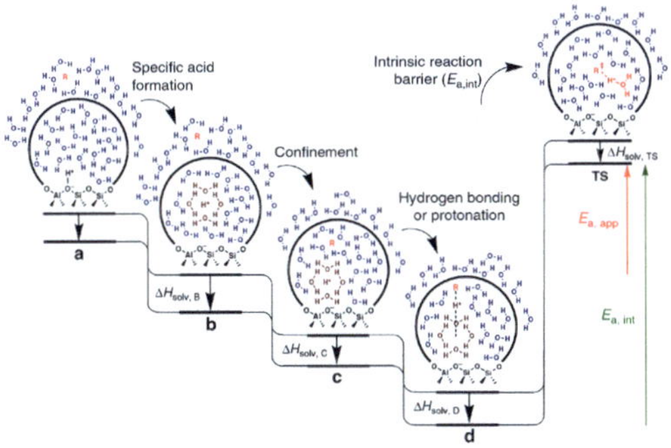

Figure 8.9. *Reaction coordinate diagram for a typical zeolite-catalyzed, aqueous-phase reaction. Initially, the reactant (R) is solvated by the liquid phase outside the porous catalyst, and the proton in the catalyst is not hydrated (a). The hydration of the proton by water forms the specific acid (b), which is followed by the adsorption of R in the catalyst pore (c). The hydrogen bonding or protonation of R in the catalyst pore is accompanied by a partial or complete de-solvation of the proton (d), which is followed by the acid-catalyzed chemical transformation by going through a transition state (TS). Water molecules in brown and red are involved in the specific acid solvation sphere and in the TS, respectively. The solid curved lines represent the zeolite pore walls. The enthalpy changes (ΔH), apparent activation energy ($E_{a,app}$), and intrinsic activation energy ($E_{a,int}$) are dependent on interactions with the solvent environment.*

Source: https://www.researchgate.net/publication/339507694/figure/fig2/AS:958993761591296@1605653338022/Reaction-coordinate-diagram-for-a-typical-zeolite-catalyzed-aqueous-phase.png.

One of the most important fields of understanding reaction mechanisms is reaction coordinate diagram study, which graphically illustrates the change in energy with the progression of reaction. These diagrams show the free energy complexity of landscapes, marking transition states and intermediates for chemical reactions. In enzyme-catalyzed reactions and generally, these plots can bridge the gap between theoretical and real, adding depth to our insight into the generation of force and molecular motion by biological motors. For instance, a reaction coordinate plot would illustrate how energy barrier variation affects reaction kinetics, thereby influencing the rate and the equilibrium constant for the reaction. This perspective, argued by Scholey et al. (2013), is important in bringing attention to the role of reaction mechanisms in chemical systems as well as in biological systems, providing an elaborate framework upon which molecular dynamics can be addressed. In addition to this, the ability to tailor reaction mechanisms at the molecular level is imperative in creating catalytic technologies. By tuning reaction conditions—temperature, acidity, and the presence of other guest molecules—scientists can tune reaction paths toward target products. It is important especially in catalytic reactions with zeolites since structural parameter change can induce dramatic changes in reaction dynamics. The synergy of these frontier modeling approaches provides access to a finer grasp of how certain factors affect the kinetic and thermodynamic profiles of reactions. Thus, this molecular regulation enables the production of new catalytic systems that can efficiently cure alternative feedstocks with limited unwanted byproducts, as noted by Bailleul et al. (2016). Systematic exploration of reaction mechanisms ultimately results in more efficient and sustainable chemical processes in academia and industry (Figure 8.9).

8.4. KINETICS AND THERMODYNAMICS

The connection between thermodynamics and kinetics is central to learning about reaction dynamics because it helps describe how reactions occur and reach equilibrium. Kinetics is concerned with the rates of chemical reaction by examining what influences how fast reactants convert to products. Thermodynamics is concerned with the energy change in chemical reactions and whether a reaction will be energetically favorable or not. This kind of correspondence is actually exposed in mass action kinetics when the process rate is proportional to the concentration of reactants. Complex-balancedness, as a conjecture on reaction networks, simplifies analysis considerably, which has very deep implications for statements about stability and characterization of equilibria, both being extremely crucial for thermodynamics and kinetics (Jayawardhana et al., 2015). Through the nuances of such interactions, researchers can gain insights into the directionality and efficiency of chemical reactions under different applications. In addition, advances in computational modeling have redefined the way researchers approach the interaction between kinetics and thermodynamics, particularly in biological systems. The buildup and organization of high-throughput data sets have unveiled the intricacies of networks that regulate biological reactions. These networks may be reconstructed using kinetic modeling, which can translate reaction parameters well into thermodynamic terms. In genome-scale models, for instance, integration of metabolomic and fluxomic data provides improved insight into dynamic behavior of metabolic networks. Analytical methods focused on decomposition of the Jacobian matrix facilitate recognition of contributions of chemical, kinetic, and thermodynamic factors to the determination of steady states (Jamshidi et al., 2008). Such comprehensive frameworks are needed in order to adequately quantify reaction dynamics, discover the underlying stability, and direct subsequent experimental protocols. Lastly, combining kinetics and thermodynamics provides necessary frameworks to investigate reaction dynamics and apply to real-world applications in numerous scientific fields. Researchers can predict reaction pathways and manipulate conditions for desired outcomes using these dynamics, something that is both successful in industrial chemistry, pharmacology, and ecological science. The thermodynamic analysis of reaction networks not only provides information about potential equilibria but also sheds light on issues of relevance such as stability under various conditions. For example, the plasticity of chemical systems can be a symptom either of their robustness or of their fragility with respect to external stimuli. With improvement in techniques of integrating heterogeneous data sets, the study of reaction dynamics is the key to new strategies for simulating complex systems and tackling problems in chemistry and other fields (Jamshidi et al., 2008; Jayawardhana et al., 2015).

8.4.1. Relationship Between Kinetics and Thermodynamics

One needs to understand the intricate relationship between kinctics and thermodynamics so that one can truly understand reaction dynamics. Kinetics refers to the rate of reactions, while thermodynamics reflects energy changes and the favorability of chemical reactions. Both areas are connected because reaction rates are often dominated by the thermodynamic stability of the products and reactants. For example, reactions whose Gibbs free energy values lower tend to proceed more spontaneously, pointing out that thermodynamic favorability can control the likelihood of a reaction's occurrence. On the other hand, even a favored reaction thermodynamically can happen slowly if it possesses a high activation energy barrier, thereby pointing out that kinetic as well as thermodynamic parameters need to be observed along with each other in order to clearly understand reaction mechanisms in chemical and biological systems. The applicability of kinetic models to thermodynamic systems has come increasingly into the focus in biochemical system research. As a result of high-throughput metabolomic and fluxomic information, scientists are now able to construct genome-scale kinetic models more accurately and incorporate thermodynamic information. This incorporation facilitates a more accurate description of biochemical networks, which can be used to better understand cellular processes. By decomposing the Jacobian matrix into chemical, kinetic, and thermodynamic elements, scientists are in a better position to investigate intricate reaction networks.

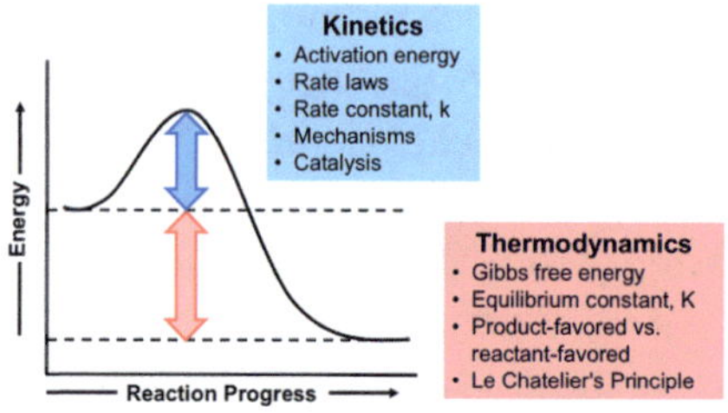

Figure 8.10. *Reaction energy diagram showing kinetic and thermodynamic relationships.*

Source: https://wisc.pb.unizin.org/app/uploads/sites/557/2020/10/104_M6Q7_kineticthermostability-768x441.png.

By employing such holistic approaches, not just our ability to model biological systems is enhanced, but the interaction between the rates of reaction and energy matters also becomes comprehensible, fostering the overall comprehension of reaction dynamics in natural and artificial systems (Jamshidi et al., 2008). Furthermore, the bond graph approach is a sound foundation for biochemical system modeling based on thermodynamic consistency in the investigation of energy transfer and storage. It enables researchers to systematically construct reaction models in accordance with the principles of thermodynamics, which are a crucial component often not explicitly included in standard kinetic models. Through the integration of stoichiometric information with simulation techniques, the bond graph approach is a scalable method for developing dependable models for biochemical cycles. Thus, for instance, breakdown at an elementary level of biochemical pathways illustrates how energy funnels through and dissipates in the system and makes kinetic behavior and thermodynamic constraints easy to understand simply. Accordingly, in deepening theoretical understanding of reaction dynamics, this novel framework also stands to have applicative implications for application in engineering and synthetic biology (Crampin et al., 2014) (Figure 8.10).

8.4.2. Gibbs Free Energy and Reaction Spontaneity

The Gibbs free energy is an important consideration for determining the spontaneity of chemical reactions, providing information beyond mere energy considerations. Gibbs free energy (G) accounts for the system's enthalpy (H) and entropy (S), leading to the equation $\Delta G = \Delta H - T\Delta S$, where T is temperature. In reaction spontaneity analysis, a negative ΔG signifies an energetically favorable process that would take place without the addition of external energy. A positive ΔG signifies non-spontaneity and the need for a shift of equilibrium or intervention to trigger the reaction. Such an association of Gibbs free energy with reaction conditions considerably impacts various scientific disciplines, particularly in studying energy behavior within complex systems, like perovskite materials whose stability is dictated by enthalpic and entropic considerations (Ciccioli et al., 2018). Furthermore, reaction spontaneity complexities arise when not only the directional thermodynamic potential is taken into account but also the kinetic barriers involving reactions. Gibbs free energy provides data on reaction profiles and through this, chemists can predict not just whether or not a reaction will proceed but also at what rate it will tend toward equilibrium. The relationship between enthalpy and entropy has a role to play in the explanation of reaction mechanisms, for example in the decomposition reactions of methylammonium halides. In this case, the contrast between kinetic feasibility and thermodynamic stability accentuates the place of Gibbs free energy in reaction kinetics. The need for stronger empirical proof is underscored in the case of competing routes in such multistep reactions, illustrating the fact that theoretical predictions must be backed by experimental data (Ciccioli et al., 2018). While the incorporation of Gibbs free energy concepts with interdisciplinarity in science continues to broaden our understanding of chemical processes, it enables connections to be established among fields like chemistry, biology, and physics. For the life sciences especially, a concentration on chemical energy's role can enhance students' knowledge about vital biochemical processes. Traditional physics courses are often dominated by mechanical energy and disregard the chemical nuances relevant to students in these disciplines. By injecting a curricular thread that addresses energy ideas from different perspectives, instructors can create a unifying framework that addresses the energy dynamics involved in life system study. Such a framework not only improves learning but also gives students the ability to address complex scientific phenomena in an effective manner so that they comprehend the important role played by Gibbs free energy in the spontaneity of reactions (Figure 8.11).

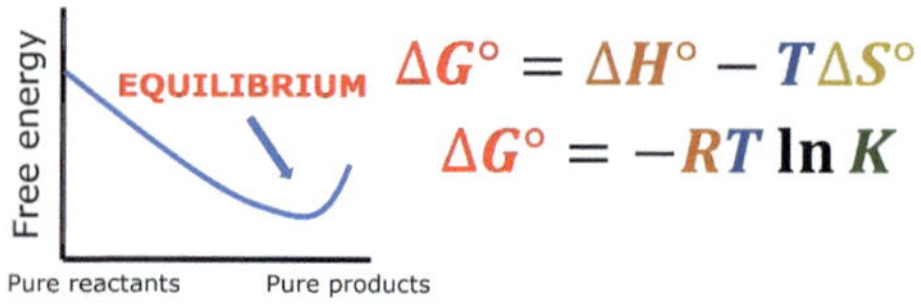

Figure 8.11. *Gibbs free energy.*

Source: https://i.ytimg.com/vi/y0YGk5oOrnw/maxresdefault.jpg.

8.4.3. Activation Energy and Its Significance

Activation energy is a basic concept to the dynamics of a reaction, being the minimum amount of energy that must be provided for reactants to experience a chemical change. The activation energy acts as the barrier that needs to be overcome for reaction to occur, dictating the rate and path of the reaction. Activation energy can be applied to different chemical reactions, such as decomposition reactions in molecular dynamics simulations. Such as the decomposition reaction of hot liquid nitromethane, which significantly varies with different compression and temperature conditions, showing that different activation energy levels are involved in different reaction paths (Goddard et al., 2011). The conclusions acknowledge that not only does activation energy dictate whether a reaction will occur but also on which products this reaction forms, thus being applicable from materials science to environmental chemistry. Understanding activation energy goes beyond just theoretical importance; it has practical applications in many scientific fields, most notably catalysis and environmental science. Catalysts work by providing alternative reaction pathways whose activation energy decreases, hence speeding up reaction rates without themselves chemically altering permanently. An example of this is in atmospheric chemistry, where the significance of polar stratospheric clouds (PSCs) is pivotal in facilitating reactions resulting in ozone destruction. PSC formation favors an activation energy to reaction involving chlorine compounds that is far lower than under standard conditions (Turco et al., 1992). This indicates the significance of the interaction between activation energy and reaction dynamics, and how variation in environment may result in extreme biochemical and climatic consequences. Besides its importance to reaction rates and mechanisms, activation energy plays a pivotal role in dictating the control of environmental variables over reaction dynamics. Pressure and temperature both influence the activation energy that accompanies chemical reactions as evidenced in various behaviors of materials under diverse physical and chemical stresses. For instance, the development in the process of decomposition of nitromethane at different pressures illustrates how much the activation energy may vary relative to extrinsic conditions and thus the application of such information in optimizing explosive efficiency and safety in real applications (Goddard et al., 2011). Besides, in conditions such as volcanic eruptions affecting stratospheric sulfate layers, an understanding of activation energy can improve the prediction of climate anomalies and demystify corresponding chemical reactions, thereby guiding effective environmental management strategies (Turco et al., 1992) (Figure 8.12).

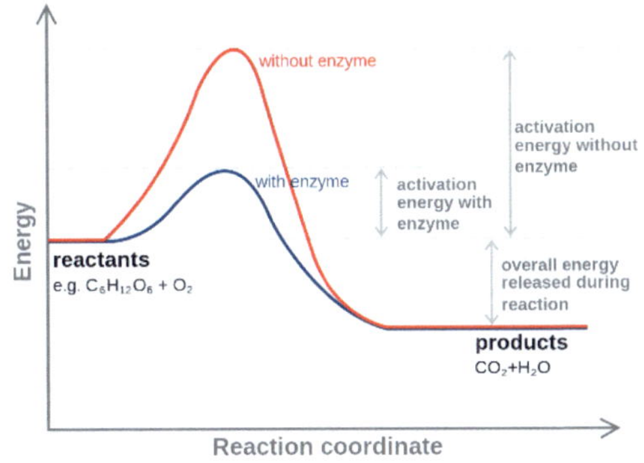

Figure 8.12. *An enzyme or catalyst lowers the activation energy.*

Source: https://sciencenotes.org/wp-content/uploads/2021/03/Activation-energy-enzyme.png.

8.5. EXPERIMENTAL TECHNIQUES IN REACTION DYNAMICS

Experiment in reaction dynamics is the basis for achieving the intricate processes occurring after chemical reactions. Experimental strategies include a range of spectroscopic and kinetic examinations that allow scientists to observe and quantify reaction pathways. Advanced techniques such as time-resolved spectroscopy and reactive kinetics are the bases for determining reactive intermediates' transient states. Using these tools, researchers can acquire real-time data charting reaction evolution paths, providing insight into mechanistic pathways as well as catalytic efficiencies. The use of computational modeling in conjunction with experimental methodologies enhances the power to predict and dissect reaction dynamics, bridging the gap between theoretical prediction and experimental observation. Thus, synergy between experimental and computational understanding is paramount to an integrated understanding of reaction dynamics. Such a key development in this field is the application of ab initio molecular dynamics, by which detailed probing of catalyst performance under real operating conditions is feasible. This technique is particularly important in the probing of zeolite-catalyzed reactions, where synergy of catalyst topology and acidity plays a major role on kinetic outcomes. As evidenced by recent research, the manipulation of these parameters has the ability to dramatically alter the mechanisms of reaction used in processes such as the conversion of methanol to hydrocarbons. Through the application of advanced molecular dynamics, researchers have the capability to mimic reaction conditions in detail, enabling them to examine key steps in such catalytic reactions. Such novel methodology not only supplements traditional experimental techniques but also allows the exploration of complex mechanistic questions heretofore inaccessible. The harmony between theoretical and experimental approaches in reaction dynamics is important to decipher the dynamic character of chemical reactions. To put it in a nutshell, the experimental approaches employed in reaction dynamics play an integral role in formulating our knowledge base for chemical processes. The marriage of high-resolution spectroscopic methods and sophisticated computational tools facilitates an elaborated understanding of the reaction mechanisms. As demonstrated by zeolite system research, the ability to finely tune reaction conditions opens up the possibility for optimization of catalytic reactions, thus improving the efficiency of chemical conversions. Ongoing improvement of such experimental methods will undoubtedly allow future work in reaction dynamics to develop more efficient catalysts and innovative processes. By integrating various experimental techniques and theoretical models, researchers are able to expand our knowledge in catalysis and reaction dynamics and ultimately drive the sustainable development of the chemical industry.

8.5.1. Spectroscopy Methods

Spectroscopic methods are leading in the investigation of reaction dynamics, providing essential data on molecular interaction and structural changes in chemical reactions. Among all such methods, time-resolved vibrational spectroscopy (TRVS) is one of the most impactful, by which researchers are able to study reaction channels on ultra-short timescales, often to the picosecond regime. This technique, discussed at the 13th TRVS conference, emphasizes the importance of old and new vibrational methods, including infrared and Raman spectroscopy, in chemical kinetics and molecular dynamics studies. The development of these techniques allows for a nuanced examination of vibrational spectra, which frees molecule- and bond-specific information essential to the unraveling of complex reactions. Through advances presented at such conferences, researchers gain the methodologies required to examine the inherent processes that control chemical reactions, making it possible to have an enhanced understanding of reaction dynamics and their applicability to a wide range of branches of science (Laubereau et al., 2007). More so, blending entanglement of experimental and theoretical methods enhances the utility of spectroscopy in shedding light on the reaction mechanisms. A very good example of such synergy can be seen in photosynthetic systems themselves, i.e., Photosystem II Reaction Center (PSII RC). Investigations using the methodology of two-dimensional electronic spectroscopy (2DES) have identified robust vibronic coherence playing an important role in solar energy conversion optimization. Visualization of interactions between electronic and vibrational dynamics by means of powerful spectroscopic tools allows observation of the competing electron transfer pathways. For instance, wavelet analysis of 2DES data reveals temporal dynamics of coherences, demonstrating the dynamical function of excitonic states in charge separation dynamics. Thus, this coupling of spectroscopic tools and theory modeling not only enriches the understanding of reaction dynamics but also holds the promise to unlock potential for innovation in energy conversion technology (Chin et al., 2017). These sophisticated spectroscopy methods are used beyond fundamental research, and commercial technology and uses evolve accordingly. The information generated by vibrational spectroscopy and other methods of similar character has significant implications in material science, biochemistry, and pharmaceuticals and can help create new materials and facilitate the drug discovery process. With more knowledge and advancements in these methods, scientists will be able to investigate more complex systems and reveal underlying principles governing reactivity and selectivity in chemical reactions. Moreover, real-time monitoring of the reaction provides promising opportunities for applications in industry, where rapid assessment is required. Lastly, ongoing research and development of spectroscopy methods will be vital in determining the destiny of reaction dynamics, facilitating deep scientific understanding and practical innovations across numerous areas (Laubereau et al., 2007).

8.5.2. Kinetic Studies and Rate Laws

Kinetic studies form the basis of information regarding chemical reactions because they indicate the rates at which reaction occurs and the determining factors for such rates. Through reaction dynamics, scientists are able to specify how concentration, temperature, and catalysts regulate reaction rates. Such investigations employ various mathematical models for describing the kinetics of reactions, including elementary steps and complex mechanisms. Rate laws actually define the relationship between reactant concentration and reaction rate in concise summary form of essential information about the nature of the reaction in question. One of the most widely applied is the Michaelis-Menten equation, which provides us with some insight into enzyme reactions that is naive but captures the essence of the trend in kinetic behavior and hence is a valuable analytical tool in biochemistry research. Application of rate laws extends beyond simple systems, where they must contend with the challenge of modeling correctly complex biochemical networks. Herein, reactions can involve numerous interacting species and necessitate the application of higher-order rate laws in order to describe the dynamics adequately. Current research points to the merit of approximate rate laws in simplifying modeling of such intricate networks. For example, Michaelis-Menten kinetics with experimentally determined enzyme parameters gives very precise predictions of system response. However, as pointed out in recent research, using simplified models gives error, particularly when parameters are approximated. The outcomes emphasize the importance of understanding the assumptions in the selected rate laws because these will make a huge impact on kinetic prediction correctness under a dynamic biochemical environment (Bordbar et al., 2016). Moreover, mechanistic bases of kinetic studies offer a close correspondence between theoretical models and biological reality. To connect the two, scientists use sets of rate equations to capture the dynamic properties of enzymic and metabolic processes. Strikingly, it has been demonstrated in studies that moderate approximation of parameter values is sufficient to obtain results consistent with actual system dynamics. Red blood cell kinetics in humans illustrates how varying reaction characteristics such as thermodynamic reversibility and substrate concentration saturation influence the choice of the appropriate rate law to use in modeling. By examining these, researchers posit that increased understanding of physiological conditions can significantly enhance attempts at kinetic modeling. Finally, the ongoing interplay between sophisticated mechanistic models and their reduced analogs steers best practice in kinetic investigation and lays out possibilities for future advancement in reaction dynamics (Pinto et al., 2017).

8.5.3. Computational Chemistry Approaches

The field of computational chemistry is vital to the determination of reaction dynamics by providing information about reaction mechanisms and pathways that are often hard to determine using experimental techniques only. Techniques such as the nudged elastic band (NEB) method are greatly valued for the determination of potential energy landscapes in relation to chemical reactions. This technique allows scientists to establish the optimal reaction pathways by study of the transition state energy change, thereby greatly advancing reaction mechanisms at surfaces. Furthermore, the integration of various computational techniques and experimental data expands the scope of reaction dynamics research to allow for the synthesis of molecular structures that are not possible with traditional techniques (Gallardo et al., 2025). In addition to quantum mechanical methods, molecular dynamics (MD) simulations have proven to be an invaluable tool in reaction dynamic studies. With QM/MM (quantum mechanics/molecular mechanics) schemes, researchers are able to simulate large systems while incorporating accurate interactions at the quantum level. This is particularly useful to explain temperature-dependent processes since this incorporates entropy as well as vibrational modes within calculations. Therefore, MD methods facilitate the access of scientists to dynamic processes that occur during chemical reactions, thereby providing an improved reaction pathway description. By having both the quantum mechanical and classical approaches, the chance of studying the reaction dynamics thoroughly is made possible, hence filling gaps between experimental results and theoretical calculations (Gallardo et al., 2025). Machine learning has increasingly been seen as a vital computational chemistry technique, offering new solutions to the challenge of high-dimensional stochastic dynamical systems. By using artificial intelligence techniques, such as variational autoregressive neural networks and reservoir computing, researchers can efficiently model and predict complex behavior in reaction dynamics. These novel approaches enable the study of free energy landscapes and dynamics of chemical reaction networks more accurately and with reduced computational effort. Additionally, application of machine learning in this regime is likely to provide meaningful analysis of noise-induced phenomena and therefore provide deeper insight into reaction dynamics. With the gradual evolution of computational chemistry in combination with machine learning, opportunities for a revolutionary breakthrough in reaction kinetics are vast, heralding a revolutionary transformation in the manner chemical processes are explored.

8.6. APPLICATIONS OF REACTION DYNAMICS

The applications of reaction dynamics extend into an enormous range of applications, to which materials science and biochemistry are a significant part. In materials science, understanding the reaction pathways and reaction kinetics involved in the synthesis of new materials is crucial. Through this understanding, researchers are able to design materials to perform in specific ways, e.g., as chemical sensors or catalysts. Reaction dynamics can be manipulated to provide tailor-made properties to these materials in a bid to optimize their performance within practical applications. Provided the mechanisms within such reactions are understood, scientists can then make accurate predictions of how materials will behave under various conditions, and thus reaction dynamics is a pivotal tool in promoting modern materials research, which requires careful planning and know-how. These views are echoed in the collaborative effort present in important scientific endeavors, for example, outlined by Bracco et al. (2010), where the need for a concerted effort to overcome challenges in this area is emphasized. Furthermore, reaction dynamics is an essential aspect of biochemistry, particularly in relation to knowing enzyme kinetics and reaction mechanisms in metabolic pathways. Through the investigation of enzyme dynamic behavior, scientists gain insight into how biological catalysts function and interact with substrates. This knowledge is paramount to drug design, where the aim is often to create inhibitors that can control enzyme activity for therapeutic ends. Sophisticated modeling of reaction kinetics allows scientists to identify how changes in biochemical conditions affect rates of reaction and provide for the development of more efficient pharmaceuticals. The significance of this realization is reflected in accomplishments awarded in a special issue dedicated to achievements of outstanding scientists engaged in fields of relevance to chemical kinetics and spin chemistry (Buntkowsky et al., 2017), reflecting presence of rich research heritage of exploratory nature based on appreciation of reaction dynamics at the foundation. Apart from biotechnology and materials science, reaction dynamics is also part of environmental chemistry. Investigating mechanisms of pollutant degradation reactions within various environmental matrices is important for effective design of remediation measures. For instance, reaction kinetics between greenhouse gases such as methane and carbon dioxide is most significant in indicating how they will interact in the atmosphere and impact climate change. With the use of reaction dynamics, environmental scientists are able to model the transformation of these compounds and contribute to improved environmental legislation and technology to reduce emissions. These instances reveal the utility of reaction dynamics in fixing some of the most pressing problems of the day, demonstrating its importance in both fundamental research and applications as also recognized in interdisciplinary scientific work (Bracco et al., 2010).

8.6.1. Industrial Applications in Synthesis

Industrial applications in synthesis are most significant to the ongoing development in chemical manufacturing and process design. Incorporation of various chemical compounds into synthesis reactions significantly enhances the efficiency and success of manufacturing. Nitromethane (NM) is a prime example, as a key compound used as a solvent and reactant in a broad variety of chemical reactions. Due to high polarity, strong exothermic reactions are achievable, which play a key role in production of energy-dense products. However, the complexities of using NM in industrial processes are accentuated by the need for maintaining safety measures due to its capacity to explode. Understanding the dynamics of the reaction of NM, like its interaction with water vapor and molecular oxygen, is critical in increasing operation methods, thereby boosting product synthesis of various chemicals to the maximum while maintaining detonation hazards to an absolute minimum (Chaban et al., 2014). Another important consideration for chemical synthesis is the concept of design-for-operability, which integrates operating effectiveness into preliminary process design.

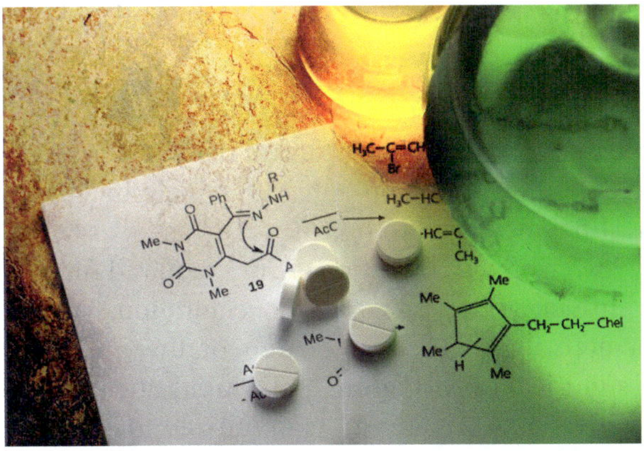

Figure 8.13. *Industrial applications in chemical synthesis.*

Source: https://venture-lab.org/wp-content/webpc-passthru.php?src=https://venture-lab.org/wp-content/uploads/2022/05/Chemical-Synthesis-e1654010693139.jpg&nocache=1.

These addresses growing chemical production complexity, particularly in relation to stringent economic and safety requirements. By addressing operability early on, chemical engineers can develop more durable and adaptable processes that can fit changing business requirements. This integration is especially significant since security and environmental concerns have become increasingly core. Discussion of the subject emphasizes that early interventions can significantly reduce the time and costs associated with making current systems retrofittable for use, hence making industrial utilization in synthesis easier (Bandoni et al., 2004). Promoting thus a culture of operability in chemical engineering practice makes it easier to create processes that are not only efficient but thrive under regulatory scrutiny. Furthermore, the implications of reaction dynamics play an important role in the sustainability of industrial synthesis processes. With industries looking for greener alternatives, research into greener reagents and pathways of reaction has been an interest. Synthesis of novel products with lower environmental footprints, such as biodegradable plastics and renewable chemicals, relies on a sophisticated understanding of reaction mechanisms. Actually, the interaction between product stability once formed, reaction kinetics, and thermodynamics must be efficiently handled to allow meaningful industrial processes. Consequently, enhancement of simulation technologies such as reactive molecular dynamics is of vast use in the examination of such complex interactions. With the use of such information, industries can innovate and streamline their synthesis steps ultimately towards more sustainable and economically viable chemical production (Chaban et al., 2014) (Figure 8.13).

8.6.2. Biological Reactions and Enzyme Kinetics

The intricate processes involved in biological reactions lie deep within enzyme kinetics, which determine reaction rates at which biochemically catalyzed reactions occur. Enzymes as biological catalysts also significantly reduce reaction rates by lowering the activation energy required for a reaction. Enzyme activity can be influenced by substrate concentration, temperature, pH, and by inhibitors or activators. An appreciation of the interplay between these influences is of significant bearing on applications such as biotechnology and pharmacology, where enzyme activity is optimized to yield enhanced product yield and therapeutic response. A kinetic investigation into enzyme activity gives insight into their mode of operation, showing the relationship between enzyme structure and function. Such an understanding not only assists in experimental designs but also assists in the design of enzyme inhibitors, which may be absolutely essential in controlling metabolic pathways under diseased conditions. Under contemporary research, the phenomenon of dynamic disorder presents new avenues for describing enzyme kinetics at the molecular level. Recent studies suggest that a majority of enzymes exhibit fluctuations in catalytic activity intrinsically linked to conformational dynamics over a range of timescales. Such a dynamical phenomenon is antithetical to the long-standing view of enzymes as structurally static molecules and underscores the importance of considering such fluctuations in kinetic theory. With improvements in experimental techniques, particularly in single-molecule enzymology, interrogation of these stochastic fluctuations has been made more feasible. This allows researchers to see the catalytic activity of an individual enzyme molecule in real time, shedding light on the previously hidden elusive complexity of enzyme reactions. Thus, dynamic disorder analysis gives important information regarding conditions under which enzyme kinetics operate in living things, an indication of the need to update existing kinetic models to accommodate this complexity. The implications of fluctuating enzyme activity are widespread in the sense that they are capable of affecting significantly the biological consequences. For instance, in cell conditions where enzyme populations are dilute, rate fluctuations generate emergent changes in substrate levels, which lead to extreme deviations from the values expected with sole use of deterministic models. The relationship between the rate of fluctuation of enzyme activity and the mean levels of substrates illustrates the mechanisms through which biological signals may be amplified or attenuated. This insight is central to predicting how enzymes will function under varying conditions and informs experimental methods that probe the role of enzyme kinetics in living cells. Armed with hypotheses grounded in the dynamical behavior, researchers can explore the biological significance of stochastic enzyme activity, unlocking its potential role in the regulation of metabolism and specification of phenotypic traits of organisms, and considerably expanding our image of biological reaction dynamics (Figure 8.14).

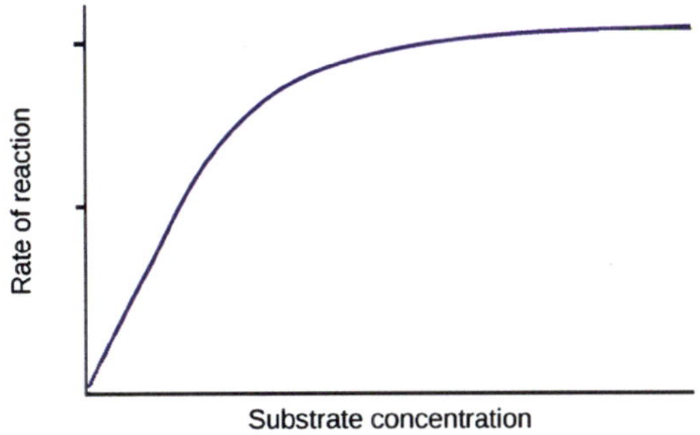

Figure 8.14. *Basics of enzyme kinetics graphs.*

Source: https://cdn.kastatic.org/ka-perseus-images/e3a3422f253ec-134f28a7a7aca27d9e08623b755.png.

8.6.3. Environmental Implications and Reactions

Multidimensional exchange among ecosystems underscores the profound environmental relevance of reaction dynamics in numerous biogeochemical processes. The hyporheic zone is one such region of particular significance, a major ecosystem interface between river channels and bordering sediments. Contemporary studies underscore how this zone acts as a dynamic metabolic modulator of stream processes, with significant influence on water quality by enhancing contact time with reactive environments. This extended interaction allows for retention of nutrients and pollution, including microplastics, and thus is central to the sustenance of aquatic ecosystem integrity. Insight from this domain calls for preserving and restoring hyporheic zones since they offer potential for enhancing ecosystem services and sustaining biodiversity along river corridors (Arnon et al., 2019). Furthermore, the understanding of nitrogen cycling in oceanic systems incorporates other scales of reaction dynamics with profound environmental consequences. Denitrification, an essential process that immobilizes fixed nitrogen into inert N_2, occurs predominantly in coastal sediments with high organic content. Recent interpretations demonstrate that traditional conceptual models of nitrogen dynamics are still incomplete, hinting at alternative pathways responsible for nitrogen removal in oceanic systems. These include processes such as anaerobic ammonium oxidation, implying that the relationships between various biogeochemical cycles such as carbon, sulfur, manganese, and iron are more coupled than ever before acknowledged. These results oppose previous conventions of nitrogen cycling and suggest the delicate interaction necessary for maintaining ecosystem stability and functionality (Aller et al., 2005). With these findings, the interplay between different ecological processes is of primary significance, particularly in aquatic environments' restoration and management. Management ought to be done with consideration of the intricacy due to such reaction dynamics, that alterations in one aspect, such as hyporheic restoration or modification of nitrogen cycling, can trickle throughout the entire ecosystem. By having a better grasp of the ecological processes of both the hyporheic zone and the marine nitrogen cycle, scientists and environmentalists are more capable of predicting the consequences of interacting interventions. Such a systems-focused strategy emphasizes the value of combining scientific understanding with practical action, finally aiming for enhanced ecosystem resilience and sustainability as we contend with the challenges of human-caused pressures on the environment (Aller et al., 2005; Arnon et al., 2019).

SUMMARY

In combining the reaction dynamics and their consequence, of particular note is the radical innovation in the establishment of microfluidics, by which tighter control over chemical process can be attained. Microfluidic systems bring along higher efficiency and promise to deliver more efficient mixing of the reactants. However, forces such as viscous resistance and surface tension can complicate these processes, potentially leading to undesirable effects such as bubbling at reaction sites. The motions of slugs of reactants at various pressures show the importance of having the forces in place in order to attain an optimum output in microfluidic systems. Through an examination of these dynamics in depth, scientists have been able to understand how to maintain stability and reduce unwanted byproducts in the process of vital reactions, which points towards the importance of ongoing innovation and research at this level (Abouali et al., 2009). Its demonstration in reaction-diffusion systems also illustrates the complexity involved in parameter estimation in the context of inherent variability. With a Bayesian inference approach, researchers can easily circumvent these challenges since it presents an estimation framework for states and parameters of chemical processes. This method not only facilitates the possibility of having a comprehensive understanding of the underlying dynamics but also identifies the potential for enhanced experimental design via credibility interval determination. The findings from synthetic as well as empirical data, e.g., those on morphogen diffusion in *Drosophila melanogaster*, confirm the value in applying stringent statistical methods when studying reaction dynamics. These findings underscore the significant role that complex modeling techniques can play in enhancing experimental design and facilitating deeper understanding within the field (Dewar et al., 2010). In sum, reaction dynamic research combines microfluidic processes and stochastic modeling, which indicates the necessity for the introduction of new methods into contemporary research. The interaction between theoretical models, e.g., slugs' dynamics in microfluidic devices, and practical implementations, e.g., parameter estimation in reaction-diffusion systems, indicates the multi-faceted nature of problems that researchers in the field face. As reaction dynamics becomes more advanced, increased research needs to refine existing models and identify novel applications. Support for coordination across distinct scientific disciplines will be needed to transcend these challenges and provide new possibilities for research, policy development, and technological innovation. Lastly, from this current work, the understanding achieved will contribute significantly towards optimizing and improving the efficiency of chemical reactions in practical applications (Abouali et al., 2009; Dewar et al., 2010).

Reaction dynamics is a generic discipline of study that describes numerous processes of describing reactant-to-product growth over time. Interpretation of complex reaction mechanisms is one of the fundamental issues of reaction dynamics, usually by means of an augmentation of simulation. Simulation constructs dynamic models that replicate the past evolution of chemical reactions and generate narrative-like insight into molecular interactions. By examining simulations of well-studied reactions such as the Diels-Alder reaction, researchers can connect theoretical models and laboratory findings. This is in reference to the significance of temporal progression in chemical reactions, where ideals of followability and contingency, as pointed out by Gallie, are significant in improving our understanding of steps in reactions and their effects, allowing scientists to visualize the delicate dance of atoms in their efforts to create new compounds (Wise et al., 2017). Recent advances in observation techniques have transformed reaction dynamics research, particularly to observe rapid molecular dynamics during chemical reactions. Femtosecond electron diffraction, e.g., has been demonstrated as a powerful method to observe atomic motion directly with high time resolution. The technique offers the prospect to observe directly real-time structural variations undergone instantaneously after excitation, e.g., like in the photoinduced ring-closing reaction of diarylethenes. Observation has shown that the torsion of the thiophene rings induces deep conformational transformations necessary to yield the closed-ring product on timescales as short as a few picoseconds. The ability to observe such dynamic processes not only confirms theoretical prediction but also provides insight into mechanistic origins of reactions. Empirical evidence here highlights the complementary nature of experimental and theoretical methods in the study of reaction dynamics (Asano et al., 2013). Recap of the essence of reaction dynamics points to the fact that multi-disciplinary knowledge is required to enhance knowledge on this topic. Integration of simulation knowledge with cutting-edge experimental methods provides insight into molecular change occurring. The narrative character of simulations, combined with the definitive information obtained from high-resolution observations, demonstrates the interactive interplay between theory and practice. Furthermore, the development of solid methodologies for real-time analysis strengthens our ability to manipulate and predict reaction pathways, ultimately leading to new applications in fields of materials science and pharmaceuticals. The investigation of these diverse methods enhances our awareness of the complexities of reaction dynamics, inviting ongoing examination of mechanistic subtleties that dictate chemical behavior and its relevance to broader scientific questions (Asano et al., 2013; Wise et al., 2017).

REFERENCES

1. Abouali, M., Breward, C., Chan, I., Ellis, G., et al. (2009). The dynamics of liquid slugs forced by a syringe pump. https://core.ac.uk/download/pdf/6183035.pdf (accessed on 10th September 2025).
2. Abou-Kandil, M., Abramov, A., Aldous, M., Anderson, D., Anderson, E., Anderson, J., Andrieux, P., et al. (2012). Markovian dynamics on complex reaction networks. http://arxiv.org/abs/1205.5524 (accessed on 10th September 2025).
3. Alberts, B., Phillips, A., Blossey, R., Bundschuh, R., Cherry, J. L., De Jong, H., Fall, C. P., et al. (2007). Compositionality, stochasticity and cooperativity in dynamic models of gene regulation. http://arxiv.org/abs/q-bio/0702016 (accessed on 10th September 2025).
4. Aller, R. C., Canfield, D. E., Dalsgaard, T., Engström, P., et al. (2005). Nitrogen removal in marine environments: Recent findings and future research challenges. https://core.ac.uk/download/12043452.pdf (accessed on 10th September 2025).
5. Arnon, S., Banks, E., Batelaan, O., Betterle, A., et al. (2019). Is the hyporheic zone relevant beyond the scientific community? https://core.ac.uk/download/288113973.pdf (accessed on 10th September 2025).
6. Asano, Y., Bertarelli, C., Boggio-Pasqua, M., Morrison, C. A., Lu, C., Chergui, M., Clark, S. J., et al. (2013). Ring closing reaction in diarylethene captured by femtosecond electron crystallography. https://core.ac.uk/download/28974479.pdf (accessed on 10th September 2025).
7. Aschaffenburg, D. J., Chen, X., & Cuk, T. (2019). Selecting between two transition states by which water oxidation intermediates on an oxide surface decay. http://arxiv.org/abs/1903.01675 (accessed on 10th September 2025).
8. Avez, B., Simenel, C., & Wakhle, A. (2012). Quantum microscopic approach to low-energy heavy ion collisions. https://core.ac.uk/download/52686907.pdf (accessed on 10th September 2025).
9. Bailleul, S., De Wispelaere, K., & Van Speybroeck, V. (2016). Towards molecular control of elementary reactions in zeolite catalysis by advanced molecular simulations mimicking operating conditions. https://core.ac.uk/download/55855326.pdf (accessed on 10th September 2025).
10. Bandoni, J. A., & Blanco, A. M. (2004). Diseño para operabilidad: Una revisión de enfoques y estrategias de solución. https://core.ac.uk/download/288803439.pdf (accessed on 10th September 2025).

11. Bell, A. T., Head-Gordon, M., Janda, A., Lin, L., et al. (2018). Understanding Brønsted-acid catalyzed monomolecular reactions of alkanes in zeolite pores by combining insights from experiment and theory. https://core.ac.uk/download/158345582.pdf (accessed on 10th September 2025).
12. Bjork, B. J., Bui, T. Q., Changala, P. B., Nguyen, T. L., et al. (2017). Direct measurements of DOCO isomers in the kinetics of OD + CO. https://core.ac.uk/download/211372083.pdf (accessed on 10th September 2025).
13. Bordbar, A., Dräger, A., Flamholz, A., Joshi, A., Kinoshita, A., Canelas, A. B., Dräger, A., et al. (2016). Evaluation of rate law approximations in bottom-up kinetic models of metabolism. https://core.ac.uk/download/323065299.pdf (accessed on 10th September 2025).
14. Bracco, A., Chomaz, P., Gaardhøje, J. J., Heenen, P. H., et al. (2010). *Perspectives of Nuclear Physics in Europe: NuPECC Long Range Plan 2010.* https://core.ac.uk/download/19502483.pdf (accessed on 10th September 2025).
15. Branco Pinto, D., Constales, D., Marin, G., Yablonsky, G., et al. (2017). Kinetic intersections as a source of information on rate coefficients. https://core.ac.uk/download/91288153.pdf (accessed on 10th September 2025).
16. Buntkowsky, G., Ivanov, K., & Vieth, H. M. (2017). From free radicals and spin-chemistry over spin-dynamics and hyperpolarization to biology and materials science. https://core.ac.uk/download/228064970.pdf (accessed on 10th September 2025).
17. Burrage, K., & Tian, T. (2004). Effective simulation techniques for biological systems. https://core.ac.uk/download/81387.pdf (accessed on 10th September 2025).
18. Cercignani, C., De Donder, T., Hill, T. L., Esposito, M., Polettini, M., Nicolis, G., Prigogine, I., et al. (2014). Irreversible thermodynamics of open chemical networks I: Emergent cycles and broken conservation laws. https://core.ac.uk/download/31208058.pdf (accessed on 10th September 2025).
19. Chaban, V. V., Fileti, E. E., & Prezhdo, O. V. (2014). Exploding nitromethane *in silico*, in real time. http://arxiv.org/abs/1408.5519 (accessed on 10th September 2025).
20. Chang, C. E. A., Huang, Y. M. M., Mueller, L. J., You, H., et al. (2016). Investigation of structural dynamics of enzymes and protonation states of substrates using computational tools. https://core.ac.uk/download/323067820.pdf (accessed on 10th September 2025).
21. Chin, A. W., Morgan, S. E., Novoderezhkin, V. I., Plenio, M. B., et al. (2017). Quantum-coherent dynamics in photosynthetic charge separation

revealed by wavelet analysis. http://arxiv.org/abs/1705.07435 (accessed on 10th September 2025).

22. Ciccioli, A., & Latini, A. (2018). Thermodynamics and the intrinsic stability of lead halide perovskites *$CH_3NH_3PbX_3$*. https://core.ac.uk/download/159268116.pdf (accessed on 10th September 2025).
23. Crampin, E. J., & Gawthrop, P. J. (2014). Energy-based analysis of biochemical cycles using bond graphs. http://arxiv.org/abs/1406.2447 (accessed on 10th September 2025).
24. Dewar, M. A., Kadirkamanathan, V., Opper, M., Sanguinetti, G., et al. (2010). Parameter estimation and inference for stochastic reaction-diffusion systems: Application to morphogenesis in *Drosophila melanogaster*. https://core.ac.uk/download/49747.pdf (accessed on 10th September 2025).
25. Dreyfus, B. W., Geller, B. D., Gouvea, J., Redish, E. F., et al. (2013). Chemical energy in an introductory physics course for the life sciences. http://arxiv.org/abs/1308.3667 (accessed on 10th September 2025).
26. Furusawa, C., & Kaneko, K. (2000). Theory of robustness of irreversible differentiation in a stem cell system: Chaos hypothesis. http://arxiv.org/abs/nlin/0009007 (accessed on 10th September 2025).
27. Gallardo, A., Pérez Elvira, E., Biswas, K., Barragán, A., Björk, J., Jelínek, P., Urgel, J., et al. (2025). Computational characterization of the on-surface synthesis of low-dimensional nanostructures. *ECS Meeting Abstracts*. https://www.semanticscholar.org/paper/28a11700b4e8b6cf0921316d41a3597050104768 (accessed on 10th September 2025).
28. Glowacki, D. R., Harvey, J. N., & Lightfoot, R. (2013). Non-equilibrium phenomena and molecular reaction dynamics: Mode space, energy space and conformer space. https://core.ac.uk/download/29026941.pdf (accessed on 10th September 2025).
29. Goddard, W. A., III, Han, S., Strachan, A., et al. (2011). Thermal decomposition of condensed-phase nitromethane from molecular dynamics from ReaxFF reactive dynamics. https://core.ac.uk/download/4889377.pdf (accessed on 10th September 2025).
30. Goddard, W. A., III, Katz, G., Kosloff, R., et al. (2011). Density-dependent liquid nitromethane decomposition: Molecular dynamics simulations based on ReaxFF. https://core.ac.uk/download/4890657.pdf (accessed on 10th September 2025).
31. Gupta, A., Khammash, M., & Milias-Argeitis, A. (2017). Dynamic

disorder in simple enzymatic reactions induces stochastic amplification of substrate. https://core.ac.uk/download/518754491.pdf (accessed on 10th September 2025).

32. Hoessly, L., & Mazza, C. (2019). Stationary distributions and condensation in autocatalytic CRN. http://arxiv.org/abs/1809.07361 (accessed on 10th September 2025).
33. Jacques, D., Mallants, D., Simunek, J., van Genuchten, M. T., et al. (2018). The HPx software for multicomponent reactive transport during variably-saturated flow: Recent developments and applications. https://core.ac.uk/download/286361774.pdf (accessed on 10th September 2025).
34. Jamshidi, N., & Palsson, B. Ø. (2008). Formulating genome-scale kinetic models in the post-genome era. https://core.ac.uk/download/323077184.pdf (accessed on 10th September 2025).
35. Jayawardhana, B., Rao, S., & van der Schaft, A. (2015). A network dynamics approach to chemical reaction networks. https://core.ac.uk/download/518707600.pdf (accessed on 10th September 2025).
36. Kominis, I. K., & Mouloudakis, K. (2017). Quantum information processing in the radical-pair mechanism: Haberkorn theory violates the Ozawa entropy bound. http://arxiv.org/abs/1607.03071 (accessed on 10th September 2025).
37. Laubereau, A., Mantel, K. H., & Zinth, W. (2007). *Proceedings of the Thirteenth International Conference on Time-Resolved Vibrational Spectroscopy*. https://core.ac.uk/download/12162307.pdf (accessed on 10th September 2025).
38. Mebel, M., Gladstone, R., Gu, X., Guo, H., Jones, B., Kislov, V., Landera, A., et al. (2009). Chemical dynamics of triacetylene formation and implications to the synthesis of polyynes in Titan's atmosphere. https://core.ac.uk/download/4882996.pdf (accessed on 10th September 2025).
39. Proppe, J., & Reiher, M. (2018). Mechanism deduction from noisy chemical reaction networks. http://arxiv.org/abs/1803.09346 (accessed on 10th September 2025).
40. Reiher, M., & Vaucher, A. C. (2016). Molecular propensity as a driver for explorative reactivity studies. http://arxiv.org/abs/1604.06748 (accessed on 10th September 2025).
41. Scholey, J. M. (2013). Compare and contrast the reaction coordinate diagrams for chemical reactions and cytoskeletal force generators. https://core.ac.uk/download/323071498.pdf (accessed on 10th September 2025).

42. Simenel, C. (2014). Challenges in description of heavy-ion collisions with microscopic time-dependent approaches. http://arxiv.org/abs/1403.3246 (accessed on 10th September 2025).

43. Tang, Y. (2025). Machine learning stochastic dynamics. *Scientia Sinica Physica, Mechanica & Astronomica*. https://www.semanticscholar.org/paper/4857760191a2ed07bac5dae51936772cb704e90e (accessed on 10th September 2025).

44. Turco, R. P. (1992). *Upper-Atmosphere Aerosols: Properties and Natural Cycles*. https://core.ac.uk/download/pdf/10437198.pdf (accessed on 10th September 2025).

45. Wise, M. N. (2017). On the narrative form of simulations. https://core.ac.uk/download/323310842.pdf (accessed on 10th September 2025).

CHAPTER 9

PHOTOCHEMICAL KINETICS

Contents

Introduction 331

9.1. Fundamental Principles of Photochemistry 336

9.2. Mechanisms of Photochemical Reactions 342

9.3. Factors Influencing Photochemical Kinetics 347

9.4. Experimental Techniques in Photochemical Kinetics 352

9.5. Applications of Photochemical Kinetics 357

9.6. Challenges and Limitations in the Field 362

Summary 366

References 368

INTRODUCTION

Photochemical kinetics is the analysis of processes and mechanisms wherein light interacts with matter, and particularly how these interactions bring about chemical change. Photochemical kinetics is crucial to an understanding of a wide range of phenomena, for example, photosynthesis, phototoxicity, and photovoltaic cell behavior. At the core of the discipline is the investigation of excited electronic states, which play crucial roles in initiating photochemical processes. When molecules absorb photons, they are pushed into higher energy levels, leading to a range of potential reactions. How well these reactions work depends mostly on the kinetics at hand, like rate constants that are relevant to various photochemical processes. Being in a position to know the kinetics allows scientists to manipulate conditions so as to optimize beneficial outcomes, and as such, it is a very important area of study in both research and industrial use. At the heart of photochemical kinetics lies the concept of molecular activation and how crucial this is in determining pathways of reaction. Temperature, pressure, and solvent character may all be significant forcing functions on activation barriers and rate of reaction.

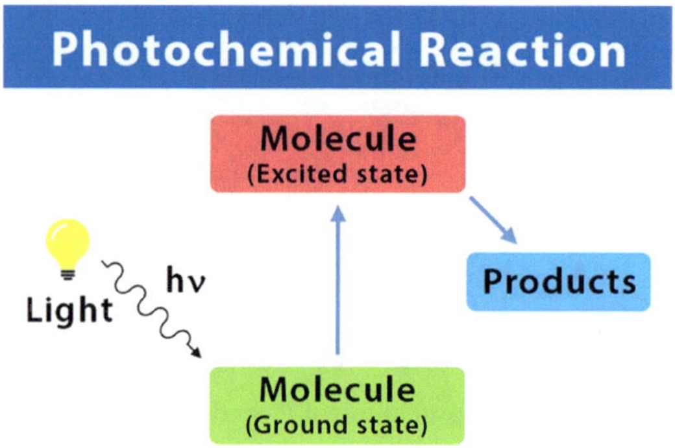

Figure 9.1. *Photochemical reaction.*

Source: https://www.chemistrylearner.com/wp-content/uploads/2020/09/Photo-chemical-Reaction.jpg.

By studying such dynamics, researchers have made available data on the photochemical response of systems and the intricate relationship between reactivity and molecular structure. For example, Penrose graph research in reaction kinetics illustrates how particular molecular structures allow for or block the transition states needed for reactions to take place. Such understanding is further nuanced by advanced spectroscopic techniques that allow the monitoring of real-time molecular development and intermediate states. Integrating information from molecular studies enhances photochemical reaction predictability, a significant goal in further establishing the discipline. Recent research of specialist molecular systems has also illuminated further the complexities involved in photochemical kinetics. Study of mutant proteins, such as the PAS-GAF-PHY sensory module of phytochromes, illustrates the significance of each intramolecular interaction on photoreaction efficiency. For instance, the mutants experiment illustrated how modifications of molecular structure can lead to reduced photoconversion efficiency with an emphasis on the subtle interaction between molecular geometry and energy landscapes (Essen et al., 2014). Similarly, kinetic experiments on surface modification of materials like polytetrafluoroethylene (PTFE) illustrate the dynamical nature of reactions in radical formation in plasma interactions that are of utmost significance in most industrial procedures (Kono et al., 2012). These analyses at length emphasize the multidisciplinary character of photochemical kinetics, illustrating the interdependence both of basic principles and of special molecular phenomena in advancing knowledge of light-activated chemical processes (Figure 9.1).

Definition of Photochemical Kinetics

Photochemical kinetics is the investigation of the mechanisms and rates of the photochemically initiated chemical reactions. Photochemical kinetics is the foundational chemistry employed to explain a lot of natural and artificial phenomena, including photosynthesis and the breakdown of air pollution. Photochemical kinetics is simply the examination of how illumination energy influences molecular interactions, leading to the formation or breakdown of chemical entities. The understanding of these processes is important in various fields including environmental science, atmospheric chemistry, and chemical engineering. As scientists probe the complexities of these reactions, they use mathematical descriptions to measure the rate of occurrence of photochemical processes. This not only permits predictions of atmospheric constituents' behavior under different conditions of light but also aids in developing solutions to mitigate environmental issues caused due to photochemical reactions. Moreover, the intricacy of photochemical kinetics is highlighted by the interaction between light and molecular systems. The photons' energy is captured, forming excited electronic states, which relax to the ground state or are involved in chemical transformations. The mechanism is critical in finding the contribution that pollutants such as chlorofluorocarbons make towards ozone depletion in the stratosphere. As outlined in recent reviews of studies, intricate chemical kinetics models provide input parameters vital in atmospheric modeling (Kurylo et al., 2025). Such models evaluate how variations in wavelength and light intensity impact reaction rates, which are variables to a full understanding of the upper atmospheres chemical composition. It is particularly relevant in environmental evaluation where accurate predictions of ozone change are needed to determine effective regulatory policy. Apart from its relevance in atmospheric science, photochemical kinetics also serves as a foundation for the evaluation of various chemical species' reactions to light. The intercomparisons and workshops of models and measurements of the stratosphere identify the necessity for accurate kinetic data in the simulation of chemical reactions in environmental diagnostics (Prather et al., 2025). The usefulness of such models is based on their ability to simulate observed phenomena in the stratosphere with high accuracy, for example, photochemistry of ozone and other trace gases. Understanding temporal and spatial behavior of photochemical kinetics is crucial because it allows one to forecast how conditions in the atmosphere vary, thereby guiding policies to address climate change and air quality issues. The ongoing disclosures of photochemical kinetics are not only crucial to scientific advancement but also to practical applications that uncover implications in public health and environmental sustainability.

Importance in Chemical Reactions

The significance of photochemical kinetics via chemical reactions is that they can manage the photochemical conversion of light to chemical energy. This is very important in most atmospheres, such as those in gas giants, where photochemical reactions serve as facilitators of complex chemical processes. As an illustration, studies of the chemistry of Jupiter's troposphere indicate that the formation and interconversion of nitrogen-bearing species are strongly temperature- and light-sensitive. The intricate mechanisms behind these processes, such as transport-induced quenching of hydrogen cyanide (HCN), demonstrate how environmental conditions dictate chemical behavior and regulate the subsequent atmospheric composition. These results indicate the essential role of chemical kinetics to understand planetary atmospheres and the dynamics of their components, with the full understanding of the sophisticated interaction between photochemistry and physical processes in a range of extraterrestrial environments (Achterberg et al., 2010). Furthermore, the understanding of photochemical kinetics extends beyond the understanding of planetary atmospheres in terms of influencing many facets of chemical research and applications. The photophenomenon of photochemistry is of great relevance in organic synthesis and materials science, where it is used to develop complex structures via photoinduced conversions. With emerging technologies gravitating increasingly towards photoactive compounds, their reactions can be regulated to yield particular outcomes when synthesizing multifunctional compounds. The overlap of light absorption, electronic excitations, and the formation of radicals explains how light is able to change molecular structures so that researchers can probe new ways of reaction engineering. Understanding these mechanisms across different contexts enhances our ability to engineer new materials, such as advanced polymers and photonic materials, which rely on specific photochemical reactions in order to gain optimal performance (Marley et al., 2016). Ultimately, the importance of chemical reactions in photochemical kinetics is essential to the creation of theoretical as well as applied knowledge within the area of chemistry. As scientists work to learn about various processes undergirding these reactions, it is ever clearer that each interaction is a puzzle piece in a larger scientific puzzle. The improvement in kinetic models allows quantification of rates and pathways for light-sensitive reactions, leading to better prediction of chemical behavior under varied conditions. In addition to this, the data can guide the design of methodologies employing photochemical processes towards solar energy conversion and reduction of greenhouse gases using sustainable approaches. Thus, the comprehension of these intricate reactions not only makes us better understand fundamental chemical ideas but also serves as a catalytic force to promote innovation in solving present-day scientific challenges, bridging the gap between theory and practice in chemistry.

Historical Context and Development

The history of photochemical kinetics is centrally embedded in the history of atmospheric chemistry, with a special emphasis on the study of ozone and its role in atmospheric chemistry. The second half of the 20th century saw a turning point in scientific understanding, particularly following the breakthrough discoveries about ozone depletion in Antarctica. Initial measurements in the 1980s showed periodic seasonal variations in ozone, which stimulated an emerging revision of stratospheric chemistry. Scientists identified polar stratospheric clouds (PSCs) as the prime suspects in the chemical processes of ozone loss, and the scientific debate shifted towards a more comprehensive understanding of complexities of pollutant interactions with atmospheric processes (Cynthia et al., 1989). The subsequent focus on these clouds and their part in ozone loss not only redefined the field of atmospheric science but also underscored the importance of photochemical kinetics for explaining these phenomena. The world science communities response to the crisis of ozone depletion led to significant policy changes and international agreements, such as the Montreal Protocol. This seminal treaty aimed at phasing out ozone-depleting substances, exemplifying the critical intersection of scientific discovery and environmental policy. Additional research into the photochemical processes was now required to better appreciate the implications for such treaties. Chemists were able to investigate the interactions of chlorofluorocarbons (CFCs) with atmospheric conditions and learn about their degradation pathways under photochemical processes to identify the driving forces for ozone loss. This linking of science with policy aided not only in the minimization of ozone depletion effects but also marked a higher degree of public engagement with atmospheric issues (Cynthia et al., 1989). All these thus created opportunities for more studies on the effect that anthropogenic processes have on atmospheric chemistry. Photochemical kinetics developments have been increasingly motivating environmental science advances, particularly as air and climate issues have grown. Progress in this field has led to enhanced tools with which to study chemical kinetics in all sorts of atmospheric conditions. Scientists have progressively drawn upon laboratory experiments and field measurements to break down the complexity of chemical reactions within tropospheric and stratospheric environments. The studies of various pollutants and their photochemical character have offered useful facts concerning oxidative processes that impact human health and ecosystems. Specifically, an understanding of how ozone is involved in working as a ground-level pollutant has warranted attention given its harmful effects, demonstrating the perfect balance required in atmospheric chemistry (Grose et al., 2025). Overall, the path of photochemical kinetics over history mirrors the relentless quest for knowledge, finally for solving the pressing environmental problems provoked by human activity.

9.1. FUNDAMENTAL PRINCIPLES OF PHOTOCHEMISTRY

The fundamental principles of photochemistry are in the domain of the interaction of light and chemical reactivity, illuminating the manner molecules absorb photons and undergo electronic change. When a molecule absorbs light, it stimulates an electron into an excited state from a ground state, a process that is important in the majority of photochemical processes. The efficiency of this energy conversion and subsequent reactions can be quantified by photochemical kinetics, establishing the rates of such chemical transformations. These principles are essential to characterize processes such as photosynthesis and photopolymerization, where light absorption triggers significant changes in chemical content. The pathway and effectiveness of these reactions depend upon various factors, including molecular form, solvent effect, and wavelength of light, and hence photochemical kinetics is a sophisticated discipline that requires meticulous examination of kinetic models and spectroscopic data. In examining the kinetic properties of photochemical reactions, scientists concentrate on the impact of molecular interactions in reaction pathways. The potential for mutation of some molecules to significantly alter photochemical efficiency makes this area of study further illuminating. For instance, a study of the *Cph1* phytochrome details the effect of mutations on the Pr-to-Pfr photo-conversion efficiency, highlighting the significance of the role of particular bonds and structural conformations at various points in the reaction pathway (cf. the yield decreases from 0.13 to 0.05–0.07 in mutants) (Essen et al., 2014). This specificity highlights the relevance of certain molecular characteristics, such as hydrogen bonds and steric hindrance, which are involved in the definition of activation barriers for photoreactions and are also responsible for our understanding of mechanisms of energy transfer and storage of photochemical systems. General uses of photochemical kinetics include industrial processes, where the understanding of light-induced reactions can be so dramatic as to produce breakthroughs. A straightforward kinetic model of photothermal oxidation of polypropylene demonstrates the approach, explaining how thermolysis and photolysis both contribute to radical formation and rates of reaction (Colin et al., 2014). Temperature sensitivity is key to this, with rate constants for thermolysis exhibiting Arrhenius dependence, while photolysis is not dependent on temperature variation. These models describe the photochemistry mechanisms of oxidation reactions in polymers, as functions of UV light intensity and wavelength. Collectively, these descriptions of the basic principles of photochemistry guide subsequent investigation and enable more effective chemical processes and materials that utilize light for transformation (Figure 9.2).

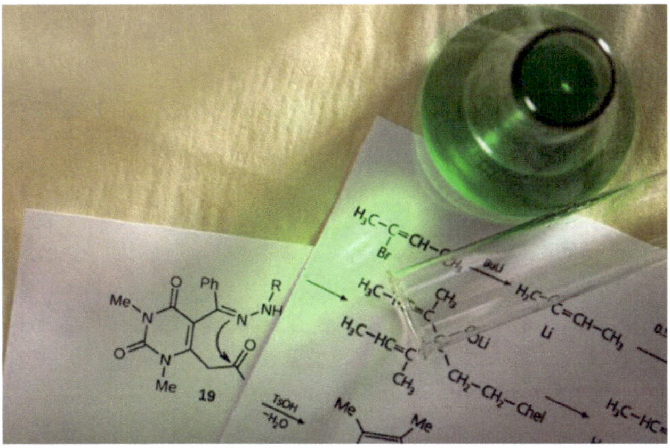

Figure 9.2. *Fundamentals of photochemistry.*

Source: https://microbiozindia.com/wp-content/uploads/2023/11/photochemistry.jpg.

9.1.1. Interaction of Light with Matter

The interaction of light with matter is crucial in the study of photochemical kinetics since such an interaction initiates most chemical reactions that cause various processes in atmospheric and biological systems. When light falls on matter, it can either be absorbed, reflected, or transmitted and hence excites the electrons in molecules with the subsequent initiation of chemical changes. The absorption spectrum of the substance, showing how much light is absorbed at different wavelengths, is an important tool for predicting chemical response to exposure to light. Such interaction would naturally be a function of factors such as temperature and state of the material. Second, the complex dynamics of these interactions are dictated by the structural properties of molecules, which dictate the specific energy levels that electrons will occupy, and therefore how efficient absorption of light can be in initiating photochemical reactions. Transition metals serve as facilitators in facilitating interactions between matter and light, particularly in atmospheric chemistry. For instance, differing oxidation states of irons render it able to catalyze numerous photochemical reactions, which might lead to the formation of reactive oxygen species upon UV-visible light exposure. It has been noted by Al-Abadleh et al. (2015) that redox conditions and solubility largely control the reactivity of irons, and this shows that light not only affects chemical reactions but also that the chemical environment affects light-matter interaction.

Figure 9.3. *Interaction of light with matter.*

Source: https://www.biolinscientific.com/hs-fs/hubfs/images/Biolin%20Scientific/Blog/Light-matter%20interaction.jpg?width=5000&name=Light-matter%20interaction.jpg.

Additionally, the presence of aerosols and organic chemicals enhances the richness of these interactions, with varying levels of light absorption and reactivity. Multi-component systems such as these possess extensive photochemical products, with effects on atmospheric chemistry and the driving of secondary organic aerosol formation, among others. Light-matter interaction is significant apart from Earth's application in the research of exoplanets, particularly in enlightening atmospheric conditions on distant planets such as 51 Eridani b. The research highlighted by Marley et al. (2016) reports the processes by which photochemistry can transform such substances as H_2S into elemental sulfur under specific conditions of the atmosphere, a process that is highly dependent on light interactions. Dynamics in such transformations indicate the need to determine the conditions under which various atmospheric species absorb and transform light energy. Such encounters contribute to the formation of haze and cloud shapes, shape the planets' climates, and may even hold the key to understanding their potential for life. With photochemical kinetics as a prominent component of astrochemistry, future work will increasingly uncover the light-matter interactions in an immense array of celestial environments (Figure 9.3).

9.1.2. Absorption Spectra and Energy Levels

It is fundamental knowledge in photochemical kinetics to know how absorption spectra and energy levels are related. Absorption spectra are the result of light-matter interactions in which molecules capture photons and change between discrete energy levels. These transitions are controlled by quantum mechanics, which allow for the absorption of some wavelengths of light corresponding to the energy between the ground state and excited states of the molecules. The determination of an absorption spectrum at the same level provides critical information on the structure of the molecules and electronic configurations, thus affecting theories of photochemical reactions. It is at these transitions that the rate of photochemical processes, i.e., dissociation or isomerization, can be understood and quantified, paving the way to kinetic modeling of chemical reactions. The relevance of energy levels can be surmised from the viewpoint of a variety of photochemical phenomena. For instance, light absorption will in general lead to population of the electronic excited states, where the energies of each of the states define the reaction pathways accessible to the system in a given, uniquely defined manner. When photochemical reactions are investigated, account must be taken of the activation energies of the reactions, which are a function of the conical intersections of the potential energy surfaces governing the transitions. The treatment is also extended to the effect of structural molecular changes, such as mutations in photoreceptor proteins, which alter not just the absorption character but also photoconversion efficiency. This is reflected in activation barriers established via state changes as observed in experiments for specific molecular mutations, thus linking absorption spectra to useful photochemical kinetics (Essen et al., 2014). Recent advances in photochemical and thermochemical kinetics have put greater focus on absorption spectra in more complicated systems, including atmospheric analysis of exoplanets. For instance, variations in molecular species concentration in atmospheres such as those of HD 189733b and HD 209458b can be correlated to absorption characteristics of light by different molecular constituents. The absorption spectra also reveal interactions of light with major atmospheric constituents, modulating their molecular concentration via processes such as photochemical reactions and transport quenching. Under these conditions, the absorption spectra play a crucial role in determining thermal profiles and spectral signatures that take into consideration the obtained atmospheric conditions. Familiarity with these spectra is critical for understanding not only the chemical makeup of exoplanetary atmospheres but also for forecasting their stability and evolution in the future (Friedson et al., 2011) (Figure 9.4).

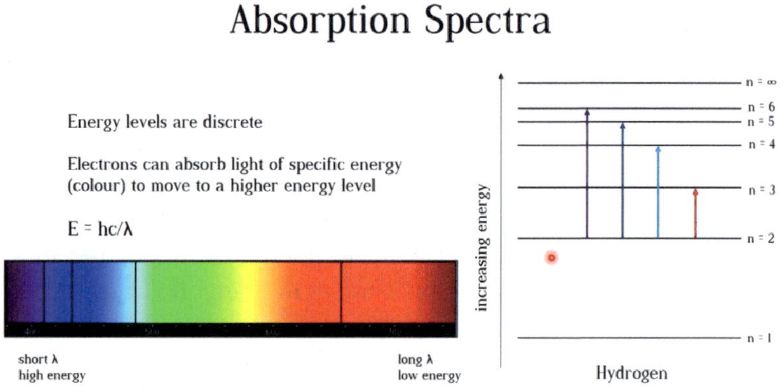

Figure 9.4. *Absorption spectra and energy levels.*

Source: https://i.ytimg.com/vi/lRx_-x2hpcM/maxresdefault.jpg.

9.1.3. Quantum Yield and Its Significance

Quantum yield is a very important parameter in photochemical kinetics since it determines the efficiency of photon absorption leading to a particular chemical transformation or change. Since it defines the ratio of the number of reacting molecules to the number of photons absorbed, quantum yield is an excellent metric for assessing the performance of various photochemical processes. Quantum yield variation among different reactions arises from factors such as the stability of the intermediate radicals and the efficiency of reactive oxygen species formation, which can significantly influence reaction pathways. Quantum yield plays a critical role in the optimization of processes in a vast array of applications ranging from environmental cleaning to energy conversion where precise control of light-induced reaction is of critical importance to efficiency enhancement. Further, the significance of quantum yield expands to utility applications, namely in the area of engineered nanomaterials (ENMs) and their implications with long-term environmental contaminants.

For instance, photochemical conversion efficiencies of perfluoroalkyl acids (PFAAs) precursors can be significantly enhanced with the inclusion of ENMs that are capable of generating reactive oxygen species upon light illumination. Experiments reveal that such improvements in rates of transformation, typically doubling to six-fold, accentuate the quantum yield for the reactive channels. It is, however, notable that all ENMs do not facilitate such transformational effects; others may even inhibit reactions due to competing effects such as adsorption or photo-shielding effects. Consequently, qualitative understanding of quantum yield directs nanomaterial design and implementation in environmental cleanup (Dai et al., 2020).

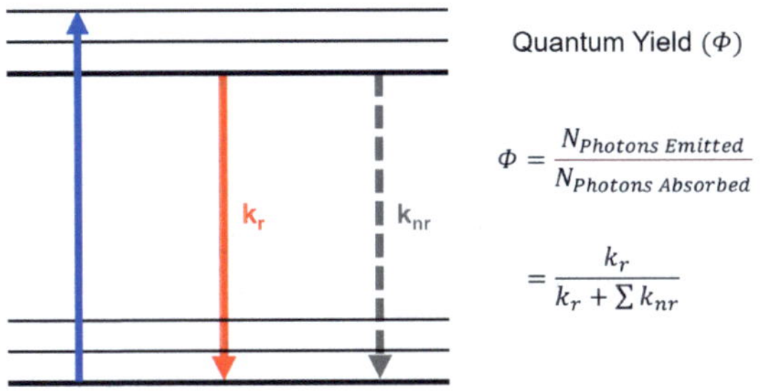

Figure 9.5. *Quantum yield and radiative and non-radiative decay processes.*

Source: https://edinst.com/wp-content/uploads/2021/09/Quantum-Yield-1024x566.png.

Apart from applications in the environment, quantum yield is essential in explaining photothermal oxidation of polymers because it directly impacts rate constants in reaction mechanisms. Specifically, in polypropylene, a kinetic model derived predicts that photolysis initiated by absorption of UV light produces oxidative species of very varying quantum yields. The unimolecular nature of photolysis simplifies observable absorbance-reaction rate relationship, and thus it becomes easy for researchers to easily predict concentration changes of various oxidation products accurately. This interaction extends understanding of the behavior of quantum yield with light intensity and wavelength, hence gaining understanding of the broader implications for material stability under light irradiation. Quantum yield is thus a significant parameter in the study of photochemical processes, acquiring knowledge not only in basic scientific investigation but also in applied material science (Colin et al., 2014) (Figure 9.5).

9.2. MECHANISMS OF PHOTOCHEMICAL REACTIONS

The mechanisms of photochemical reaction are intricate processes that include the absorption of light and consequential chemical modifications. The mechanism of photon absorption by molecules results in the molecule transitioning from the ground state to the excited state, leading to varying paths of reaction. Excitation of this type alters the energy distribution of the molecule, allowing the destruction of chemical bonds as well as bond creation. The efficiency of these reactions relies to a large extent on numerous parameters, including the wavelength of the used light, the nature of the absorbing species, and the environment.

These processes must be well understood, particularly in atmospheric chemistry, where photochemical reactions are ubiquitous. They decide the shape of planetary atmospheres, for example, the creation of hydrogen cyanide (HCN) in Jupiter's atmosphere that is intensely influenced by transport and photolytic processes (Achterberg et al., 2010). The interaction between light and matter also gives details regarding kinetic properties of photochemical processes. Reaction rates also greatly depend on what sort of photochemical reaction is occurring and whether it involves a straightforward absorption, a complicated dissociation, or even the formation of transient intermediates. The kinetics are also typically controlled by both thermal and non-thermal processes, which include such processes as internal conversion and intersystem crossing. The unwilling dynamics in those systems containing nitrogen or hydrocarbon species mimic processes observed on Earth.

For instance, studies reveal cyclic cohesive and production limits of nitric and ammonia compounds in extraterrestrial systems, and they refer to the significance of knowing connected photochemical networks (Yung et al., 1990). Data derived from such kinetic models offer key insights to ensure accurate prediction of planetary atmospheric behaviors. Analysis of the photochemical reaction mechanisms refers to their extensive implications in various fields of science. For example, Venusian and Jovian planet photochemistry reveals broader trends that may be extrapolated to Earth's atmosphere as well, giving clues about how these planets evolved and formed.

The prevailing view is that while some chemical pathways have ubiquitous characteristics, others assume a shape that is shaped by the specific conditions present within various atmospheres, including temperature and particle density. Moreover, heterogeneous reactions, of considerable interest to aerosols and the formation of haze, introduce a further complication in the interaction of light with chemically active species. With the development of science, data on such processes will find their place in advancing the knowledge on not only extraterrestrial atmospheres but also reactive processes associated with Earth's environment (Achterberg et al., 2010; Yung et al., 1990) (Figure 9.6).

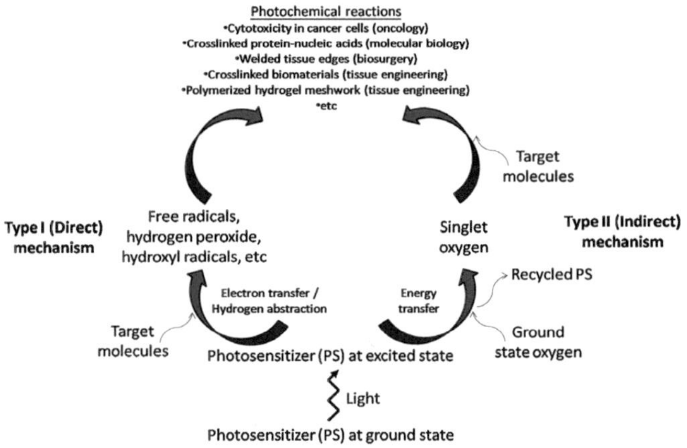

Figure 9.6. *Type I and type II mechanisms of photochemical reactions.*

Source: https://www.researchgate.net/profile/Barbara-Chan/publication/43180479/figure/fig3/AS:601790777155591@1520489510385/Type-I-and-type-II-mechanisms-of-photochemical-reactions.png.

9.2.1. Excited States and Their Formation

Excited state formation is an important phenomenon in photochemical kinetics because it decides the fate and reaction of molecules when they are excited by light. When a molecule is excited with a photon, electrons get excited from the ground state to higher energy levels and form excited states. The excited states are vastly different in energy and stability and decide what course of action the molecules will take next. Interestingly, intramolecular interactions and molecular conformations could influence the photochemical efficiency of these transitions. For instance, mutations of specific residues of photoreceptors, as resolved in the PAS-GAF-PHY sensory module of phytochromes, alter the energy landscape of excited states. This alteration, in turn, affects the quantum yield and the efficacy of photoconversion, defining the intricate relationship between photochemical behavior and molecular structure (Essen et al., 2014). Illustrating excited state dynamics also involves sophisticated experimental techniques that provide insights into their formation and relaxation mechanisms. Sophisticated techniques like time-resolved laser gain-versus-absorption spectroscopy provide evidence for the yield of excited photofragments, while time-resolved Fourier transform infrared (FTIR) spectroscopy enables detection of transient species formed in photochemical reactions. These techniques have emerged as essential to investigating molecular dissociation kinetics and the accompanying behaviors of excited states. Of special significance, these experiments have proven the role of these excited states in being of critical significance in broader applications like atmospheric chemistry. They help in the explanation of process concerning polycyclic aromatic hydrocarbons, with implications for planetary atmospheres and implications for fundamental photochemical kinetics (Leone et al., 1990). An understanding of excited states is central to the development of our current knowledge concerning photochemical reactions, in particular energy storage and energy conversion. The deactivation channels for excited states, like those between different conformers, reflect the complexity of molecular kinetics. In particular, the manner in which excited states occupy different energy surfaces could define how efficiently a molecule can accumulate energy or transform into another species. As an example, studies of the Pr → Pfr transition in phytochromes illustrate the concept of energy barriers and quantum yields in different conformations. Chemical alteration of the key functional groups, as identified in mutation research, will also regulate these dynamics, with effects on biological function and also on synthetic use. Excited states are thus a bridge of utmost significance between excited light and consequent chemical function in photochemical systems, guiding the focus of future work in the area (Essen et al., 2014) (Figure 9.7).

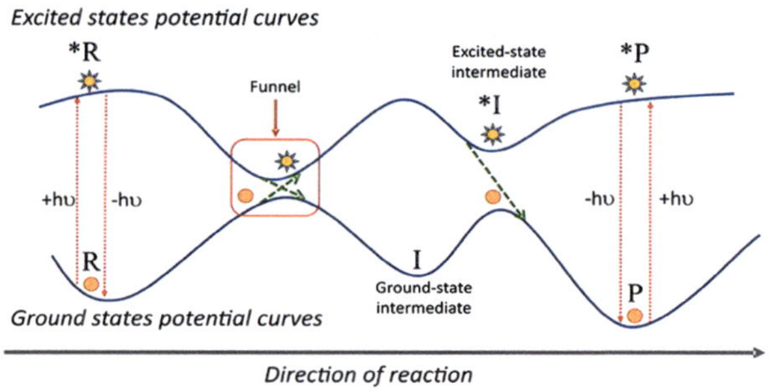

Figure 9.7. *Potential curves of ground and electronic excited states of a photochemical reaction. (R) is the ground-state reactant; (*R) is the excited-state reactant; (I) is the ground-state intermediate; (*I) is the excited-state intermediate; (*P) is the excited-state product; and (P) is the final product of the overall process.*

Source: https://cdnintech.com/media/chapter/56771/1512345123/media/F7.png.

9.2.2. Role of Intermediates in Reactions

The role of the intermediary in photochemical reactions is a key component in reaction mechanism and reaction kinetics understanding. Transient species are formed under specific conditions within these reactions, forming the reaction significantly faster. For instance, the availability of engineered nanomaterials (ENMs) in water can facilitate photochemical conversion of perfluoroalkyl acids (PFAAs) precursors to generate harmful byproducts like perfluorooctanoic acid (PFOA) and perfluorooctane sulfonate (PFOS) (Dai et al., 2020). These intermediates could have varying stabilities and reactivities that are significant in establishing the degradation process of pollutants. Their formation and interaction with other living organisms may decide the overall efficiency of photochemical reactions, i.e., intermediates' existence is required not only for further process development but also for environmental effects of photochemical kinetics. Intermediate mechanisms are described further by describing electron transfer (ET) reactions, which are usually a component of various biological processes. In respiration and other contexts, electron transfers are more than one, and this underlines the necessity of intermediates in bringing about such processes.

Quantum tunneling, with electron transfer distances larger than usual limits, finds a place in elucidating how intermediates are able to span gaps between redox cofactors within elaborate structures of proteins (Dempsey et al., 2010). These transition states have a tremendous impact on the rates of reactions, revealing that they are not transitory side products but integral components that allow efficient energy transformation. Their existence allows rapid transition states that govern reaction kinetics and stability, illustrating their central function within concerted biochemical pathways. Additionally, the importance of intermediates extends to the field of application in photochemistry.

The ability of ENMs to generate reactive oxygen species (ROS) implies that intermediates can be catalysts or promoters in speeding up rates of reactions of pollutants upon irradiation with light (Dai et al., 2020). The connection between intermediates, their stability, and their ability to adsorb on surfaces reveals how photochemical processes can be optimized and hence pave the way for new strategies in environmental remediation.

Identification of such intermediates' dynamics allows scientists to predict results in complex reaction environments and targets and establishes their relevance in basic research and applied science. This refers to the important function of intermediates in the overall kinetics of photochemical reactions and how they control strategies for more effective chemical conversions (Figure 9.8).

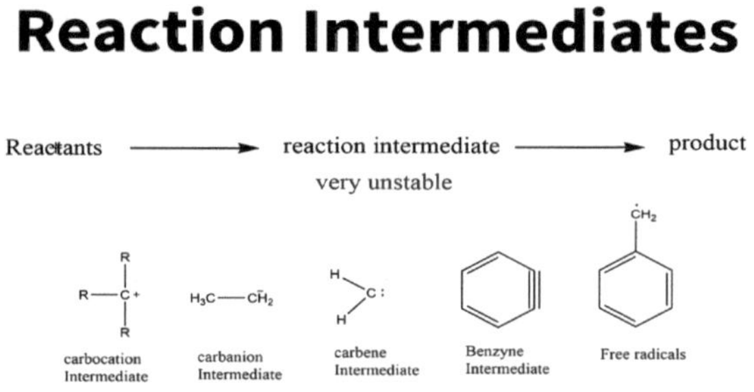

Figure 9.8. *Reaction intermediate.*

Source: https://chemistnotes.com/wp-content/uploads/2021/10/Reaction-Intermediates-1.jpg.

9.2.3. Pathways of Energy Transfer

The exploration of energy transfer pathways is focal in the analysis of photochemical kinetics, particularly for the process governing chemically initiated reactions by light. Core to such pathways are electron transfer processes that are attainable by direct excitation or sensitization. For instance, in N-(selenomethyl)alkyl-phthalimides, intramolecular electron transfer from the selenium center to the phthalimide functionality is responsible for energy transfer via signaling complex reaction pathways. The destiny of such energy transfers depends not only on the excitation state but also molecular architectures and substituents that can direct the reaction towards specific products. This hidden understanding of energy pathways supplements the predictive power toward photochemical system behavior and facilitates more nuanced approaches in synthetic and environmental chemistry. As we delve deeper into energy transfer mechanisms, reactive intermediates become increasingly central to the picture. The dynamics of proton transfer reactions, for example, are immensely important for the formation of many biradicals, which are critical precursors to cyclization reactions to yield selenium-heterocycles. The length of the alkyl spacer between the phthalimide and selenium groups determines which direction of proton transfer is favored, to expose researchers to the challenge of not only the efficacy of the energy transfer, but also its mechanistic implications. This intersection of energy dynamics and molecular structure is a manifestation of the delicate ballet of chemical change that photochemical kinetics strives to explain. Understanding these pathways is crucial for the synthesis of new materials and strategy for harnessing light for efficient chemical transformation and synthesis. When understanding energy transfer pathways is applied in the context of environmental problems, such as pollutants like perfluoroalkyl acids (PFAAs) and precursors, understanding of energy transfer pathways becomes crucial. Their persistence in the environment despite reductions in their production is evidence of the significance of photochemical processes to their degradation. Most importantly, the advent of engineered nanomaterials (ENMs) can significantly increase photochemical conversion of precursors for PFAAs by indicating the influence of reactive oxygen species on reaction rates. These processes mirror how engineered systems are able to remodel energy pathways to create more toxic byproducts. The interaction between photochemical kinetics, energy transfer processes, and environmental factors brings to the fore the complex interrelationships that need to be dealt with in scientific study and in applications, reflecting the polyvalent character of energy transfer in photochemistry (Figure 9.9).

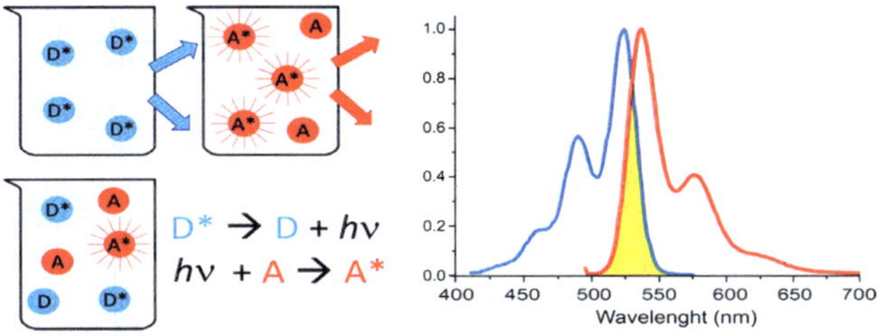

Figure 9.9. *Scheme of the trivial mechanism of energy transfer.*

Source: https://cdnintech.com/media/chapter/56771/1512345123/media/F3.png.

9.3. FACTORS INFLUENCING PHOTOCHEMICAL KINETICS

The kinetic photochemical dynamics is determined by many factors sensitively, which may have a significant influence on the rates of chemical reactions when exposed to light. One important factor is the light source intensity, as it determines what energy is available for photochemical processes. Increasing light intensity will tend to cause maximum excitation among reactant molecules, thereby making more instances of energy absorption and favoring rapid chemical conversions. In addition, the spectral composition of the light plays a critical role in determining reaction pathways. Different wavelengths are linked with different energy levels, and hence, selectively trigger specific photochemical reactions. Hence, understanding light intensity and spectral characteristics is crucial in order to make predictions about products of photochemical processes, particularly in intricate atmospheres when multiple reactive species may compete for energy and react concurrently. Temperature also emerges as a central determinant in photochemical kinetics. Increased temperatures have a tendency to accelerate the rates of unimolecular and bimolecular reactions, with ease of overcoming activation energy barriers. This is particularly in instances where gaseous reactants or volatile molecules are involved, where increased kinetic energy can lead to more frequent and effective collisions between molecules. Besides, temperature influences gas solubility in liquids and also the spatial distribution of reactants in heterogeneous systems, thereby the overall photochemical efficiency. For example, ambient temperatures during hot summer seasons can impact the concentration of pollutants, thereby altering local rates of ozone production, as observed from the evidence of the studies for $P(O_3)$ under conditions of high-NOx. Proper monitoring and regulation of temperature are thus required for optimization of photochemical processes. Finally, the chemical makeup of the reaction environment significantly influences photochemical kinetics.

The occurrence of reactants, intermediates, or inhibitors may either enhance or degrade reaction pathways. For example, volatile organic compounds (VOCs) co-emitted with nitrogen oxides (NOx) can easily influence the rate of ozone formation. Summer observations indicate discrepancies between model and measured rates of ozone formation, suggesting other contributing factors like peroxy organic radicals that remain to be fully investigated. This suggests an ongoing necessity for improvement of those chemical models predicting photochemical behaviors. Having an understanding of how environmental chemistry interacts with light and the absorption of energy is important to create reliable predictions and strategies towards the reduction of ozone and other photochemically initiated processes, which enables a more responsive approach towards addressing air quality issues in different geographic regions (Apel et al., 2017; Ebenhöh et al., 2010).

9.3.1. Wavelength and Intensity of Light

The correlation between wavelength and intensity of light is a significant domain of photochemical kinetics that determines the direction of reactions that are light-induced. Different wavelengths imply different energies, and therefore the photochemical effects are diverse. For instance, ultraviolet (UV) light, commonly characterized by its reduced wavelength, possesses sufficient energy to initiate reactions that cannot be induced by visible light. This is especially true for materials like polypropylene in which exposure to UV causes oxidative mechanisms. The discrete nature of energy carried by different wavelengths has a direct relation with reaction rate constants, and this means there must be a clear relationship between light properties and reaction kinetics. It is important to discern this relationship in the development of models that predict the behavior of photochemical systems at varying light intensities, hence enhancing our knowledge of material degradation and aging. The intensity of light is also an important parameter in photochemical reaction. Higher intensity not only increases the quantity of photons to interact with but can also enhance the activation of more than one reaction pathway. In accordance with the kinetic model of photothermal oxidation, photolysis is independent of temperature and proportional to absorbed energy from photosensitive species. The correlation demonstrates the significance of light intensity as a driving force of photochemistry.

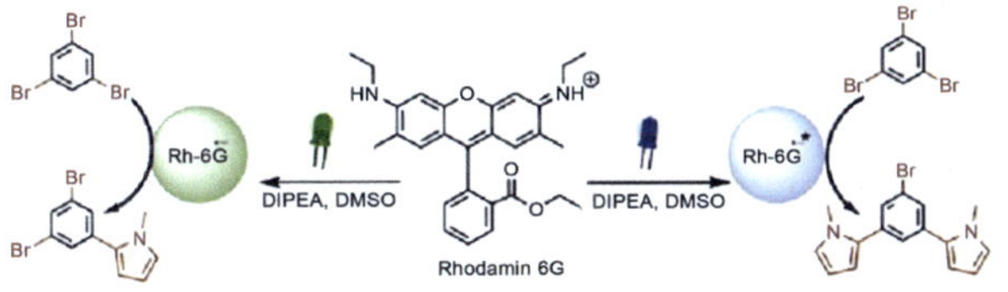

Figure 9.10. *Catalyst selection and wavelength of light.*

Source: https://www.perfectlight.com.cn/Uploads/UserFile/Image/1/20221129/638 57a09bc314.jpg.

Deviations from calculations might also be encountered at elevated irradiances due to multiphotonic excitations. Thus, a delicate consideration of intensity's interaction with wavelength is essential to proper photochemical kinetics modeling, particularly in estimating sunlight exposure degradation of polymers and other materials. Models integrating both intensity and wavelength are powerful tools to increase our understanding of photochemical kinetics. According to recent studies with the help of quantum theory, energy absorption-based criteria were developed to explain the initiation of oxidation reactions. These theoretical models have advanced the understanding of maximizing the efficiency of light-induced reactions, particularly in the process of polymer degradation processes.

For example, while reciprocity laws are useful guidelines for certain scenarios, more sophisticated kinetic models developed from elaborate mechanistic schemes turn out to be of superior utility in a broader set of conditions. The findings suggest the limitations of simpler models and verify that there is a need for stringent analysis of the manner in which wavelength and intensity interact in complex photochemical reactions, providing improved predictions of material response under illumination, as outlined by Colin et al. (2014a, b) (Figure 9.10).

9.3.2. Temperature and Pressure Effects

In photochemical kinetic experiments, pressure and temperature play key roles within the dynamics of chemical reactions, particularly in planetary atmospheres. Thermodynamic characteristics at various temperatures can potentially influence the rates of reactions as well as stability in chemical species. For instance, higher temperatures can facilitate the acceleration of photochemical reactions to yield complex molecules. On hot exoplanets like 51 Eridani b, the temperature limit which differentiates the activity of certain substances like sulfur is significant. The effective temperature on which sulfur condenses to solid particles is between 280 K and 320 K. Thus, temperatures higher than this can sustain gases that cause photochemical activity, leading to a dynamic and sulfur compound-filled atmosphere, according to Marley et al. (2016). Pressure dynamics also play an immense role in determining the chemical makeup of planetary atmospheres, though. Pressure fluctuations can change the concentration of the molecules and can cause departures from equilibrium conditions, particularly in the case of hot Jupiter. The experiments with HD 189733b and HD 209458b illustrate the way in which greater pressures can enrich mole fractions of certain gases, such as methane and ammonia. This phenomenon, known as transport-induced quenching, occurs when eddy diffusion alters vertical gradients of the molecules to generate concentrations greater than predicted by equilibrium theory. Therefore, both pressure and temperature are crucial to understanding how chemical species behave under non-equilibrium conditions, as elucidated in the work of Friedson et al. (2011). In addition, the temperature-pressure interaction is the source of the generation and processes of molecular radicals and other atmospheric constituents. Hot temperatures can facilitate the formation of reactive radicals such as OH, which are central to numerous photochemical processes. At the same time, differences in pressure can lead to the loss or creation of large atmospheric constituents, influencing the overall spectral signatures observed from exoplanets. For instance, although CO_2 will augment initial photochemical production, it ultimately becomes a casualty to destruction at greater altitudes due to the intense ultraviolet radiation characteristic of these environments. Therefore, the nuanced matrix of pressure and temperature influences basically delineates our understanding of photochemical kinetics and atmospheric chemistry of distant planets in concordance with the results from Marley et al. (2016) and the models suggested by Friedson et al. (2011).

9.3.3. Solvent Effects on Reaction Rates

The solvent is important in the regulation of rates of reaction in photochemical kinetics due to its interactions with the reactants and transition states of chemical reactions, mainly. Owing to the fact that solvents possess individual polarity and viscosity properties, they stabilize or destabilize some reaction intermediates. For example, charged transition states are stabilized in polar solvents, and this gives way to low activation energy, thereby faster rates of reaction. On the other hand, in non-polar solvents, the lack of solvation can lead to a detrimental energetic profile for charge species, which can slow down the reaction rate. Further, the solvent dynamics also govern the direction of the reaction; the solvent relaxation time may be at the same timescale as that of certain photochemical reactions, therefore influencing reaction efficiency and yield, a scenario applicable in electron transfer reactions described in protein redox machine reports like photosystem II and ribonucleotide reductase (Dempsey et al., 2010).

In addition, the use of additives or co-solvents can regulate reaction rates to a large degree by adjusting solvent characteristics, establishing a tailor-made environment for some reactions. For instance, co-solvent addition may tailor dielectric constants and thereby affect the extent of ionic species solvation. Modulation by this changes reaction kinetics by accelerating or hindering the formation of the transition state or intermediate. In photochemical systems, in which electron dynamics and energy transfer are of prime relevance, having information regarding the role of the solvent becomes even more crucial. The functioning of fluorophore-labeled probes in membrane models, for instance, illustrates how the solvent effects can intrude into antioxidant actions' kinetics and interpreting the intricacies of molecular interactions in heterogeneous media (Gramlich et al., 2004). Thus, the insight into the solvent effects is essential to the rate prediction of the reaction and the optimization of photochemical processes. Lastly, one must consider the reactivity trends developed in special solvent environments. Solvents can form an effect known as solvent polarity inversion, in which an enhancement of the solvent polarity enhances the reactivity of certain molecules in an unexpected manner.

This is demonstrated in the example of fluorophore-labeled membranes and antioxidants, where the reversal of the expected behavior was observed under various conditions of pH. The findings in such examples show the complexities involved behind solvent effects, leading towards the fact that reactivity largely depends upon the character of the solvents as well as its interaction with solute species. With continued research, employing refined techniques to study these effects will continue to enlighten us about solvent dynamics in photochemical kinetics and refine our control of reaction rates further, thereby increasing the range of application in fields ranging from materials science to biochemistry (Dempsey et al., 2010; Gramlich et al., 2004) (Figure 9.11).

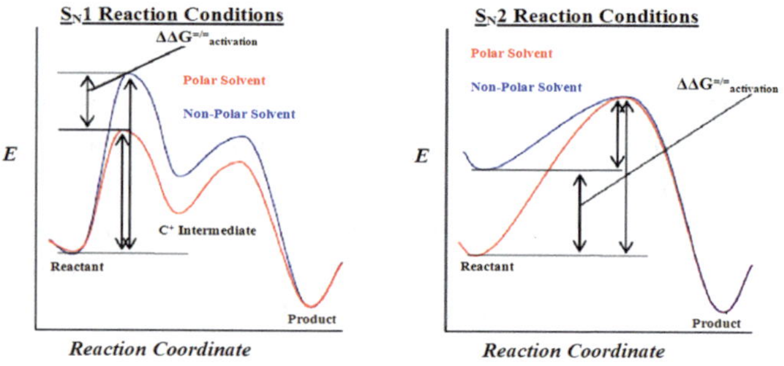

Figure 9.11. *Solvent effects on S_N1 and S_N2 reactions.*

Source: https://upload.wikimedia.org/wikipedia/commons/1/13/SN1_SN2_comparison.png.

9.4. EXPERIMENTAL TECHNIQUES IN PHOTOCHEMICAL KINETICS

Experimental techniques for photochemical kinetics must be investigated in an attempt to decide the mechanisms and dynamics of light-powered chemical reactions. Experimental techniques are numerous, from traditional spectrophotometry to advanced laser-induced techniques. Spectrophotometry, for instance, enables researchers to trace the light absorbed by a sample, providing knowledge on the time-dependent nature of different photochemical reactions. These, though, with progress in the field have been substituted by techniques such as laser flash photolysis, which have attracted global attention due to the generation and observation of reactive intermediates in real time. This acceleration in speed for data acquisition allows precise kinetic analysis to be obtained, giving vital insight into transient species pivotal to photochemical reactions, thereby increasing the overall understanding of such complex reactions. In addition to spectroscopic methods, utilization of femtosecond laser techniques is a great advancement in the study of photochemical kinetics. By the focusing of high-intensity laser pulses in liquid media, plasmas rich in reactive species like hydrated electrons and hydroxyl radicals can be generated. These radicals play a vital role in chemically facilitating changes, such as the reduction of metal ions. Recent research has identified that the introduction of specific additives, including liquid ammonia, can potentially impact the dynamics of such reactions significantly, controlling the formation of reactive intermediates. As an example, the concentration of ammonia will control the rates of metal ion reduction, as evidenced in Ag^{+} reduction to form silver nanoparticles with various properties. This study of radical chemistry illuminates the fine interplay between photochemistry and conditions of reaction, facilitating greater insight into complex reactions in solution. Furthermore, significant findings from recent studies highlight the importance of insight into photochemical mechanisms on a larger scale, as regards their relevance for atmospheric chemistry and prebiotic processes. The inclusion of nitriles within photochemical aerosols, initiated by ultraviolet wavelengths, shows the effect of light-matter interactions on complex systems. As detailed, the reaction with CH radicals generated by photolysis of methane provides insights into how reactive intermediates contribute towards the generation of molecular complexity. The findings underscore the significance of experimental approaches in understanding mechanisms of photochemical reactions on the planet and in extraterrestrial systems. By employing sophisticated experimental methods, researchers can gain important information about mechanisms of reactions and the rules governing them and, as a result, become better at understanding photochemical kinetics and its application in various fields of science (Ahsan et al., 2018; Jimenez et al., 2012).

9.4.1. Spectroscopic Methods

Spectroscopic methods form the core of photochemical kinetic studies, providing useful information about molecular transitions and reaction dynamics. Fluorescence spectroscopy, ultraviolet-visible (UV-Vis) absorption spectroscopy, and nuclear magnetic resonance (NMR) spectroscopy are some of the most important methods through which researchers can explore the energy states and activities of excited molecules. An example is the fluorescence spectroscopy, which is useful in the uncovering of the photochemical nature of proteins and other large molecules. Low-temperature fluorescence spectroscopy analysis of phytochromes underscored activation barriers and quantum yields of various mutations, representing the significance of spectroscopic approaches in elucidating the correlation between molecular structure and photochemical activity (Essen et al., 2014). By enabling researchers to detect and quantify transient species, the methods dramatically enhance the understanding of the kinetics of light-induced reactions. Ultraviolet-visible spectroscopy (UV-Vis) is also a critical tool in photochemical kinetics, as it allows observation of the absorption characteristics of a molecule under illumination. By tracking changes of absorbance with time, researchers can infer rates of reaction and mechanisms of photochemical reactions. For instance, N-(selenomethyl)alkyl-phthalimides experiments demonstrated how photolysis and intramolecular electron transfer yield characteristic products based on reaction pathways initiated by light absorption. Acetone-sensitized or direct excitation highlighted the significance of spectroscopic assessments in visualizing these processes (Argüello et al., 2015). Such observations not only advance our rudimentary understanding of photochemical reactions but also provide a window for new applications in materials science and photobiology, where reactions photochemically driven must be controlled. Furthermore, compatibility with spectroscopic methods provides broad latitude to investigate the photochemical kinetics of different systems and so allow the probing of both intrinsic molecular dynamics and extrinsic environmental influences.

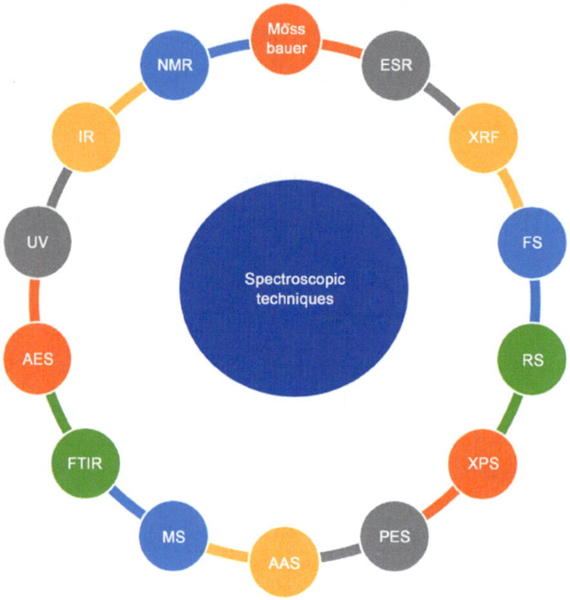

Figure 9.12. *Types of spectroscopic techniques.*

Source: https://scienceinfo.com/wp-content/uploads/2023/08/image-160.png.

The capacity to observe reaction intermediates in real-time holds profound implications for the development of more effective photochemical systems. For example, the above research describes the crucial role played by molecular interactions and structural properties in determining reaction routes and efficiencies, as illustrated in the investigation of Pr-to-Pfr photoconversion in phytochromes (Essen et al., 2014). Such results indicate that structural modification at individual sites on the molecule can dramatically influence photochemical efficiency and indicate the delicate interdependence of function and structure in photochemical kinetics. Overall, the collaborative application of multiple spectroscopic methods makes significant contributions to the establishment of the field and towards a better appreciation of light-induced chemical processes (Figure 9.12).

9.4.2. Time-Resolved Techniques

Time-resolved techniques play a pivotal role in enhancing understanding of photochemical kinetics by enabling scientists to observe fast molecular dynamics as they occur. The most significant among these are time-resolved laser gain-versus-absorption spectroscopy and time-resolved Fourier transform infrared (FTIR) emission studies. By applying these techniques, scientists can capture transient species generated under photochemical reactions and obtain valuable information on the dynamics of excited states. For instance, tunable diode and F-center lasers are employed in laser gain-versus-absorption technique, which yields precise photofragment yield measurements and their kinetics. These observations are vital in investigating complex reactions where sudden events may have an impact on the overall kinetics and product. Moreover, the data obtained serve as a starting point for extrapolating kinetic models that are significant in the study of photochemical processes both in laboratory conditions and planetary atmospheres, such as studies on polycyclic aromatic hydrocarbons and cometary emissions (Leone et al., 1990). The use of time-resolved methodologies in photochemical studies has immensely enhanced our knowledge of pathways of mechanisms following light-induced reactions. For instance, time-resolved spectroscopy-obtained behavior makes it possible for scientists to monitor intermediate processes like electron transfer and proton transfer. During studies involving N-(selenomethyl)alkyl-phthalimides, scientists applied such new methods to check that reactions proceeded through direct excitation and subsequent interfacial electron transfer between phthalimide moiety and selenium. This is significant since it provides a model for the explanation of the ensuing behavior of formed biradicals, as well as folding into cyclization products. Surprisingly, the efficiency of electron transfer pathways can be influenced by the structure of the molecule involved, so that careful substrate selection in photochemistry can lead to diverse kinetic response (Argüello et al., 2015). These results show the importance of applying time-resolved measurements to develop sound kinetic models that actually describe photochemical processes. As further advancement is made on time-resolved approaches, their application will be spread to other areas aside from traditional photochemistry, for example, planetary science and materials science. The possibility of measuring transient states with caution creates room to explore systems with non-stationary processes, particularly in complex environmental situations. This aspect is particularly important in research on planetary atmospheres, where the unique molecule-molecule interactions can give rise to diverse kinetic pathways. The new information provided by time-resolved techniques not only contributes to basic molecular dynamics information but also holds promise for practical applications, e.g., in designing photochemical material for maximum efficiency in energy conversion or in pollution abatement. As scientists continue to refine and implement these techniques, their research is to remain key in bridging new windows within the broader field of photochemical kinetics and thereby completing gaps in understanding and spurring creativity (Leone et al., 1990).

9.4.3. Computational Modeling Approaches

The growth of computational model techniques has significantly enhanced our understanding of photochemical kinetics, with complex chemical processes simulated and analyzed by robust frameworks. Computational techniques allow one to investigate reaction mechanisms that propel light-induced chemical conversions, and scientists can foresee the outcome of a set of photochemical reactions under various environmental conditions. With the implementation of sophisticated algorithms and simulation models, researchers are able to effectively simulate molecular behavior in a system, thus developing correlations between molecular interactions and reaction rates. Not only does the deployment of computational analysis reduce experimental data analysis to ease, but it also allows the identification of theoretical conditions that may not even be available experimentally in the first instance. In planetary science, computational modeling methods have been invaluable when trying to determine the atmospheric state of exoplanets, particularly hot Jupiter. An example is the SCARLET model, which is already being employed to forecast transmission spectra of such planets and obtain their elemental compositions. Such models incorporate statistical treatment of observational errors along with advanced atmospheric modeling in order to achieve more precision in measuring important parameters, like carbon-to-oxygen ratios. The findings show that these ratios are always less than some threshold value of 0.9 for some examined exoplanets, pointing to the robustness of computation models in describing atmospheric chemistry pertinent to photochemical processes. Such modeling results in greater understanding of planetary atmospheres as well as the broader astrophysical significance of photochemistry in diverse environments (Benneke et al., 2015). Also, the use of computational modeling is in fields such as the interstellar medium, wherein it assists in comprehending how various molecular species are produced through gas-phase reactions and surface chemistry on interstellar dust. New surveys are highlighting how accurately modeling rate coefficients for important reactions influencing molecular abundances is important. Sensitivity studies in the range of environments, such as cold dense clouds, point out what processes must be improved with regard to modeling accuracy. Adding gas-grain interactions to models provides a higher predictive capability, making it possible to build a more comprehensive picture of chemical evolution in astrophysical systems. As researchers continue to develop these frameworks, the impact of emerging laboratory experiment data will continue to refine our theoretical predictions, closing the gap between computational models and observations (Agundez et al., 2010).

9.5. APPLICATIONS OF PHOTOCHEMICAL KINETICS

The implications of photochemical kinetics are far-reaching and profound in various disciplines such as medicine, environmental science, and industrial processes. Photochemical kinetics plays a significant role in the creation of treatment methods in medicine, particularly for extremely malignant tumors like glioblastoma multiforme (GBM). As traditional therapies fail to remove tumor cells that infiltrate the surrounding tissues, new approaches such as photodynamic therapy (PDT) are being explored. PDT employs photosensitizers activated by light to cause cytotoxicity in cancer cells, thus providing a targeted therapy. Experimental research has proven that the combination of PDT with chemotherapeutic agents can immensely amplify therapeutic efficacy as evident in *in-vitro* and *in-vivo* models that demonstrate synergistic outcomes more than using individual modality, as noted in recent studies involving glioma cell lines, showing the promise of photochemical kinetics towards cancer treatment (Singh et al., 2018). Outside the health sciences, photochemical kinetics is of paramount importance in the science of the environment and particularly in understanding atmospheric processes. A particular and pertinent example of such processes is the action of chlorofluorocarbons (CFCs) and airplane emissions on stratospheric ozone. Stratospheric models, based on photochemical kinetics, attempt to simulate and forecast the actions of such chemicals on ozone depletion and global warming. By comparing model predictions with observations collected over the past 15 years, researchers can verify the precision and reliability of such models. The data obtained from such comparisons are vital in informing environmental policy and constructing a more comprehensive portrayal of climate-chemistry interactions, placing a high priority on accurate photochemical modeling in the reduction of global environmental concerns (Prather et al., 2025). Photochemical kinetics finds application in process chemistry in the development of more energy-efficient, sustainable processes. One of the leading fields is in the synthesis of fine chemicals, where photochemical processes yield unique pathways to product formation under milder conditions than conventional thermal pathways. Light-initiated initiation and propagation of chemical reactions provide a mechanism of lowering energy consumption and preventing toxic byproducts. Besides, research into the enhancement of photochemical reactions, such as photocatalysis, confirms practical applications of photochemical kinetics in the production of cleaner fuels and materials. With companies transitioning towards the adoption of greener processes, the principles of photochemical kinetics are expected to become increasingly vital in contributing towards the achievement of sustainable development goals in various industries.

9.5.1. Photovoltaics and Solar Energy Conversion

The development of photovoltaics (PV) is a critical step in the path to sustainable conversion of energy. Sunlight's colossal energy potential, which exceeds human annual usage within an hour, has encouraged researchers to emphasize pushing the conversion efficiency beyond traditional limits. The Shockley–Queisser (SQ) limit offers a theoretical limit to single-junction PV devices' efficiency, typically around 30%. Nevertheless, the manufacturing of high-efficiency crystalline silicon (c-Si) solar cells, whose efficiency approaches this limit, poses challenges to cost-effectiveness and material sustainability. As solar energy production expands, affordability and accessibility of such technologies are crucial to their adoption in the world's energy mix (Cheng et al., 2012). Advanced techniques are being searched for to bridge the SQ barrier and reduce the cost-to-efficiency ratio of PV devices. Among the most hopeful techniques is hydrogenated amorphous silicon (a-Si:H), which can attain high efficiencies with much thinner film layers, enhancing charge carrier extraction. However, a-Si:H has its drawbacks too; its high optical threshold of 1.7–1.8 eV and light-induced degradation sensitivity impose necessary performance limitations. In order to achieve maximum efficiency, the absorber layer thickness needs to be optimized because thin films can enhance stability but also contribute to increased losses in absorption, particularly in the red region of the solar spectrum. This contradiction manifests the complexity of material selection and design strategies in photovoltaic research (Cheng et al., 2012). Alternative materials research, such as dye-sensitized and organic cells, further complicates the field of solar energy conversion. Although these materials are flexible and potentially less expensive to produce, their inherent disadvantages also replicate those of traditional technologies, namely the inability to effectively collect low-energy photons. As the solar energy field continues to mature, an integration of photochemical kinetics becomes necessary. As an example, knowledge of the kinetics of charge separation and recombination in most materials could be used to advance more effective photovoltaic systems. As researchers continue challenging conventional wisdom with revolutionary advances in solar technology, adding efficient photochemical processes will be key to shaping energy conversion paths in the future (Figure 9.13).

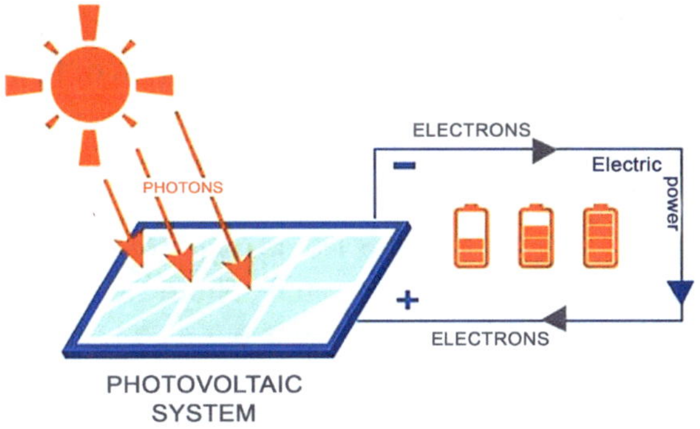

Figure 9.13. *Solar energy conversion.*

Source: https://cdn.prod.website-files.com/65b217df08a6f1f6aba3d4ce/681b30f43c936079d3ea0002_pasted-image-0--5--compressed--8-.jpeg.

9.5.2. Phototherapy in Medicine

Phototherapy in medicine has become a groundbreaking method, employing light to create specific biochemical responses within the body. With carefully controlled wavelength, intensity, and light exposure duration, clinicians are able to induce beneficial responses that target the treatment of a vast array of health conditions, including skin diseases, mood disorders, and even certain cancers. This treatment technique is utilizing the fundamental principles of photochemistry as the interaction of light with biological tissue gives rise to photochemical reactions, whose effects can be to drive or suppress cellular processes. Furthermore, advances in phototherapy have also been enhanced through novel photoactivatable molecules, allowing selective and safe in-clinic use. The use of such techniques is an emphasis of the interdisciplinary nature of modern medicine in bringing together chemistry, biology, and medicine under one discipline aimed at new therapeutic methods. Metal complexes of phototherapy indicate the potential of photochemical kinetics in medical interventions. Such complexes, e.g., ruthenium and iridium-based, possess favorable luminescent properties that confer precise imaging and treatment of targeted tissues. Recent research has highlighted the use of such metal-based complexes to bring about photorelease of biologically active molecules with enormous potential in site-directed treatment, particularly in cancer treatment. With the help of ultrafast-pulsed radiation techniques and advanced computational methods, scientists are unraveling the photochemical process mechanism at short time scales. This early understanding is the basis for the development of future phototherapeutics that may provide improved and less invasive treatments and thus revolutionize the management of disease. Successful design of phototherapeutic agents depends on an in-depth understanding of the interaction of light with biological systems by combining concepts of medicinal chemistry and structural biology. As scientists unravel complex biological mechanisms and determine novel targets for therapy, the need for interdisciplinarity becomes more and more crucial. Based on existing knowledge from a variety of disciplines—such as molecular pharmacology, physiology, and photochemistry—scientists can design molecules that not only react to light but also have therapeutic effects with specificity. Furthermore, the synergy of these advances with conventional therapeutic modalities promises a shining future to personalized medicine, in which phototherapy can be tailored to specific patient needs. The convergence of these emerging trends is symptomatic of the ongoing advancement in the evolution of phototherapy as a medical intervention, in light of its broader place in the subject of photochemical kinetics (Bruijnincx et al., 2009; Sadler et al., 2013) (Figure 9.14).

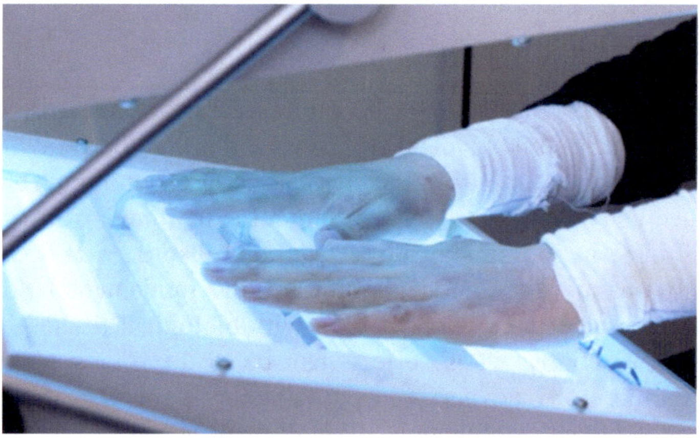

Figure 9.14. *Phototherapy treatment on hands.*

Source: https://emedia1.nhs.wales/CAVUhb/cache/file/0FB9E995-B2C6-4A8A-A150E8BCDA5E0D05_featureblockimage.jpeg.

9.5.3. Environmental Photochemistry

The environmental photochemistry is indicative of its critical role in the atmospheric chemistry and ecosystems on our planet. Photochemical reactions are initiated when solar radiation interacts with atmospheric components and creates various chemical species that influence air quality and climate. Among such significant processes, there is photolytic destruction of pollutants like nitrogen oxides, volatile organic compounds, and particulate matter. The role of hydroxyl radicals in this process can be exceptionally important, since they are generated by UV radiation interacting with water vapor and play a significant role as an oxidant in the atmosphere. As such, it is necessary to assess these processes, specifically when establishing the impact of human activities that cause air pollution and global climate change. Such considerations emphasize the importance of studying these kinetics in an effort to effectively model and mitigate environmental impacts.

Moreover, the effect of some pollutants like fluorides on vegetation illustrates the overall relevance of environmental photochemistry. Research indicates that the dry deposition of fluorine compounds disrupts plant metabolism and eventually contributes to ecosystem health. This disturbance is typically discussed in the context of gaseous hydrogen fluoride and particulate phase, with a focus on the complexity of such interaction. When looking at how precipitation affects fluoride deposition, one should consider how its mechanism for elution of deposited pathogenic fluoride from leaves can be considered. Such degradation reactions show how photochemical kinetics is able to account for plant response to environmental stressors, with implications that can help develop environmental protection schemes. Understanding the intricate interaction of pollutants with biological systems is necessary to support sustainable practices and assure food security. Photochemical kinetics extends beyond Earth, as seen under remarkable atmospheric conditions on other worlds such as Titan. Nitriles added to Titan aerosol result from complex hydrocarbon chemistry triggered by extreme UV wavelengths, a fascinating analogue to Earth's photochemistry.

Notably, the reaction of methane radicals produced by methane photolysis can be significant to nitrogen activation within Titan's atmosphere. The interplanetary comparison reveals the significance of photochemical kinetics within the general framework of astrobiology and the potential for prebiotic chemistry elsewhere in the universe. By observing these extraterrestrial processes, scientists can gain precious insights into the beginnings of life and atmospheric evolution on planetary bodies, further enhancing the importance of environmental photochemistry in unveiling both terrestrial and extraterrestrial contexts.

9.6. CHALLENGES AND LIMITATIONS IN THE FIELD

The photochemical kinetics is hampered by enormous challenges that can drag advances in research and clinical application. One of the chief limitations arises from the inherent complexity of the photochemical reactions in question. Light interaction with various chemical substrates can give rise to a number of pathways that yield a variety of products making quantitative analysis challenging. Furthermore, the effectiveness of such reactions is often highly reliant on extrinsic factors, such as temperature, pH, and solvation interactions. Understanding these nuanced dynamics is crucial to optimizing photochemical processes, particularly in biomedicine applications, like photodynamic therapy (PDT). As with recent reviews, novel laser technologies, such as quantum-dot laser diodes, are of enormous promise for addressing some of these challenges. Notwithstanding this, quantifying the whole kinetic profile of singlet oxygen generation and its bioactivity remains a challenge in the area (Bragina et al., 2017). Modulation of light absorption and penetration by various mediums, particularly biological tissues, remains a main challenge in photochemical kinetics. Light attenuation due to absorption and scattering limits the effectiveness of therapeutic applications, a concern that takes on special significance in PDT, whose singlet oxygen effectiveness in targeting cancer cells is decisive. The variations in optical properties among different tissues can result in inconsistent therapeutic action. The discovery of new laser systems focusing on the near-infrared (NIR) region offers a solution to enhance penetration of light. However, in spite of the technological progress, the optimization of the wavelength and the duration of exposure remains complexity due mainly to the effect of the surrounding biological environment on absorption coefficients. Constant optimization and research of the mentioned parameters are required in order to realize the maximum clinical effect of photochemical approaches (Bragina et al., 2017). Lastly, conversion of the volumes of scientific information and utilization as useful applications is also a series of complexities. The combination of interdisciplinary knowledge, from molecular biology to medicinal chemistry, is crucial for the successful design of new drugs that take advantage of photochemical kinetics. The challenge is not just to design molecules that will benefit from interaction with targeted biological processes, but also to ensure that they do so without losing therapeutic effectiveness and exacerbating toxicity. As is emphasized in current literature, successful drug discovery is highly reliant on a deep understanding of the governing chemical and biological processes at work. Such a sophisticated network of required expertise requires cooperation between disciplines, useful as that may be, while also with it comes logistical and communications problems. The further researchers explore these waters of interdisciplinary study; there continues to be a demand for innovative solutions to link theoretical constructs and practical application in the drug development process.

9.6.1. Complexity of Reaction Mechanisms

The complexity of reaction mechanism in photochemical kinetics is apparent in the multifactorial processes controlling chemical reactions in various environments. It takes an appreciation of such mechanisms to consider cautiously the intricate interaction among diverse species in sequences of reaction. In the Earth and extraterrestrial atmospheres, for instance, the underlying chemical cycles reveal a cosmopolitan nature connecting different planetary systems. The well-established earth chemistry of hydrogen, oxygen, nitrogen, and chlorine serves as a yardstick to compare the kinetics of more complicated systems on other planets. Still, there is less defined chemistry in terms of sulfur because there are considerably larger uncertainties regarding our understanding of complicated reaction mechanisms that can be crucial in atmospheric chemistry. The impact of heterogeneous reactions arrives as a further complication because such interactions have the potential to create aerosols and to the ozone hole phenomenon, indicating the need for extensive research in chemical kinetics (Yung et al., 1990). Further research on photochemical processes occurring in other planetary atmospheres arrives as a witness to the complexities involved in modeling these reaction schemes. The case of 51 Eridani b illustrates how sulfur photochemistry plays a significant role even when early studies focused on carbon compounds. This finding indicates that there must be models explaining more species and their interaction during photochemical reactions. The modeling indicates that various parameters, such as temperature and composition of the atmosphere, play an important role in the creation and characteristics of inorganic compounds like elemental sulfur. Since temperatures control the phase of sulfur—vapor to solid—the findings illustrate the fine nuances of reaction mechanisms and consequences under varying environmental conditions. Hence, there is a need to understand such dynamics in making predictions for atmospheric behavior between various planetary systems, thus widening the overall scope of planetary atmosphere research (Marley et al., 2016). Implications of complex reaction mechanisms extend beyond the common chemical reactions; they have far-reaching implications on our understanding of planetary evolution and habitability. Whereas chemical kinetics illustrates how minor variations in conditions can cause a reaction to take different directions, this knowledge becomes important in astrobiological studies. For example, the potential for organic hazes on other worlds and their importance to atmospheres necessitates an advanced level of understanding how light, chemical compounds, and temperature interact. These are complications that demonstrate that the evolutionary paths of planets may be based on such subtle chemical balances in creating their atmospheres and hence their capacity for life. The sustained modeling and research in photochemical kinetics continue to unveil these complexities, making connections between chemical mechanisms and astrophysical phenomenology in general, thus highlighting the importance of intense studies in the field (Marley et al., 2016; Yung et al., 1990).

9.6.2. Limitations of Current Experimental Techniques

In photochemical kinetic studies, limitations of current experimental techniques are the principal hindering factor in increasing knowledge of molecular interaction and reaction. One of the basic issues is that there is difficulty in obtaining strict control on reaction conditions, such as temperature, pressure, and light intensity. Changes in these factors make results unsystematic and thereby difficult to interpret reaction behavior. Besides, traditional methods tend to be insensitive to detect species formed in photochemical reactions that are fleeting. This is vital when carrying out rapid reactions' studies where ordinary methods will not capture the transitory nature of the intermediates, and hence an incomplete picture is given of the mechanism involved. It is thus apparent that such experimental limitations in terms of precision and sensitivity are serious setbacks to extensive study work in photochemical kinetics. A second major limitation derives from the very inherent character of biological systems, so that experimental approaches find it challenging to reproduce the *in vivo* situation. Application of laser technology, even as revolutionary, is limited with regards to tissue penetration and specificity in addressing biological phenomena such as singlet oxygen generation. For instance, as lasers become more widely used in medicine, their potential is hampered by limited penetration depth of light and photosensitizer toxicity, and the development of more efficient methodologies becomes a necessity (Bragina et al., 2017). Added to that is the complexity of interaction between photochemical reactions and cellular structures, which cannot be accurately modeled in regular laboratory conditions. Consequently, most current experiment approaches give results that cannot be applied directly with a correlation to practical use, hence leaving an implication of limited application on broader research outcomes in photochemical kinetics. Finally, rate coefficient uncertainties of important reactions, especially in more complex chemical settings, represent another challenge in experimental approaches. Recent evaluations have identified that current models are often incomplete in data, particularly if reactions under astrophysical environments are being examined, such as in the interstellar medium (Agundez et al., 2010).

Insufficient experimental confirmation of the models might lead to erroneous calculations for molecular abundances and reaction kinematics. Additionally, the difficulty of simulating the interstellar conditions in laboratory setups only compounds this issue, where much is still poorly understood. Consequently, the limitations of today's experimental techniques restrict not only the accuracy of kinetic models but also the range of drawing informative conclusions about the chemical processes in complex systems. As photochemical kinetic studies proceed, it will be important to overcome these limitations for their expansion with the purpose of enhancing experiment results' reliability and practicality.

9.6.3. Theoretical Challenges in Modeling

One of the most pressing theoretical issues in photochemical kinetics is the adequate representation of complex chemical reactions in varied atmospheric conditions. Designing good models is impeded by the scarcity of the chemical kinetics database, particularly in planetary atmospheres, where a great deal remains poorly understood. For instance, the photochemistry of light hydrocarbons is relatively well mapped in the atmospheres of Jovian planets and Titan, but higher hydrocarbon and nitrogen compound activities remain enigmatic. The gaps in our understanding prompt scientists to rethink the structure used for modeling and emphasize necessity in refining methods to include heterogeneous reactions. This will ultimately further our knowledge of the way that gas-phase chemistry results in aerosol formation and other atmospheric effects (Yung et al., 1990). The synthesis of molecules in the interstellar medium is also a significant theoretical problem for which modeling is both conceptually and practically challenging. A number of gas-phase reactions that contribute to the richness of molecules within different astronomical environments have been identified by researchers, such as dense molecular clouds and circumstellar envelopes. However, the inaccuracies in rate coefficients for primary reactions cast a ripple effect on these models' predictive capability. Conducting sensitivity analyses of these reactions can enlighten us as to which processes are most critical in establishing molecular abundances. In addition, the presence of interstellar dust introduces a level of complexity as it alters molecular formation and transformation dynamics. Weighing in the physical interactions of dust into chemical models requires new approaches that both incorporate gas-phase and gas-grain interactions (Agundez et al., 2010). Finally, challenges in photochemical kinetics modeling also include environmental variables influencing the reaction result. The existing models will often fail in capturing the impact of outside parameters such as temperature, pressure, and radiation intensity on reaction kinetics. This deficit is particularly pertinent in planetary atmospheres, where conditions can be wildly variable.

Moreover, the need for adequate information on reaction mechanisms under various conditions necessitates experimental confirmation as well as theoretical models. There is then a resulting need not only for increasingly sophisticated modeling techniques but also for interdisciplinarity in order to elicit empirical data that can hone predictive algorithms. Resolving these theoretical issues could be a solution to comprehending not only photochemical kinetics, but also more general chemical dynamics in numerous surroundings (Agundez et al., 2010; Yung et al., 1990).

SUMMARY

In concluding our survey of photochemical kinetics, it is important to note the significant consequences such processes have on our knowledge of planetary atmospheres. The interaction between thermochemical processes and photochemistry has been demonstrated to determine the atmospheric makeup of exoplanets like HD 189733b and HD 209458b. As in our research, disequilibrium chemistry ensuing from transport-induced quenching profoundly changes the vertical profiles of molecular species. These methanate and ammonia upgrades to equilibrium levels point toward the introduction of kinetic models with transport as well as photochemical impacts. Subsequent research, therefore, should focus on the importance of these interactions in a bid to develop better prediction models for atmospheric composition of other bodies, particularly under conditions like hot Jupiter (Friedson et al., 2011). Furthermore, conclusions gathered from investigation of nitrogen-containing species, particularly in planetary atmospheres like Jupiter, reveal complexity that photochemical dynamics introduce into chemical equilibria. Through the analysis, it is observed that traditional models that draw expectations from equilibrium assumptions have the potential to miss significant pathways that control the abundance of vital molecules like hydrogen cyanide (HCN). As revealed by our findings, nitrogen species coupling mechanisms are significantly influenced by thermal and photochemical separation, which means observational constraints effectively assist in improving our kinetic models. This underscores the need for interdisciplinary approaches that synthesize observational data with photochemical kinetics to harmonize differences in predicted species abundances (Achterberg et al., 2010). Lastly, the broader implications of photochemical kinetic knowledge extend beyond the atmospheres of the individual planets and into planetary habitability and astrobiology. By delineating the intricate chemistry involved under diverse atmospheres, scientists can better conjecture concerning potential biosignatures and the processes that sustain life. As our models become ever more precise and we continue to gather additional data, it is clear that the study of photochemical kinetics will remain significant in theoretical and experimental work. Ultimately, broadening our knowledge in this field could result in the discovery of new, Earth-like exoplanets and increase our understanding of the fundamental processes by which atmospheric evolution occurs throughout the universe. Photochemical kinetics gives the dramatic revelation of the interactions between chemical reactions and light, particularly in atmospheric contexts. Modeling stratospheric chemistry is among the areas of significance, where empirical findings over the past 15 years have been pivotal. These observational findings serve the purposes of reference in calibrating and tuning the credibility of various stratospheric models simulating the dynamics of ozone depletion and the impact of chlorofluorocarbons. As noted by the Workshop on Stratospheric Models and Measurements, the comparisons have brought to the fore the necessity for accurate stratospheric observation to determine predictive models' reliability, a requirement essential for environmental analysis (Prather et al., 2025). Therefore, understanding these chemical processes in the stratosphere is essential to addressing big environmental problems, like climate change and preservation of the ozone layer. Besides the verifications of models, there has been enormous progress in comprehending the upper atmosphere of Earth through continuous research funded by NASAs Upper Atmosphere Research Program. The report series of the agency condense goals and accomplishments and are new findings related to stratospheric composition and structure. Of interest, the assessments provide quantitative estimates of the manner in which ozone levels are expected to alter due to anthropogenic emissions. The inclusion of reviews of future atmospheric loading of chlorine and bromine serves only to underscore the relevance of photochemical kinetic observation and investigation into the future. Such research not only facilitates a sophisticated comprehension of stratospheric chemical processes but also informs policy and regulatory response to mitigate environmental impact (Kurylo et al., 2025). Further, current research has recognized the intricate interplay of photochemical kinetics and emerging technologies, such as the stratospheric supersonic aircraft. With the evolution of environmental studies, the potential implications of introducing new aircraft into the stratosphere come under keen examination. These technologies have been shown to have an interactive function of altering the stratospheric chemical surroundings, particularly concerning mechanisms involving ozone destruction and greenhouse gas release. Such assessments indicate an understanding of photochemical reactions in depth to be at the fore of informing future environmental and aerospace policy developments. With the integration of understanding from international assessments and ongoing research, stakeholders are adequately positioned to intervene in the complex relationships between technological advancement and air health, focusing on environmental concerns amidst innovation (Kurylo et al., 2025).

REFERENCES

1. Achterberg, R. K., Allen, M., Allen, J. E., Atkinson, D. H., Atkinson, D. B., Atkinson, R., Atreya, S. K., et al. (2010). On the abundance of non-cometary HCN on Jupiter. http://arxiv.org/abs/1006.3508 (accessed on 10th September 2025).
2. Agúndez, M., Cuppen, H. M., Geppert, W., Herbst, E., et al. (2010). Reaction networks for interstellar chemical modelling: Improvements and challenges. http://arxiv.org/abs/1011.1184 (accessed on 10th September 2025).
3. Ahsan, S., Frias Batista, L. M., John, M. G., Meader, R., et al. (2018). Radical chemistry in a femtosecond laser plasma: Photochemical reduction of Ag^+ in liquid ammonia solution. https://core.ac.uk/download/215485623.pdf (accessed on 10th September 2025).
4. Al-Abadleh, H. A. (2015). A review on the bulk and surface chemistry of iron in atmospherically-relevant systems containing humic like substances (HULIS). https://core.ac.uk/download/143689511.pdf (accessed on 10th September 2025).
5. Apel, E., Baier, B. C., Blake, D., Brown, S. S., et al. (2017). Higher measured than modeled ozone production at increased NOx levels in the Colorado Front Range. https://core.ac.uk/download/287622642.pdf (accessed on 10th September 2025).
6. Argüello, J. E., Heredia, A. A., Oksdath Mansilla, G., Peñeñory, A. B., et al. (2015). Photochemistry of *N*-(selenoalkyl)-phthalimides: Formation of N,Se-heterocyclic systems. https://core.ac.uk/download/159300430.pdf (accessed on 10th September 2025).
7. Benneke, B. (2015). Strict upper limits on the carbon-to-oxygen ratios of eight hot Jupiter's from self-consistent atmospheric retrieval. https://core.ac.uk/download/216243642.pdf (accessed on 10th September 2025).
8. Bragina, O. A., Gekaluyk, A. S., Goltsov, A., Rafailov, I., et al. (2017). Laser-induced generation of singlet oxygen and its role in the cerebrovascular physiology. https://core.ac.uk/download/228178032.pdf (accessed on 10th September 2025).
9. Bruijnincx, P. C. A., & Sadler, P. J. (2009). Controlling platinum, ruthenium, and osmium reactivity for anticancer drug design. https://core.ac.uk/download/44701.pdf (accessed on 10th September 2025).
10. Cheng, Y. Y., Clady, R. G. C. R., Crossley, M. J., Danos, A., et al. (2012). Improving the light-harvesting of second-generation solar cells with photochemical upconversion. https://core.ac.uk/download/199433580.pdf (accessed on 10th September 2025).

11. Colin, X., Desnoux, E., François-Heude, A., Richaud, E., et al. (2014a). Influence of temperature, UV-light wavelength and intensity on polypropylene photothermal oxidation. https://core.ac.uk/download/29137377.pdf (accessed on 10th September 2025).
12. Colin, X., Desnoux, E., François-Heude, A., Richaud, E., et al. (2014b). A general kinetic model for the photothermal oxidation of polypropylene. https://core.ac.uk/download/29137950.pdf (accessed on 10th September 2025).
13. Dai, Y., Guo, X., Hoffmann, M. R., Wang, Y., et al. (2020). Photochemical transformation of perfluoroalkyl acid precursors in water using engineered nanomaterials. https://core.ac.uk/download/323085206.pdf (accessed on 10th September 2025).
14. Dempsey, J. L., Gray, H. B., & Winkler, J. R. (2010). Proton-coupled electron flow in protein redox machines. https://core.ac.uk/download/4887683.pdf (accessed on 10th September 2025).
15. Ebenhöh, O., Houwaart, T., Lokstein, H., Schlede, A., et al. (2010). A minimal mathematical model of nonphotochemical quenching of chlorophyll fluorescence. https://core.ac.uk/download/177276.pdf (accessed on 10th September 2025).
16. Essen, L. O., Feilke, K., Hughes, J., Kopycki, J., et al. (2014). Tyrosine 263 in cyanobacterial phytochrome Cph1 optimizes photochemistry at the prelumi-R→lumi-R step. https://core.ac.uk/download/30729823.pdf (accessed on 10th September 2025).
17. Friedson, A. J., Showman, A. P., Agol, E., Ayers, T., Baraffe, I., Barman, T., Beaulieu, J. P., et al. (2011). Disequilibrium carbon, oxygen, and nitrogen chemistry in the atmospheres of *HD 189733b* and *HD 209458b*. http://arxiv.org/abs/1102.0063 (accessed on 10th September 2025).
18. Gramlich, G. (2004). From 2,3-diazabicyclo[2.2.2]oct-2-ene to Fluorazophore-L, a membrane bound fluorescent probe for antioxidants. https://core.ac.uk/download/18233474.pdf (accessed on 10th September 2025).
19. Grose, W. L., Jones, R. L., McCormick, M. P., Molina, M. J., et al. (2025). Polar ozone. https://core.ac.uk/download/pdf/42814590.pdf (accessed on 10th September 2025).
20. Jimenez, J. L., Tolbert, M. A., Toon, O. B., Trainer, M., et al. (2012). Nitrogen incorporation in CH_4–N_2 photochemical aerosol produced by far ultraviolet irradiation. https://core.ac.uk/download/4895062.pdf (accessed on 10th September 2025).

21. Kono, A., Carbone, E. A. D., Clark, D. T., Fridman, A., Haupt, M., Horibe, H., Kondo, H., et al. (2012). Synergistic formation of radicals by irradiation with both vacuum ultraviolet and atomic hydrogen: A real-time *in situ* electron spin resonance study. http://arxiv.org/abs/1206.6920 (accessed on 10th September 2025).
22. Kurylo, M. J., Ormond, F. M., Prather, M. J., Watson, R. T., et al. (2025). Present state of knowledge of the upper atmosphere 1990: An assessment report. https://core.ac.uk/download/pdf/42821958.pdf (accessed on 10th September 2025).
23. Leone, S. R. (1990). Novel laser gain and time-resolved FTIR studies of photochemistry. https://core.ac.uk/download/pdf/10419420.pdf (accessed on 10th September 2025).
24. Marley, M. S., Morley, C. V., Moses, J. I., Zahnle, K. J., et al. (2016). Photolytic hazes in the atmosphere of *51 Eri b*. https://core.ac.uk/download/pdf/76423622.pdf (accessed on 10th September 2025).
25. McCune, D. C., & Weinstein, L. H. (2025). Assessment of effects on vegetation of degradation products from alternative fluorocarbons. https://core.ac.uk/download/pdf/42814572.pdf (accessed on 10th September 2025).
26. Prather, M. J., & Remsburg, E. E. (2025). The atmospheric effects of stratospheric aircraft. *Report of the 1992 Models and Measurements Workshop. Volume 1: Workshop Objectives and Summary*. https://core.ac.uk/download/pdf/42807416.pdf (accessed on 10th September 2025).
27. Ravirajan, P., & Ravirajan, P. (2004). Characterization and optimization of hybrid polymer/metal oxide photovoltaic devices. https://core.ac.uk/download/1591619.pdf (accessed on 10th September 2025).
28. Sadler, P. J., & Smith, N. A. (2013). Photoactivatable metal complexes: From theory to applications in biotechnology and medicine. https://core.ac.uk/download/16663845.pdf (accessed on 10th September 2025).
29. Singh, K. (2018). Efficacy of combined 5-fluorouracil and photodynamic therapy in glioma spheroids. https://digitalscholarship.unlv.edu/cgi/viewcontent.cgi?article=4388&context=thesesdissertations (accessed on 10th September 2025).
30. Whittman, C. L., Grosjean, D., Cass, G. R., Druzik, J. R., Whitmore, P. M., & Nazaroff, W. W. (1989). Protection of works of art from atmospheric ozone. https://core.ac.uk/download/71354084.pdf (accessed on 10th September 2025).

CHAPTER 10

SURFACE REACTIONS AND ADSORPTION KINETICS

Contents

Introduction.. 371
10.1. Fundamentals of Surface Chemistry.. 374
10.2. Adsorption Phenomena 380
10.3. Kinetic Models of Adsorption 385
10.4. Experimental Techniques for Studying Adsorption 390
10.5. Factors Influencing Adsorption Kinetics 395
10.6. Applications of Surface Reactions and Adsorption Kinetics 400
Summary.. 406
References... 408

INTRODUCTION

Investigation of adsorption kinetics and surface reactions lies at the heart of understanding many processes in materials science and chemistry. Such processes are crucial for applications ranging from catalysis to clean technologies. Surface reactions, characterized by adsorbate-substrate interactions, determine the effectiveness of chemical processes. This is followed by a similar discussion of adsorption kinetics, such as how fast these processes occur. Adsorption kinetics directly influences how effectively systems such as catalysts work. Here, understanding how mass transport and chemical reaction occur at one surface and in porous material becomes crucial in designing the correct means of leveraging the processes for industrial applications. This background highlights the complexity of reaction rate and mass concentration prediction, which warrants a critical examination of the underlying mechanisms and methodologies in these processes. Recent developments in theoretical treatments have exposed the intricacy of adsorption processes. The generalized Fick-Jacobs equation serves as a good foundation for the study of diffusive mass transport along porous geometries, especially when chemical reactions occur both on the solid surface and in the bulk fluid. The approach emphasizes local geometric features such as pore length and curvature as key drivers that control mass concentration within the pore system. Having an understanding of these interdependencies, researchers can accurately simulate stationary and transient conditions for various equilibrium as well as nonequilibrium conditions. These findings bring about more precise predictions on concentration profiles, directly relating to actual applications in the fields of catalysis and gas sensing technology, thereby furthering the characterization of surface phenomena in confined geometries (Hernández-Hernández et al., 2016). Also, advancements in computation techniques, especially statistical mechanics, have transformed the ability to predict and simulate the outcome of surface reactions. Using density-functional theory together with transition-state theory, researchers can calculate energetics of elementary steps that occur in heterogeneous catalysis. This statistical mechanical methodology, as embodied by the kinetic Monte Carlo scheme, addresses the correlations and fluctuations of surface chemical systems and thus allows for detailed simulations under various conditions—high pressure and temperature, for example. These methodologies have returned significant findings, such as the dynamic and disordered nature of active surfaces, which is imperative in optimizing catalytic efficiency. These advances consolidate the connection between surface structure and reactivity, positioning studies in surface reactions and adsorption kinetics at the forefront of materials science and catalysis research (Gross et al., 2005).

Definition of Surface Reactions

Surface reactions are important in numerous fields, including catalysis, materials science, and environmental chemistry. Surface reactions are the processes that occur at the boundary between two phases, one of which is predominantly a solid surface and the other gaseous or liquid reactants. Reactants adsorb on the surface here, where they are converted into products. The significance of surface reactions is because they have the ability to control rates and mechanisms of chemical reactions in heterogeneous systems. Such phenomena need to be understood in order to optimize catalytic processes, as surface reactions have the tendency to control efficiency and selectivity. The intricate balance of molecular interactions at the surface largely influences reaction dynamics, hence the development of sophisticated models of predicting behavior and outcome under varying conditions. Kinetic Monte Carlo (kMC) simulations have been found to be a particularly powerful tool for the study of surface reactions, allowing for a detailed exploration of the processes occurring at catalytic surfaces. Adopting a formalism that deals with the site-selective nature of elementary steps, kMC simulations are capable of dissolving the complexity present in the lattice approximation of reactions onto discrete active sites. This approach facilitates the investigation of various geometries and interactions that govern surface reactions, highlighting the significance of surface composition, temperature, and partial pressures of reactants in defining catalytic performance. By enabling the visualization of model geometries and possible pathways, kMC enables researchers to understand mechanistic details of surface reactions more effectively and, as a consequence, assist in the development of more efficient catalysts and materials (Hoffmann et al., 2014). Secondly, first-principles statistical mechanics methods complement kMC simulations by providing fundamental knowledge regarding the energetics of relevant elementary reactions. Both of them enhance the understanding of heterogeneous catalysis under realistic conditions as they take into account fluctuations, correlations, and spatial distributions of reactants on the catalyst surface. To cite an example, application of these principles to catalytic CO oxidation reveals variations in atomic surface structures and composition with dynamic surroundings. These studies indicate that catalytic behavior is a consequence of a synergistic relationship of numerous reactions with an assortment of active sites on the catalyst. The understanding achieved with these methods leads to tremendous advances in catalysis science, demonstrating how surface reactions have a tremendous influence on the effectiveness and efficiency of chemical processes (Gross et al., 2005).

Importance of Adsorption Kinetics in Chemical Processes

Adsorption kinetics plays a critical role in the wide variety of chemical processes, as it regulates reaction rates and the efficiency of material interaction. The rates of molecule adsorption onto surfaces influence the way that systems function, varying from catalysis in industrial systems to environmental remediation processes. Knowledges of such kinetics not only assist in predicting reactant behavior under different conditions but also assist in enhancing chemical reactor design and operation. Perhaps most importantly, factors like surface area, pore structure, and temperature affect the rates of adsorption and thus highlight the importance of understanding such factors at a fine-grained level in order to enhance reaction efficiency. Furthermore, as illustrated in various research works, the existence of a generalized Fick-Jacobs equation in describing adsorption is evidence of vitally important information on the intricacies of mass transport and surface reaction kinetics, which in essence are obligatory in maximizing chemical processes in academic research and industrial processes (Hernández-Hernández et al., 2016). The significance of adsorption kinetics also goes beyond theoretical importance to extend into practical applications with regards to energy storage and conversion technologies. For instance, in electrochemical devices, rapid ion adsorption on the surfaces of electrodes can significantly influence the efficiency and performance of fuel cells and batteries. Breakthroughs in material science, particularly the discovery of carbon nanomaterials, identify the need for a critical analysis of how the kinetics of adsorption affect electrochemical properties. Electrochemical imaging of high-resolution permits scientists to examine reaction dynamics on a micro-scale to see localized differences in adsorption behavior that can yield remarkably better performance outcomes (Aleix et al., 2016). The tools of understanding these nuanced kinetics permit scientists to predict and control the manner materials respond in real-life applications, providing the potential for new energy technologies and efficiency of chemical processes. Moreover, the adhesion of adsorption kinetics with surface reaction emphasizes the importance of developing proper models to model these processes. As adsorption involves complex molecular-level interactions, the application of proper predictive models can help significantly in the field of chemical engineering, particularly in the area of catalysis. The interplay between adsorption kinetics and reaction processes calls for sophisticated modeling approaches, typically involving the inclusion of various cross-disciplinary methodologies to assess conditions under both equilibrium and nonequilibrium states. Convergent methodologies in recent studies prove that electrochemical behavior of carbon materials can be manipulated by their features, successfully linking rates of adsorption with general performance (Aleix et al., 2016). As developments are realized in theoretical and experimental models, the phenomenon of adsorption kinetics will always be improving, providing insight into ways to better optimize chemical processes in a wide range of applications.

10.1. FUNDAMENTALS OF SURFACE CHEMISTRY

The interaction between the adsorbate and surface is central to understanding the richness of surface chemistry, particularly in surface reactions and adsorption kinetics. Adsorption is the phenomenon where atoms, ions, or molecules from the liquid or gas phase deposit on a solid surface, forming a film. This is a phenomenon of significant consequence in catalysis, environmental science, and materials science, in which the yield and rate of adsorption can have a significant impact on overall reaction dynamics. Basic parameters governing adsorption processes include surface area of the material, surface energy, and the chemical nature of both adsorbate and adsorbent. Thus, an understanding of these variables allows researchers to maximize materials for specific applications, either catalytic performance or adsorption capability for environmental cleanup.

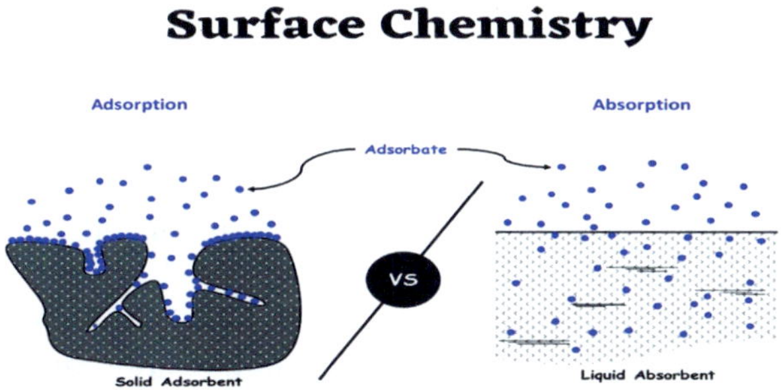

Figure 10.1. *Surface chemistry.*

Source: https://scienceinfo.com/wp-content/uploads/2023/08/Surface-Chemistry.jpg.

Moreover, the adsorption kinetics is controlled by many factors involving temperature, pressure, and adsorbate concentration. An investigation into these factors provides an understanding of surface reaction mechanisms. One of the kinetic models describing the adsorption is the Langmuir adsorption isotherm, which is monolayer adsorption on a surface with a finite number of equivalent sites. In reality, systems tend to be more complex, such as multilayer adsorption or cooperative effects of adsorbate molecules. Using techniques like the Cavity-MicroElectrode (CME) technique has been helpful in quantifying such interactions, particularly in systems where ion size and solvation states are critical. For instance, recent research has shown how ion size affects capacitance behavior in systems like TiC-derived carbon, indicating the relevance of such properties to adsorption kinetics (Chmiola et al., 2009). Surface reactions will often involve electron transfer effects, and these can significantly influence the efficiency of adsorption processes. Biochars, for example, possess large redox capacities for electron transfer, and thus influence their adsorption behaviors. The dependency of electron accepting and donating capacities on pyrolysis temperature of biochar indicates the importance of optimizing material properties to enhance performance. To the surprise of many, studies have revealed that such abilities were likely underestimated in the past owing to slow kinetics of electron transfer, providing new avenues for research in surface chemistry. By the rigorous investigation of such electron transfer kinetics and their effects on adsorption, there can be enhanced understanding of the relationship between surface chemistry and kinetic behavior across a broad number of applications (Boeckx et al., 2016). This knowledge not only serves the formulation of theories but also real development, providing new avenues for innovation in science and engineering fields (Figure 10.1).

10.1.1. Characteristics of Solid Surfaces

The solid surface properties also determine the behavior of the adsorbate, regulating most surface reactions and adsorption kinetics. A solid surface may possess special atomic packing and energy levels giving rise to varying reactivity and adsorption characteristics. The atomic packing of dissimilar crystal structures, for instance, may create high-energy sites favorable for chemical reaction. This can be closely noted in carbon materials such as graphite or graphene where the carbon atoms' orientation and nature significantly determine their electrochemical characteristics. The electrochemical activity of highly oriented pyrolytic graphite basal plane has been found to be equivalent to noble metals, stressing structural aspects in determining a solid's function in surface reactions (Aleix et al., 2016). Other than atomic structure, defects or chemically active functional groups on solid surfaces also greatly influence adsorption behavior. Vacancies, edge sites, and chemical functionalization are examples of such defects that could form reactive centers for adsorption and subsequent chemical reaction. Defective sites, for instance, in carbon nanotubes have been identified as sources of enhanced catalytic activity through localized charge distribution in such defective sites. Besides, theoretical understanding of mass transfer in solid-liquid systems recognizes that solid particles can enhance gas-liquid mass transfer reactions by means of physical adsorption and reactive interactions. These factors illustrate that the relationship between surface properties and defects is critical to the determination of the adsorption kinetics and the mode of surface reaction (Beenackers et al., 1993), which consequently guides the development of more effective catalytic materials. Moreover, the methods used to characterize solid surfaces are at the core of what advances our understanding of their properties and behavior in surface reactions. Advanced methods like electrochemical imaging and microscopy allow researchers to observe and characterize the surface features on the nanoscale. The methods allow the correlation of activity and structure through elucidation of the active site spatial distribution and the dynamics of adsorption. For instance, advances in electrochemistry illustrate that basal plane activity of carbon materials can be superior to common expectations, indicating structural nuances play a pivotal role in electrochemical performance. Consequently, precise characterization of solid surfaces not only affords greater understanding of underlying adsorption phenomena but also provides access to the creation of new materials with defined properties tailored for targeted reactions in catalytic and electrochemical systems (Aleix et al., 2016) (Figure 10.2).

Figure 10.2. *Single crystalline form of solid insulin.*

Source: https://upload.wikimedia.org/wikipedia/commons/8/8c/Insulincrystals.jpg.

10.1.2. Types of Surface Reactions

Study of surface reactions is essential in order to understand heterogeneous catalysis, an industrial process of prime significance. Surface reactions are generally classified into three types: adsorption, reaction, and desorption. Adsorption is the process of deposition of molecules over the surface of a catalyst in order to form a stable interface. This is the initial step, which is very much important because the nature of the adsorbed species significantly influences subsequent chemical transformations. Various adsorption phenomena, like physisorption and chemisorption, dictate the manner in which molecules react on a catalyst's surface. Physisorption is governed by weak van der Waals forces for binding molecules, while chemisorption involves the formation of firm covalent bonds. Catalytic performance is not just a function of the surface area but also a function of the distinctive adsorption dynamics typical under reaction conditions. Following adsorption, the next step in surface reactions is the chemical reaction between adsorbed species. The successive elementary steps of reaction taking place at the surface can greatly vary based on the chemical nature of the reacting species and the catalyst used. For instance, under catalytic oxidation, the adsorbed reactants convert via a series of steps leading to the formation of products. The kinetics of such surface reactions can be complicated with the reaction following more than one pathway and having intermediates.

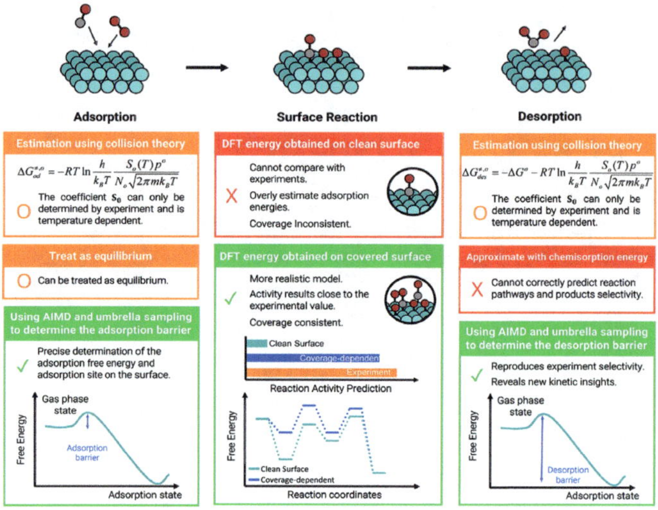

Figure 10.3. *Schematic representation of the reaction steps in a general reaction in heterogeneous catalysis and the three key approaches our group adapted (green) in building a microkinetic model, including solving the issues of reaction barriers in the adsorption/desorption steps and coverage effects in surface reaction steps. Some microkinetic results from different approaches are listed.*

Source: https://www.researchgate.net/publication/360087348/figure/fig2/AS:1153502290153472@1652027785054/Schematic-representation-of-the-reaction-steps-in-a-general-reaction-in-heterogeneous.ppm.

One of the most common mechanisms employed in explaining such reactions is the Langmuir-Hinshelwood mechanism, which assumes the reaction to occur among two adsorbed species. This model emphasizes the competitive character of adsorption sites and shows how variations in reactant concentrations impact reaction rates as in experiments on selective oxidation of ethene over silver catalysts (Borman et al., 1996). The final step of surface reactions is desorption, whereby reaction products escape from the catalyst surface, thereby regenerating accessible active sites for repeated reaction cycles. Desorption is also a key feature since it controls not just the rate of the reaction as a whole but also the stability of the catalytic activity. Desorption effectiveness is subject to variation by the reaction conditions, e.g., temperature and pressure, in accordance with the complexity of the system. Dynamically stable surface phases with compositional fluctuations can provide insight into high catalytic activities, e.g., in catalytic CO oxidation over RuO_2(110) (Gross et al., 2005). Such altered conditions can drive the product generation rates faster while keeping the catalysts regenerative capability. Knowledge of such various surface reactions, such as adsorption, reaction, and desorption, is important in order to optimize catalytic processes as well as to improve the operations of various industrial catalysts (Figure 10.3).

10.1.3. Role of Surface Energy in Reactions

The role of surface energy is of main concern in explaining the mechanisms of surface reaction and adsorption kinetics. Surface energy contributes to the determination of the thermodynamic stability of adsorbates on specific catalytic surfaces through defining the amount of adsorption as well as the reactivity of adsorbed species. When a molecule adsorbs on the surface, surface energy changes must be added into the total free energy of the reaction. Higher surface energy will enhance the likelihood of occurrence of reactions due to lower barriers to adsorbates adsorbing onto the surface, thus enhancing catalytic activity. This interdependency between surface energy and activation energies gives rise to an intimate relationship between material properties and catalytic activity, shaping the design of more efficient catalysts. Another complexity arises from interdependency between surface coverage and energetic compensation effects, which is a significant feature of heterogeneous catalysis. Specifically, an increase in surface coverage results in the effect of the surface energy on the apparent activation energy to generate compensation effects, where changes in activation energy are compensated by modifications in configurational state-related entropy. This effect is explained by research that determines measurable surface interactions, which also corroborates changes in surface properties to directly affect reaction kinetics. As described by Farra et al. (2012) results, understanding this equilibrium enables researchers to manage surface properties to optimize catalytic activity and selectivity, thus optimizing reaction efficiency in real-world applications. Methodological progress in computational methodologies over the recent past, particularly those centered on density-functional theory and kinetic Monte Carlo simulations, has allowed deep insight into the dynamics of surface energy during reactions. These methods enable observation of change and correlation between species at the catalyst surface and are more complete in describing catalytic oxidation processes, as illustrated in the research performed with $RuO_2(110)$ surfaces by Gross et al. (2005). It makes it possible to make predictions on the formation of surface energy under varying conditions and the potential design of optimal states for catalytic activity. By marrying theoretical models with experimental results to a complete degree, researchers can establish an advanced understanding of how surface energy governs reaction directions and eventually catalytic process efficiency and sustainability.

10.2. ADSORPTION PHENOMENA

Adsorption is an important aspect of surface kinetics and reactions, particularly in heterogeneous catalysis. Adsorption is the process in which molecules of a gas or liquid are drawn to a solid surface and adsorb onto it to create a film on the catalyst. Adsorption can have a very strong influence on reaction rates because the amount of active sites as well as concentration of adsorbed material has a direct bearing on the kinetics of surface reactions. The comprehension of the complex interplay between adsorption and desorption processes is extremely important for the goal of improving catalysts. The energy landscape that adsorbates experience upon undergoing such events usually obeys an Arrhenius-like temperature dependence, a general observation in a wide range of catalytic systems. This correlation shows how temperature fluctuations can influence not just the rate of adsorption but also the stability of adsorbed intermediates, which finally defines the overall reaction path. According to the underlying principles of adsorption, effects of compensation further elucidate the relationship between reaction kinetics and surface coverage. Specifically, the linear correlation between apparent activation energy and the logarithm of the pre-exponential factor plays a crucial role in the description of catalytic reactions. New studies indicate that such compensation effects are caused by the interplay between surface coverage and entropic configurations of the adsorbates on the catalyst surface. For instance, for HCl oxidation to Cl_2 over a RuO_2 catalyst, the systematic examination of surface coverage through temporal analysis has shown that variations in surface species are responsible for the determined activation energy. These findings indicate that heterogeneous reaction kinetics cannot fully be achieved without considering the dynamic nature of adsorption phenomena and influencing the reaction's energetics, as per the findings of Farra et al. (2012). Apart from theoretical concepts, practical processes of adsorption phenomena in membrane reactors for hydrogen production validate the significance of theoretical modeling in real processes. Optimization of effective membrane reactor design involves an understanding of varied adsorption properties, mass transfer, and reaction kinetics within the system. Researchers have explored these reactors in considerable detail for the purposes of optimizing hydrogen purity and production rates, with a focus on the subtle interplay between the adsorptive interactions. Theoretical modeling of such adsorption behavior is vital to achieving accurate predictions of reactor performance. However, such models can vary significantly in complexity, sometimes leading to conceptual flaws that make comprehension and application burdensome. Such flaws need to be addressed to develop solid models with the ability to characterize the transport and reaction processes Annesini et al. (2018) introduced. Ultimately, future research on adsorption kinetics will also advance catalytic efficiency and the production of high-performance materials (Figure 10.4).

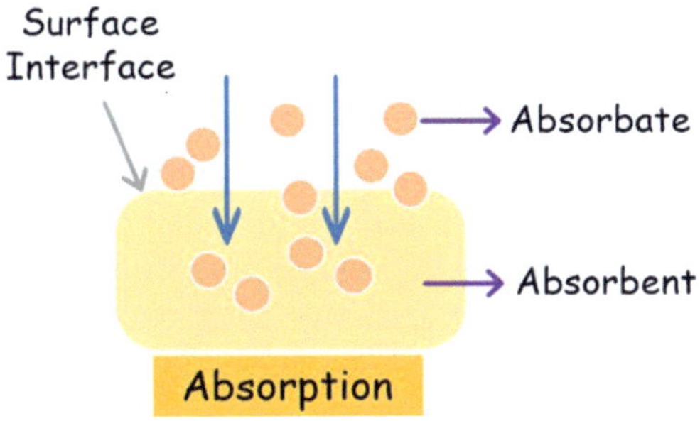

Figure 10.4. *Adsorption.*

Source: https://biodifferences.com/wp-content/uploads/2018/03/Absorption-image.jpg.

10.2.1. Definition and Types of Adsorptions

In the context of adsorption kinetics and surface chemistry, understanding the adsorption definition and types is essential. Adsorption is a process by which atoms, ions, or molecules of a dissolved solid, liquid, or gas adhere to a surface. Adsorption can be broadly categorized into two types: physisorption and chemisorption. Physisorption is accompanied by feeble van der Waals forces that allow for a comparatively low energy interaction between the surface and the adsorbate and is, therefore, reversible. Chemisorption, apart from the creation of strong chemical bonds, leads to a more stable attachment of the adsorbate on the surface. Both have important consequences in catalysis and surface science, with an impact on reaction mechanisms and adsorption energies, which are of concern when optimizing industrial catalytic reactions. The conditions that affect adsorption are complex, including the properties of the surface, temperature, and pressure levels. The level to which a material gets adsorbed onto the surface usually depends on the type of interaction between the adsorbent and adsorbate. For example, under lower temperatures, physisorption takes place due to the fact that there is no available energy to desorb chemical bonds. On the other hand, with rising temperature, kinetic energy may improve chemisorption. The process is of great importance in kinetic studies where a clear determination of the nature of adsorption will provide much more accurate predictions concerning reaction rates and product selectivity. Notably, modeling methods such as kinetic Monte Carlo (kMC) simulations have been useful in explaining such interactions at the molecular level, giving insight into how changes in adsorption kinetics might influence overall reaction dynamics (Hoffmann et al., 2014). Lastly, understanding adsorption processes is the cornerstone of most applied applications in catalysts and materials science. For instance, gas adsorption characteristics of solid heterogeneous catalysts determine their effectiveness and durability. The degree of rate control (DRC) policy even further enhances the comprehension of such adsorption processes with the identification of the most influential steps for reaction rates in dense networks. By simulating various conditions, researchers are able to ascertain which aspects of the adsorption process require extensive modeling and what can endure simplifications (Baranski et al., 2008). This analysis not only maximizes kMC calculation criteria but also helps maximize adsorbent material for a specific catalytic reaction. The intricate interaction of adsorption types and kinetic processes highlights the ongoing revolution in the science of surface reactions, advancing the frontiers of catalytic technology and materials (Figure 10.5).

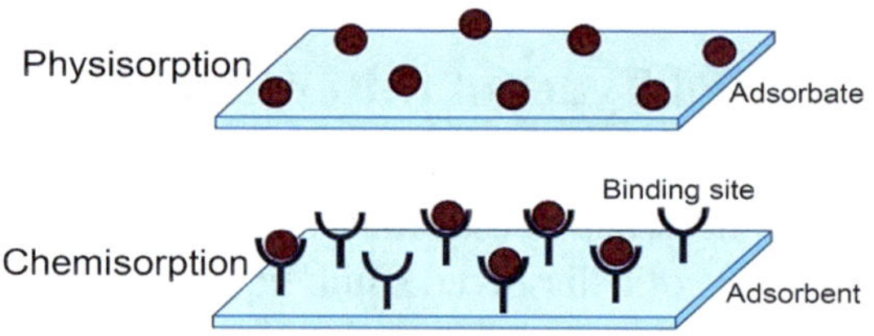

Figure 10.5. *Types of adsorptions.*

Source: https://biologyreader.com/wp-content/uploads/2022/08/types-of-adsorption.jpg.

10.2.2. Factors Affecting Adsorption

The mechanism of adsorption is controlled by several critical factors that determine efficiency and scale of surface interactions. Physical and chemical nature of adsorbate and adsorbent, temperature, concentration, and nature of the solution are some of the most critical ones. Surface area is one of them, for instance; increased surface areas correspond to higher numbers of adsorption sites, thus enhancing overall adsorption capacity. Besides, chemical features of the adsorbate, such as molecular size, polarity, and functional groups, have significant roles to play in impacting interactions. Some interactions, for instance, hydrogen bonding and van der Waals interactions, are accountable for the phenomenon of adsorption.

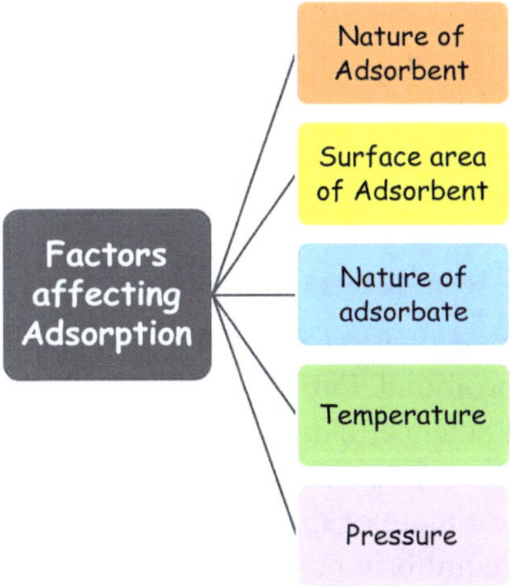

Figure 10.6. *Factors affecting adsorption.*

Source: https://biodifferences.com/wp-content/uploads/2018/03/Factors-affecting-adsorption.jpg.

This is why understanding these elementary factors is vital for enhancing adsorption procedures in varied applications, for example, catalysis and removal of contaminants. Temperature is also another critical parameter of the adsorption process as it influences the kinetic energy of participating molecules. Higher temperature will lead to higher mobility of adsorbate molecules, which in turn can result in a greater rate of adsorption. The effects are complicated, however; while kinetic energy will promote adsorption, excessive heat energy will bring about desorption, thereby counteracting the beneficial effects of higher temperatures. In addition, the adsorption isotherms that describe how the amount of adsorbate on the adsorbent varies with pressure at constant temperature may be influenced by variations in temperature. These nuances emphasize the importance of conducting extensive experiments to ascertain optimum temperature conditions for various systems. Lastly, being aware of the delicate aspects of temperature influences in the study of surface reactions and adsorption kinetics is important. The medium of adsorption also plays a critical role in the overall process. For instance, in heterogeneous systems such as two-phase interfaces, adsorption dynamics can be worlds apart from those in homogeneous systems. As demonstrated in the study of DNA adsorption onto a water-phenol two-phase system, interfacial molecule behavior can lead to new mechanisms of interaction, i.e., particular patterns of adsorption regulated by stacking and hydrogen bonding mechanisms. In such circumstances, vigorous agitation can enhance bimolecular rates of reactions, involving that mixing effects adsorptive behavior and reactivity. These findings explain the complex dynamics of surface and interface forces, necessitating a deeper interpretation of adsorption processes in various scientific disciplines, from biochemistry to material science (Annesini et al., 2018; Bergkvist et al., 2003) (Figure 10.6).

10.2.3. Adsorption Isotherms and Their Significance

Adsorption isotherm investigation is central to knowledge of solute-surface interactions and has profound implications in numerous fields of research such as environmental science, catalysis, and material design. Isotherms are straightforward tools for solute behavior prediction in systems because they provide a quantitative estimate of the way adsorbates interact with adsorbents under conditions of equilibrium. The most common models, including Freundlich and Langmuir isotherms, explain the interaction between the solute concentration in the liquid phase and the adsorbed amount of solute per unit weight of the adsorbent. Besides providing insight into the adsorption kinetics, they also define capacity and selectivity, which are crucial for applications ranging from waste treatment to sensor development. Hence, the inclusion of adsorption isotherms in more detailed analyses increases the predictability and efficiency of surface reaction control. Beyond their theoretical significance, adsorption isotherms also have real-world value in soil science and pollution abatement. For instance, the adsorption characteristics of organic acid anions such as citrate have been discovered to control nutrient solubilization and metal detoxification reactions in soils. It is stressed by research that the knowledge of this sort is very important because it reveals the complex relationships inherent in soil contaminant behavior and fertility. Most especially, the sorption phenomena are governed both by the experimental initial conditions and the properties inherent in the soil, which means that helpful models should consider both temporal and sorption hysteresis factors. These findings indicate the complexity of adsorbate-soil systems and the need for special models to predict the adsorption behaviors of various substances in different systems, which will lead to improved agricultural management and environmental restoration

(Jones et al., 2010). More importantly, the advancement of adsorption modeling—e.g., the conceptualization of the SLAM model—demonstrates the evolving understanding of solute mobility and bioavailability in soil systems, particularly concerning heavy metals like cadmium. This combined approach facilitates the explanation of long-term impacts of contaminants introduced by processes such as sewage sludge application. Based on factors such as soil type and chemical forms within the contaminants, the SLAM model explains solubility and plant uptake of cadmiums-controlled processes. The implications are significant as they offer feedback to significant agricultural practice, risk assessment, and soil health and food safety policy decision-making. Therefore, while the significance of adsorption isotherms continues to be revealed by emerging models, they continue to be important in capturing the complexities of surface chemistry and adsorption kinetics so that there may be an overall approach towards environmental sustainability (Bergkvist et al., 2003) (Figure 10.7).

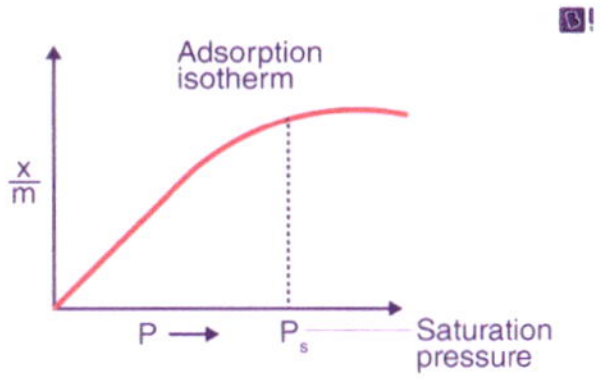

Figure 10.7. *Adsorption isotherms.*

Source: https://cdn1.byjus.com/wp-content/uploads/2018/04/Adsorption-Isotherms-1.png.

10.3. KINETIC MODELS OF ADSORPTION

An understanding of adsorption kinetics plays a fundamental role in catalytic process optimization since it allows predictive models to be developed that are able to effectively describe reactant behavior on surfaces. A prime example of such a leading kinetic model employed is the Langmuir-Hinshelwood mechanism which governs adsorbed species reaction on the surface of catalysts. The mechanism assumes that the reaction occurs between adsorbed reactants and describes how adsorption influences the overall reaction rate. In experimental practice, as for instance in selective ethene oxidation on silver catalysts, this model highlights competition between various adsorbates for active sites and influencing selectivity as well as activity. The kinetic response under such systems highlights the need for empirical data from a variety of reactor geometries, which show variances under different operation conditions and catalyst activity. As in recent studies, an understanding of such dynamics is paramount to the betterment of catalytic processes and optimization of yield (Borman et al., 1996). Kinetic Monte Carlo (kMC) simulations have proved very useful in the examination of adsorption phenomena due to the fact that they offer a microkinetic explanation of the mechanisms involved in heterogeneous catalysis. By modeling at specific sites, kMC methods give researchers the capability to describe how individual active sites coordinate with reactants on a substrate surface.

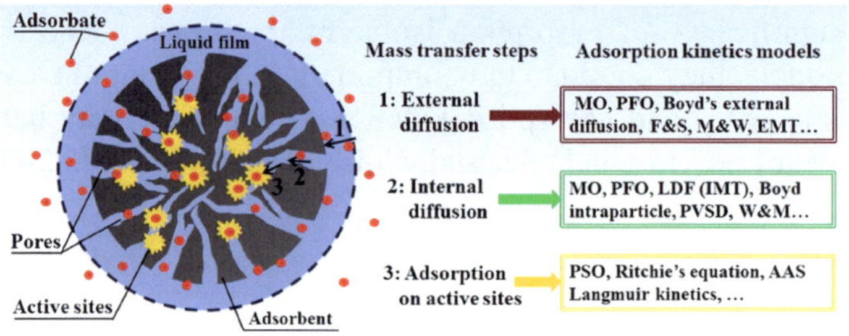

Figure 10.8. *Adsorption kinetic models.*

Source: https://ars.els-cdn.com/content/image/1-s2.0-S0304389420301448-ga1_lrg.jpg.

The versatility of tools like the *kmos* software package allows one to explore intricate networks of reactions and site-resolved interactions, which may not be represented by conventional models. These simulations can efficiently map out elementary reactions, describing how the conditions change and determine adsorption dynamics and catalyst performance. For instance, with parameters dynamically adjusted, kMC provides a responsive understanding of catalytic activity and product selectivity as influenced by temperature and partial pressures. Besides providing a better picture of the mechanism, the computational approach facilitates experimental testing of kinetic models (Hoffmann et al., 2014). The interaction between kinetic adsorption models and empirical realities and sophisticated simulation techniques is essential in the encouragement of catalysis research. The correlation between better knowledge achieved from kinetic studies and computational simulations enables more robust and efficient catalytic systems to be developed. The integration of theoretical models, e.g., the Langmuir-Hinshelwood mechanism, and computational methods, e.g., kMC, enables the more detailed study of adsorption kinetics under various conditions. This two-pronged strategy not only enhances the understanding of surface reactions but also closes the gap of challenges brought about by interactions between reaction products that are detrimental to catalytic performance. Consequently, efforts continue to refine these kinetic models so that they are robust and generalizable to other chemical systems. By continuous iteration on experimental and theoretical levels, the field of surface reaction and adsorption kinetics is poised for huge breakthroughs (Hoffmann et al., 2014) (Figure 10.8).

10.3.1. Langmuir Adsorption Model

Langmuir adsorption model is a fundamental concept of surface reaction as well as the kinetics of adsorption. Irving Langmuir formulated this model in the early part of the 20th century on which it is held that the adsorption occurs at fixed sites on a surface and these sites can either be occupied by adsorbate molecules or can be empty. One of the assumptions that underlie the Langmuir model is that the adsorption process is a reversible one in which adsorbates have access to these positions until monolayer coverage is achieved. The model is elegant and simple mathematically, and it is typically explained through the use of the Langmuir isotherm equation that allows prediction of how the concentration of an adsorbate will impact its adsorption to the surface. This fundamental knowledge is highly beneficial for understanding various applications like catalysis and decontamination. In practical applications, the Langmuir model has extensively been applied to study adsorption kinetics of diversified materials like contaminants in wastewater treatment. For instance, the most applied photocatalyst, TiO_2, operates based on principles explained by the Langmuir model under adsorption of reactive dye molecules from wastewater. The efficiency of TiO_2 is not only because of its natural adsorbing capacity; rather, it also includes further photocatalytic oxidation reactions that further decompose the adsorbed dyes on exposure to UV radiation. As indicated in corresponding research, the adsorption kinetics of dye may be properly explained by applying the Langmuir isotherm to achieve the modeling of dye removal efficiency in respect to different concentrations and operational parameters, demonstrating the model's application in practical real-world situations (Bandyopadhyay et al., 2009). However, the Langmuir adsorption theory is also not devoid of constraint because it also performs under the supposition that there exists a certain number of adsorption sites and does not take into account interactions among adsorbate molecules or surface heterogeneity changes. This may lead to deviation in predicted adsorption behavior in complex mixtures or under varying environmental conditions. Intriguingly, while testing the performance of adsorbents like TiO_2 over numerous cycles, the increase in catalyst deactivation due to the buildup of intermediate by-products can also complicate kinetics. While the Langmuir model itself is a robust framework in which to interpret adsorption mechanisms, incorporating other kinetic models and ideas, such as the L–H model, may enhance prediction more precisely for a wide range of applications, and so emphasize the need for a continued theoretical model development in the face of complexity in reality (Ibbotson et al., 1981) (Figure 10.9).

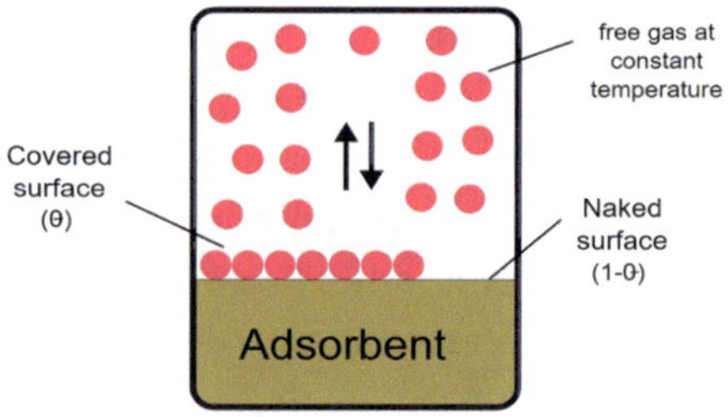

Figure 10.9. *Langmuir adsorption isotherm.*

Source: https://psiberg.com/wp-content/uploads/2022/02/langmuir-adsorption-isotherm-single-layer.svg.

10.3.2. Freundlich adsorption model

The Freundlich adsorption model is a simplified conceptual model for the adsorption of solutes on heterogeneous surfaces. Unlike the Langmuir model with a fixed number of sites available for adsorption, the Freundlich model accounts for surface energy heterogeneity and possibilities of multilayer adsorbates. This experimental equation is represented by the equation $q_e = K_{fCe}^{1/n}$, where q_e is the adsorption of solute per unit mass of adsorbent, C_e is the solute concentration in solution at equilibrium, and K_f and n are constants that express the adsorption capacity and intensity, respectively. The use of the models is especially relevant to systems whose surface properties are complex, for example, wastewater treatment processes, thus making its usage significant in environmental engineering and surface science. In addition, the Freundlich model is found to be especially useful in scenarios of use with organic compounds, including those within wastewater. As revealed in previous research, adsorption kinetics plays a critical role in eliminating contaminates within water supplies. For instance, the use of titanium dioxide (TiO_2) as an adsorbent to adsorb the dyes in wastewater confirms the applicability of the models. The Freundlich isotherm obeyed by the experimental parameters of dye adsorption confirms the heterogeneity of TiO_2's surface and the ability of it to adsorb various organic molecules in an efficient manner. Such effective adsorption strategies not only open the door to environment remediation but also give insight into the design of better materials for catalytic purposes, thereby emphasizing the relationship between adsorption phenomena in addressing contemporary environmental challenges (Bandyopadhyay et al., 2009). By the exploration of organic acid anions in the context of the Freundlich model, research has come to reveal intricate behaviors associated with the adsorption and desorption processes that deviate from the simpler models. Experiments revealed that citrate adsorption in certain soils reaches equilibrium within a very short time, spanning merely one hour, identifying rapid sorption kinetics. The process dynamics, however, show significant hysteresis, whereby conditions generated by desorption attain lower equilibration concentration levels compared to adsorption-generated conditions. This observation points toward the limitation of the models, particularly in the modeling of temporal and complex dynamics of organic processes in soil systems. As such, there is a need to rethink and redefine existing models to incorporate variable adsorption dynamics in order to more clearly articulate organic acid behavior in soils, thus filling gaps in the existing literature as well as contributing to more effective soil management protocols (Jones et al., 2010) (Figure 10.10).

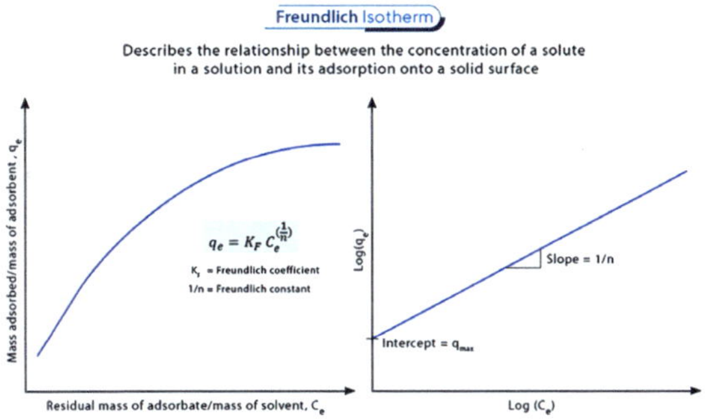

Figure 10.10. *Freundlich isotherm.*

Source: https://www.chemistrylearner.com/wp-content/uploads/2023/12/Freundlich-Isotherm.jpg.

10.3.3. Comparison of Kinetic Models

The surface reaction theory and the theory of adsorption kinetics rest on many kinetic models that are significant in unscrambling the complex interactions at solid-gas interfaces. At the center of these models is the Langmuir-Hinshelwood mechanism that believes that surface reactions involve the reaction of the adsorbed species on active sites. This model has been demonstrated to successfully apply in practice, for instance, in the selective oxidation of ethene using silver catalysts, where competition of reaction products with active sites is the primary factor controlling reaction rates. Consistent with the results reported, the kinetic expressions arising from solution using various reactor conditions differ, offering insight into the effect of catalyst poisoning on reaction kinetics across various operating conditions (Borman et al., 1996). Thus, while the Langmuir-Hinshelwood mechanism remains prevalent, its validity is contingent upon the specific conditions and the nature of the catalytic system in question. In contrast, mechanisms such as the Eley-Rideal mechanism are in conflict with the Langmuir-Hinshelwoods' assumptions to propose one of the reactants directly reacting with a surface-adsorbed compound without prior adsorption of the two. This perspective is most relevant in cases when gaseous reactants possess high kinetic energy such that they can interact directly with the surface. Comparison among these models is important in ascertaining variable catalytic behaviors in varying materials. Besides, in comparing CO oxidation over different RuO_2 facets, it was realized that while the kinetic Monte Carlo simulations had anticipated qualitative binding energetics and elementary process disparities, catalytic activity, in general, was similar under realistic circumstances (Reuter et al., 2015). This complex observation highlights the importance of context when kinetic models are used to explain surface reactions. Validity of kinetic models, apart from the mechanistic pathways that they illustrate, also relies on whether the models can be fit to various experimental conditions. One such instance is that simultaneous fitting of kinetic data sets gives rise to less accurate reaction rates, revealing the challenges involved in model selection in heterogeneous catalysis. In practical applications, a relationship between the conditions for the reaction, i.e., temperature and pressure, complicates the applicability of the models further. Analyzing the effect of by-products on reaction rates, as seen for ethene oxidation, serves to highlight the necessity of updating kinetic models so that in complex catalytic systems, predictions can be made accurately (Borman et al., 1996). As the field continues to evolve, the application of different kinetic approaches may allow greater insight into surface reactions and enable a transition from oversimplified categorization towards more complex frameworks for the incorporation of real-world variables in catalysis.

10.4. EXPERIMENTAL TECHNIQUES FOR STUDYING ADSORPTION

Understanding of adsorption phenomena is a part of the overall field of surface reactions and kinetics and requires a variety of experimental techniques for accurate measurement and characterization. One commonly applied method is temperature-programmed desorption (TPD), enabling the study of the manner in which adsorbed surface species are desorbed upon increasing temperature. By measuring amounts of species desorbing at specific temperatures, researchers are able to calculate adsorption energies and judge interaction strengths between adsorbates and surfaces. TPD also provides essential data about impacts of surface function and morphology, effects that significantly affect adsorption kinetics.

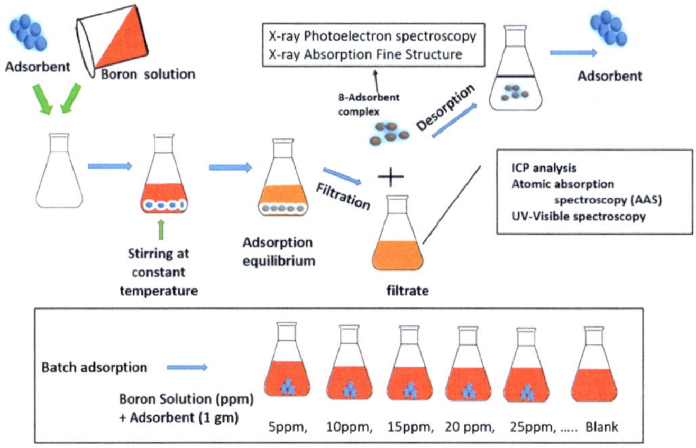

Figure 10.11. *Schematic representation for the general procedure of adsorption study.*

Source: https://www.researchgate.net/profile/Sneha-Mohan-Bhagyaraj/publication/350687190/figure/fig1/AS:1009834279325696@1617774662196/Schematic-representation-for-the-general-procedure-of-adsorption-study.ppm.

This method is complemented by techniques like reflective high-energy electron diffraction (RHEED) that can provide real-time observation of surface changes on adsorption, making a complete insight into molecular interactions on surfaces. Another useful technique in the study of adsorption is molecular dynamics (MD) simulations that allow the study of molecular-level dynamics and interactions under varied conditions. By simulating the molecules' behavior when interacting with the surfaces, scientists can estimate kinetic parameters that are otherwise difficult to measure experimentally. For instance, simulations can account for the influence of temperature, pressure, and surface properties on the rate of and amount of adsorption. Current advances in MD, combined with density functional theory (DFT), have been particularly useful in the investigation of zeolite-catalyzed reactions; these illustrate how variations in conditions at the molecular level can significantly impact mechanisms of reaction. Such simulations are of relevance to the optimization of catalyst performance and structure, especially in complex catalytic processes like the catalytic conversion of methanol to hydrocarbons, where molecular control is critical (Bailleul et al., 2016). In addition to TPD and molecular dynamics, work on non-equilibrium adsorption processes, particularly polymer solution ones, emphasizes the complexity of the kinetics of adsorption. The behavior of high molecular weight polymers when they encounter attracting surfaces presents a complicated interplay between physisorption and chemisorption with consequences for layer structure and dynamics. Recent theoretical research emphasizes that when the energy of interaction between the monomer and surface is larger than thermal energies, non-equilibrium behavior dominates and leads to layers not in a state of thermodynamic equilibrium. This phenomenon is exhibited in the structure of adsorbed polymer chains and their respective binding probabilities. By examining these non-equilibrium layers, researchers expect to isolate long-term dynamics and aging processes on adsorption stability. This branch of science carries great significance for numerous applications from material science to biotechnology since it determines the behavior of the adsorbed structures that are vital in the development of products (Ben et al., 2004) (Figure 10.11).

10.4.1. Surface Characterization Methods

The study of surface characterization methods is at the heart of our knowledge expansion on surface reactions and adsorption kinetics. A set of analytical technologies provides information on the structural, compositional, and electronic characteristics of surfaces, and provides scientists with the ability to unravel the concerns of microscopic-scale interactions. Scanning electron microscopy (SEM) and atomic force microscopy (AFM) provide resolution at high scale, providing spatially resolved information on surface morphology. In contrast, spectroscopic techniques such as X-ray photoelectron spectroscopy (XPS) and Fourier-transform infrared spectroscopy (FTIR) provide chemical composition and bonding states, respectively, which are central to analyzing the influence of surface properties on adsorption behavior. Such comprehensive characterization must be done to correlate surface properties with kinetic patterns and ultimately result in more effective adsorption phenomenon modeling. In addition to these traditional methods, emerging technologies are enhancing the study of surface reactions. For instance, the integration of electrochemical imaging techniques offers unparalleled detail and spatial resolution of electrochemical phenomena on surfaces, particularly in carbon-based materials, which are pivotal in various applications. As described in recent studies, the ability to observe surface activity at the nanoscale allows for the identification of active sites and electron-transfer pathways and significantly improves understanding of interfacial chemistry. Additionally, correlative-electrochemical microscopy combines image data with spectroscopic techniques. This provides a wider perspective on the correlation between the surface structural features and their electrochemical activity, hence deepening the insight into adsorption kinetics in more intricate and heterogeneous systems (Aleix et al., 2016). Implications of employing high-level surface characterization techniques extend beyond basic research; they are significant for applications, especially environmental and materials science applications. For example, understanding how different surface properties affect contaminants like arsenic adsorption on iron oxide minerals demonstrates an important application of these methods (Perez et al., 2020). With the identification of the specific binding sites and mechanism of interaction, researchers can design materials for the maximum elimination of toxic compounds from water. Besides, precise characterization of nanocarbon materials further enhances the performance of energy storage devices by elucidating how structural aspects influence electrochemical performance. As the methodologies advance in the future, they will undoubtedly play a critical role in creating state-of-the-art approaches to address key issues related to surface reactions as well as adsorption kinetics in various fields.

10.4.2. Kinetic Measurement Techniques

Kinetic measurement techniques are critical to understanding the dynamics of surface reaction and adsorption kinetics. These techniques provide information regarding reactant-surface interaction behavior, allowing scientists to deduce reaction mechanisms and rate constants. Various techniques like temperature-programmed desorption (TPD) and fast kinetics spectroscopy provide significant information regarding the molecular adsorption energy and effects of surface heterogeneity. For instance, TPD can measure the desorption temperatures of adsorbates, and with them, their binding strengths, helping identify potential reaction channels. With these interactions trapped, kinetic measurement techniques facilitate easier formulation of overarching thermodynamic principles that govern the systems studied, ultimately leading to enhanced efficiency in catalytic processes. In practical applications, experimental configurations range from those sustaining in-depth kinetic surface reaction studies, typically with high temporal resolution technologies. For example, integrated approaches couple mass spectrometry with surface-sensitive methods to monitor parallel desorption and reaction rates. Such research can identify underlying characteristics of coverage dynamics at the surface, which can inform the design of more effective catalysts. Reactor designs such as gradientless reactors also address flow dynamics issues and eliminate inconsistencies that are associated with traditional batch reactors. The precision of such configurations is imperative; however, it is challenging to obtain accurate images of advanced reaction networks, as was demonstrated in the examination of systems including CO and H_2 (Borman et al., 1989). In addition, computational modeling now complements kinetic measuring techniques, allowing for a predictive system that can enhance experimental data. Simulations can produce more extensive understanding of molecular level interactions and help in the adjustment of kinetic parameters from experimental findings. Computational methods, for example, can be applied to describe the impact of such variables as pressure and temperature on surface reaction rate constants. Such interfacing of experimental kinetic data with advanced computational simulations is important to bring forth increased understanding in adsorption kinetics. As can be seen in a variety of research, such as in research on thermo-oxidative stability within composite materials, results emphasize the importance of refining the efficiency of kinetic measurement techniques (Čechová et al., 2012).

10.4.3. Data analysis methods

In adsorption kinetics and surface reactions, analysis techniques are crucial in describing the complex interactions at the molecular level. With the application of various analytical procedures, scientists are able to quantify and analyze the reaction and adsorption kinetics on catalytic surfaces. For instance, measurements of reaction rates can be greatly enhanced by modeling kinetic, from where important parameters such as activation energy and rate coefficients can be extracted. Advanced data analysis tools such as linear regression and non-linear fitting enable the correlation of experimental measurements and theoretical predictions. These approaches ensure that the resultant kinetic models accurately reflect the underlying phenomena and provide a sound foundation for further research and application in catalysis. The integration of computational approaches with experimental data can yield more comprehensive insight into the mechanisms of surface reactions. One of the notable aspects of data analysis in this regard is the use of statistical methods to refine the interpretation of adsorption isotherms. The explanation of adsorption behavior is very crucial to understand how various compounds respond to reaction catalyst surfaces. Langmuir and Freundlich isotherm models are techniques that allow one to determine data on surface saturation and affinity between adsorbate and active sites. The intrinsic properties of materials being tested can be derived from the constants acquired from these models, such as K_{ads}-H^+, which is also important in linking the adsorption characteristics to the overall reaction kinetics, as postulated by Bell et al. (2018). Scientists are able to achieve better understanding of how parameters such as pressure and temperature influence the efficiency of catalysis and adsorption by adhering rigidly to these models and experimental data. Secondly, the newly emerged methods of data analysis, such as machine learning and artificial intelligence, possess significant potential for facilitating a deeper examination of surface reactions and adsorption kinetic processes. The new methods are capable of handling large sets of data collected via experiments in a way that the underlying patterns and correlations undetectable by traditional analytical methods can be discerned. For example, their application can make it possible to automate the discovery of new recipes for catalysts based on future performance from historical data. In particular, studies on catalytic reactions, e.g., hydrogenation of diphenylacetylene, have revealed that advanced data analysis can make it possible to produce more active and selective catalysts, as seen in the findings of Bonrath et al. (2015). Ultimately, the inclusion of sophisticated data analysis techniques will facilitate a better understanding of surface reaction dynamics and allow for the introduction of next-generation catalyst design.

10.5. FACTORS INFLUENCING ADSORPTION KINETICS

Adsorption kinetics—adsorbate attachment speed to the adsorbent surface—is controlled by several critical factors. Prior to them all are the surface area and porosity of the adsorbent material. The larger the surface area, the more adsorption sites there are, significantly enhancing the binding capacity and rate. Also, the chemical nature of the adsorbent plays a role in determining kinds of interactions, such as van der Waals forces, electrostatic attraction, or covalent bonding that can also define the rate of adsorbate adhesion. As exemplified in the electrocatalytic processes, different mineral phases can alter adsorption behavior and kinetics, stressing the importance of material selection for specific applications (Burggraaf et al., 1988). Therefore, an understanding of these intrinsic and extrinsic features is vital in designing effective adsorbent materials for many environmental and industrial processes. Solution conditions effects on adsorption kinetics cannot be overstated. pH, ionic strength, and co-existing competitive ions are important among them. For instance, pH can alter the charge of the adsorbent and adsorbate, hence influencing the magnitude of electrostatic attraction or repulsion at the interface. Similarly, ionic strength can facilitate or hinder ion-ion interactions that can enhance or reduce adsorption rates, respectively. In addition, the role of co-existing inorganic ions, for example, as seen in studies of arsenic immobilization, can significantly influence adsorption process kinetics. Such ions' activity is directly connected with physical and chemical stability of the adsorbent, affecting reactivity in the applied environmental conditions (Perez et al., 2020). Thus, the optimized parameters are crucial for maximizing the efficiency of adsorption processes in a large number of applications. Temperature is also a basic factor influencing adsorption kinetics. Greater temperatures often enhance molecular kinetic energy and therefore enhance rates of adsorption due to increased motion within the molecules. Temperature and adsorption, though, are closely related. While elevated temperatures facilitate faster kinetic reactions, they can lead to desorption if thermal energy overcomes the adsorbate binding forces. The dual nature of such an effect must be evaluated in deciding on adsorption efficiency and stability under varying conditions. In the case of the like of green rust, which possesses massive arsenic adsorption capacities, temperature-adsorption mechanism interactions are essential in long-term performance. Understanding these temperature influences is therefore paramount for application in processes involving pollutants and waste management and where stability in adsorbent materials determines environmental containment policy (Burggraaf et al., 1988).

10.5.1. Temperature Effects

Temperature is extremely crucial for surface reactions as well as adsorption kinetics because it has the primary role of changing the kinetic energy of involved molecules. Upon a rise in temperature, vibrational energy of the adsorbate also rises, which enhances its opportunity to overcome the energy barriers to adsorption. This is a very significant phenomenon within catalytic reactions, where the efficacy of reactants converting to products relies on the velocity with which they adsorb onto catalytic active sites. This is most apparent within reactions with the Langmuir-Hinshelwood mechanism, where introduced thermal energy accommodates the cooperative adsorption of both reactants and thereby increases the rate of the reaction. Temperature dependence illustrates why, in industrial processes, it is so important to optimize operating conditions so that reaction rates and product yield are maximized under controlled thermal conditions. Furthermore, competitive adsorption of reaction products will modify the effects of temperature on surface reactions. High temperatures can cause intermediates or products that would otherwise be using active sites to desorb, which can alter the selectivity and overall reactivity of a catalyst. For instance, in studies of selective oxidation of ethene, high temperature not only accelerates reactions but also affects adsorption dynamics, as products like carbon dioxide can hinder rates of desired reactions while enhancing selectivity towards desired intermediates like ethene oxide. This complexity of temperature effect highlights the complicated interplay between kinetics and thermodynamics involved in catalysis. Besides, the interlocking of temperature, adsorption, and inhibition by product presents challenges to overcome for improved reaction efficiencies in many catalytic operations (Borman et al.,

1996). Structural characteristics of catalysts themselves may also regulate temperature influence on the kinetics of reactions. For example, in acidic zeolite catalytic cracking operations, temperature variability impacts the alkane dehydrogenation and cracking reaction kinetics at their most elementary levels. The generation of active sites and their action with reactants are temperature-sensitive, necessitating a clear awareness of adsorption equilibria at different temperatures. This correlation underlines the significance of investigating and quantifying factors such as adsorption equilibrium constant and intrinsic rate coefficients coupled with temperature changes. These conditions are invaluable in the formulation of predictive models that quantify how temperature changes affect catalytic activity and selectivity, culminating in more advanced catalyst design and use strategies leveraging the beneficial effect of temperature on surface reaction and adsorption kinetics (Bell et al., 2018) (Figure 10.12).

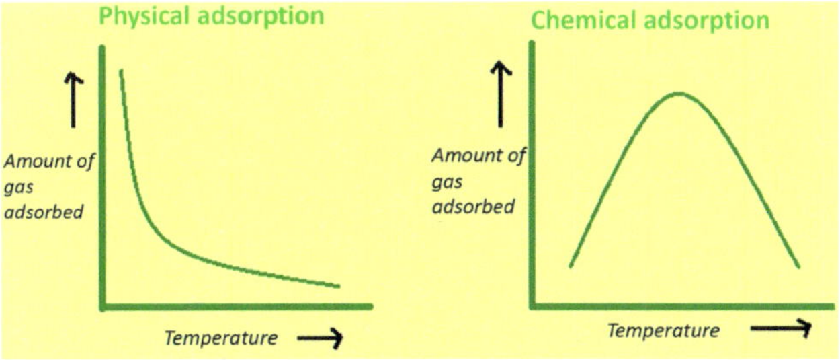

Figure 10.12. *Relationship between adsorption and temperature.*

Source: https://www.w3schools.blog/wp-content/uploads/2019/08/image-result-for-factors-affecting-adsorption.png.

10.5.2. Pressure Effects

The influence of pressure on adsorption kinetics and surface reaction is key to the understanding of catalytic reaction mechanisms and planning for reaction conditions. Increased pressure tends to make reactants denser in the gas phase, causing increased molecular collision rates with the catalyst surface. This action has a tendency to enhance adsorption rates, accelerating reactions. Pressure oscillations can have a significant influence on yield and selectivity of target products in catalytic reactions. For instance, while pressure increase is favorable to product formation, it stands an equally good chance to trigger competing adsorption, where the products of the reaction compete with reactant adsorption, as in selective ethene oxidation, where some reaction products decelerate the overall kinetic (Borman et al., 1996). Thus, pressure management is significant in industrial use to optimize the adsorption kinetics and efficiency of surface reactions. Adding to the complexity of the pressure environment are the intricate mechanisms that control adsorption phenomena. In the Langmuir-Hinshelwood model, the most widely applied kinetic representation of intricate surface reactions, both adsorbed reactants must occupy space on the catalyst surface for reaction. As pressure increases, the equilibrium governing such adsorbed species is perturbed, which can accelerate or retard reaction rates. Further, the reality that pressure-induced composition changes at the surface should not be overlooked.

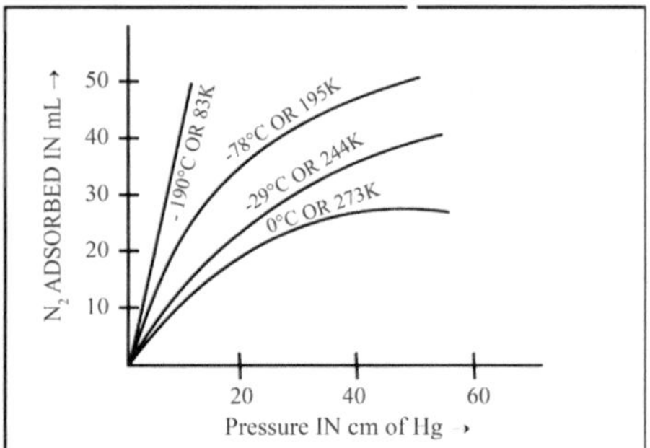

Variation of adsorbtion of N_2 on charcoal with pressure at different constant temperatures

Figure 10.13. *Variation in the adsorption of nitrogen gas on charcoal with varying pressure and temperature.*

Source: https://www.w3schools.blog/wp-content/uploads/2019/08/image-result-for-factors-affecting-adsorption-1.png.

For metal catalysts, increased pressures will tend to promote the formation of many different surface oxides or adlayer forms different from those that form at lower pressures (Rogal et al., 2006). This pressure-induced alteration affects not only the catalyst activity but also selectivity towards specific reaction pathways, making pressure an inevitable element in catalyst design and reaction optimization. In addition, pressure influence on temperature is a significant parameter in surface reactions and adsorption kinetic processes. With high pressures, temperature oscillations can generate significantly different adsorptions and reaction processes. For instance, under optimal temperature conditions, there are some catalysts that stabilize some surface structures facilitating catalysis, while high pressure may trigger unwanted interactions that destabilize them. The dynamic nature of the interplay between pressure, temperature, and catalyst composition highlights the requirement for sophisticated modeling approaches. As has been shown in multiscale modeling simulations, the co-adsorption of various surface states—like thin oxide films or CO adlayers—at high pressures can generate complex reaction dynamics that are essential to predicting catalyst performance (Rogal et al., 2006). Researchers must thus consider such interdependent factors to accurately evaluate and optimize the efficacy of catalytic processes in real-world applications (Figure 10.13).

10.5.3. Surface Area and Porosity

Surface area and porosity are intrinsic properties that play a major role in controlling the kinetics of surface reaction and adsorption processes. The effectiveness of adsorbents and catalysts is positively linked to their accessible surface area, which dictates the number of active sites for chemical interaction. Increased surface area typically implies increased adsorption capacity, which makes materials such as nanoscale metal oxides extremely valuable in applications such as photocatalysis. For instance, in halloysite-TiO_2 composites, structural organization offers sufficient porosity, enhancing diffusion and permeability of the reactants. Synthesis techniques resulting in optimal surface area to porosity ratio and consequently guaranteed active sites accessibility are essential in attaining maximum efficiency in diversified catalytic reactions. Furthermore, the synthesis pathway plays a crucial role in defining structural properties including surface area and porosity. The sol-gel method, for example, allows tunable porosity and particle size on the nanometer scale to influence the overall functionality of the materials in catalytic functions. Films of nanoparticle origin, e.g., those derived from titanium isopropoxide, can exhibit extremely distinct behavior based on the size of their aggregated crystallites. Changes in process conditions can thus yield variations in porosity, influencing the ability of the materials to facilitate surface reactions. As stipulated in the literature, stable sols could give rise to materials of close-packed structures when mishandled, minimizing their porosity and subsequent catalytic activity. These process conditions have to be properly controlled in forming catalysts that are capable of optimizing the potential of their surface area. Addition of other constituents, i.e., polymers, to these materials increases the porosity further, leading to an increased surface reaction. The incorporation of polymers, for instance, can induce a better porous structure, significantly increasing the channels for diffusion of adsorbates. This is evident in studies where post-synthesis modifications created porous nanocomposite films with improved photocatalytic activity. The interaction between the metal oxide matrix and the polymeric materials lends itself to a more effective adsorption setting, which enhances faster kinetics in the degradation of pollutants. With new synthesis and post-modification methods tailored to engineer desired porosity, researchers can build complex materials that not only meet but also exceed the demands for effective surface reactions in most applications, thus emphasizing the importance of optimizing material surface area and porosity in adsorption kinetics (Figure 10.14).

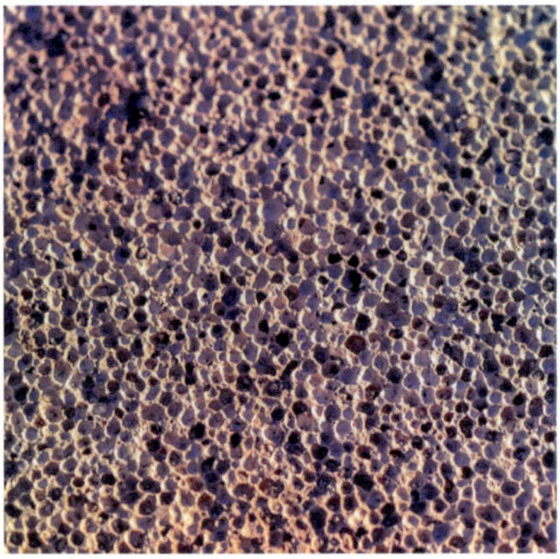

Figure 10.14. *Surface area and porosity.*

Source: https://e3ekiq82rvu.exactdn.com/wp-content/uploads/2022/03/Membranes.jpg?strip=all&lossy=0&ssl=1.

10.6. APPLICATIONS OF SURFACE REACTIONS AND ADSORPTION KINETICS

The study of surface reactions and adsorption kinetics is of critical significance in a broad spectrum of applications across numerous scientific fields. The most important one of these is in the field of materials science and the development of novel thin films and coatings. Surface reaction mechanisms determine the properties of these materials, influencing their functionality, stability, and even aesthetics. For instance, the random sequential adsorption model describes the mechanisms in which the surface coverage in coatings may be optimized. The model provides valuable information on designing desired states in surface engineering in the micro- and nanoscale, a necessity for applications ranging from electronics to biomedical implants (Cadilhe et al., 2007). Tuning these surface interactions allows researchers to engineer materials with properties designed to meet specific performance needs. In addition to materials science, surface reaction kinetics and surface adsorption are the key aspects in the area of catalysis. The efficiency of the catalytic process largely relies on reaction kinetics at the catalyst surface. An understanding of the rate of reactant adsorption and desorption on catalytic surfaces can offer a method for creating more effective catalysts and thereby enhancing the rate and selectivity of the reaction. For instance, first-principles DFT calculations have contributed to better knowledge of molecular beam epitaxy surface processes. Researchers employ these calculations for simulating essential aspects of growth, such as adsorption, desorption, and nucleation, which are invaluable when it comes to optimizing the synthesis of semiconductor materials. Through detailed study of these kinetics, researchers are able to maximize processes to achieve greater yields and fewer defects in these critical materials (Kratzer et al., 2002). Furthermore, the implications of adsorption kinetics extend into environmental science, particularly into pollution control and resource recovery. Adsorption is the principle through which contaminants are removed from air and water; understanding what these interactions accomplish, and how they do it, makes it possible to design effective filtration systems. The alteration of surface properties of adsorbents can then be utilized for optimizing removal efficiencies by promoting favorable reaction pathways. Adsorption kinetics, for example, can be regulated through pre-patterning of surfaces in order to enhance self-assembly mechanisms within adsorptive processes. While technology keeps moving forward towards reducing environmental issues, changing understanding in surface reactions and adsorption kinetics will play a fundamental role in developing innovative solutions for ensuring sustainable methods in industries (Cadilhe et al., 2007).

10.6.1. Catalysis in Industrial Processes

Surface reactions and adsorption kinetics play a crucial role in catalysis in industrial processes, significantly enhancing efficiency and product yield. Catalysts, which provide an alternative pathway with less activation energy for a reaction, are essential in the majority of chemical reactions, particularly in the petroleum and petrochemical industries. Heterogeneous catalysis, where solid catalysts react with gaseous or liquid reactants along their surface, is of utmost reliance in these industries. The efficiency of such catalytic systems depends mainly on the rate of surface reactions as well as adsorption kinetics because these mechanisms dictate how rapidly reactants can be converted into products. It is thus imperative that the kinetics of such processes be known in order to optimize industrial catalysis and enhance the general efficiency of catalytic processes. Recent advances in computational methods, such as the development of Kinetic Monte Carlo (kMC) simulations, have significantly extended our ability to study surface reactions in catalysis. The k_{mos} software, its user-friendly interface and speed, allows researchers to simulate complex catalytic processes within various lattice systems, many active sites, and interactions. By focusing on surface catalytic reactions, kMC simulations can deliver valuable information on reaction dynamics under specified operating conditions, i.e., temperature and reactant pressures. The simulations not only allow the evaluation of catalytic activity and selectivity but also deliver valuable mechanistic information on individual elementary processes in the reaction network. Therefore, kMC is an important tool in investigating and predicting the activity of catalysts under real conditions, facilitating optimization of existing industrial operations and design of novel catalytic substances (Hoffmann et al., 2014). Zeolites, with their flexibility in structure and the catalyst property, are a prime example of the complexity involved in catalysis in industrial processes. These microporous materials have been widely applied in a variety of chemical conversions, e.g., methanol conversion to hydrocarbons. Recent ab initio molecular dynamics simulations have revealed the complex synergy between zeolite topology, acidity, and presence of guest molecule on reaction kinetics. By studying representative reactions like benzene and propene methylation, researchers can distinguish between the influence of such variables on reaction pathways, determining how much a concerted or stepwise route is favored. This molecular data is invaluable for optimizing existing zeolite-catalyzed processes and developing new technology using substitute feedstocks. Lastly, improving our understanding of zeolite catalysis by means of enhanced modeling not only makes operations more efficient in current industrial applications but also inspires the building of greener catalytic solutions for the future (Bailleul et al., 2016) (Figure 10.15).

Figure 10.15. *The significance of heterogeneous catalysis in industrial processes.*

Source: https://catalysts.com/wp-content/uploads/2024/04/iStock-1447817599.jpg.

10.6.2. Environmental Applications

The intersection of surface reactions and adsorption kinetics is crucial in environmental uses, particularly water decontamination of polluted water sources. Adsorption processes utilize the spontaneous affinity between the adsorbent surface and contaminant to simplify the process of eliminating harmful contaminants from water sewage. For example, the transformation of lignin to produce an aminated cellulose-based adsorbent is a promising development in water treatment. This aminated CELF lignin has been found to exhibit a superior adsorption ability toward toxic dyes such as methylene blue and direct blue 1 with a superior maximum capacity for adsorption being 502.7 mg/g. Kinetic studies indicate the adsorption processes to adhere to pseudo-second-order kinetic models effectively, supporting the role of surface reaction in effective adsorbent use in environmental remediation (Cai et al., 2020). A further significant environmental application is the remediation of hexavalent chromium, a highly toxic contaminant in groundwater.

Nanoscale zero-valent iron nanoparticles (nZVI-p) have emerged as an effective treatment for reducing Cr(VI) content in contaminated aquifers. Experimental studies indicate that nitrate can influence the effectiveness of Cr(VI) removal by nZVI-p due to the complexity of contaminant interactions under natural field conditions. Through controlled alteration of the nitrate concentration, researchers have obtained insights into the competitive nature of interactions in the Cr(VI) reduction process. The application of a proposed kinetic model provides a better understanding of such interactions, and the path towards optimized remediation strategies in reducing environmental pollution. These findings highlight the pivotal role played by adsorption kinetics towards the development of effective remediation technologies for hazardous groundwater contaminants (Palma et al., 2017). The application of advanced adsorption materials to environmental purposes is increasingly identified as a viable measure to address worldwide pollution issues. Studies have demonstrated that the modification of natural adsorbents, like lignin, can improve their chemical characteristics to function better in the decontamination of wastewater. Not only does this address some of the toxic substances, but it also fosters the employment of renewable resources to environmental protection. Similarly, the use of engineered nanoparticles like nZVI-p boasts a twofold benefit: effective targeting of toxic species and imposing a lesser additional burden on the environment. Through an understanding of the concepts of surface reaction and adsorption kinetics, scientists are refining these processes to improve contaminant removal efficiency and stimulate the regeneration of contaminated water. Ultimately, the developments portray a favorable way forward to more efficient and more sustainable environmental operations that are compatible with the growing need for effective pollution control operations in various ecosystems.

10.6.3. Material Science and Nanotechnology

The intersection of material science and nanotechnology becomes increasingly significant, particularly within the realm of adsorption and surface reaction kinetics. The new properties at the nanolevel—increased surface area, quantum phenomena, and enhanced reactivity—remodel traditional models of the phenomenon of adsorption. For instance, when transition metal sulfides are analyzed, their unique chemical properties over metal oxides create a unique environment for investigating surface behavior and kinetics. These materials possess varied functional properties that may be exploited in applications ranging from catalysis to bioimaging. Understanding the complex surface chemistry of these materials, especially their adsorption modes, is essential for optimizing their applications. Preparation of colloidally stable suspensions of these materials involves rigorous functionalization protocols that regulate their surface properties, a cumbersome task given the limited knowledge in comparison with their oxide counterparts (Dengo et al., 2017).

Nanotechnology enables regulation of the properties of materials at the molecular or atomic level, yielding tailored nanomaterials whose surface functionality is specific. Such manipulation plays an important role in environmental treatments, such as the treatment of toxic pollutants. For example, nanoscale zero-valent iron particles (nZVI-p) have been established in research for the remediation of hexavalent chromium in contaminated groundwater. The behavior of nZVI-p reacting with different types of contaminants demonstrates the intricate dynamics of surface reactions and adsorption kinetics. Experimental evidence reveals that the presence of competing species, e.g., nitrates, plays significantly into the mechanism of Cr(VI) reduction. Here, a kinetic model assists in unfolding how parameters affect the efficacy of nZVI-p in real-world applications, especially emphasizing the role of surface chemistry in organizing effective remediation schemes (Palma et al., 2017).

The combination of material science and nanotechnology offers an opportunity for profound technological developments across every field, particularly environmental science and engineering. The building of a superior understanding of surface reactions in addition to adsorption kinetics not only enhances material performance but also extends the scope of application of nanomaterials. The synergy between experimental tools and modeling strategies enables researchers to predict how shifts in surface properties can influence reaction mechanisms and kinetics. Thus, information from the characterization of nanostructured materials can lead to breakthroughs in catalyst, sensor, and adsorbent design. As the field continues to grow, continued synergistic interaction among material scientists, chemists, and environmental scientists will give rise to novel solutions for significant global problems, in turn supporting the fundamental imperative of researching surface reactions and adsorption kinetics in nanotechnology (Dengo et al., 2017; Palma et al., 2017) (Figure 10.16).

Figure 10.16. *Material science and nanotechnology.*

Source: https://www.aiu.edu/wp-content/uploads/2024/04/nano1024-e1400502699196.webp.

SUMMARY

In anticipating the study of surface reactions and adsorption kinetics, it is worth noting the complicated interactions that control catalytic activity. Study of heterogeneous catalysis shows that surface dynamics are key to the efficiency and selectivity of chemical reactions. With advanced methods such as density-functional theory and kinetic Monte Carlo simulations, researchers can simulate these interactions under various conditions. The results indicate a strong correlation between surface structure and catalytic performance, such as in the catalytic CO oxidation at $RuO_2(110)$, where dynamic surface behavior is highly responsible for high turnover rates. This indicates that there is a need for a complete understanding of phenomena at the surface in designing next-generation catalysts and optimizing current processes to ultimately promote industrial application. In addition, research into acidic zeolites illuminates the importance of structural characteristics towards catalytic activity. The relationship between zeolite structure and Brønsted acidity provides insightful information regarding the kinetics of dehydrogenation and alkane cracking. To illustrate, the reliance of adsorption equilibrium and intrinsic reaction rates on the apparent rate coefficient highlights the intricacies of catalytic reactions. The advances in configurational-bias Monte Carlo simulations allow for scientists to better predict and control these interactions, thus making the design of improved catalysts possible. This not only improves theoretical models but also pushes experimental techniques toward new paradigms in catalysis that can address current energy and environmental challenges. Ultimately, combined data from surface reactions and adsorption kinetics present a roadmap for future research and technology. The interdisciplinary nature of the field necessitates integration between chemistry, materials science, and engineering in order to untangle the intertwined processes at play. Looking forward, coupling machine learning and data-driven approaches has the potential to enhance our predictive capabilities, allowing for the advancement of catalyst design at an accelerated rate. As the energy map evolves, application of the underlying principles of surface chemistry will be fundamental to building sustainable solutions. The relationship between theory and practice in catalysis and industrial chemistry will continue to open new doors, confirming the central role of surface phenomena in shaping the future of chemical processes.

The study of surface reaction and adsorption kinetics is at the heart of heterogeneous catalysis and materials science. A crucial aspect of such processes is the interaction of reactants and active sites on a surface, which can be effectively captured using Kinetic Monte Carlo (kMC) simulations. The simulations enable fine-grained examination of elementary reactions on a lattice of discrete active sites. The k_{mos} software package is a prime example of such an approach and provides an easy-to-use platform for the application of intricate kMC models to a range of dimensional frameworks. Especially in the investigation of catalytic processes, kMC offers essential information on catalytic activity, selectivity, and mechanistic pathways, connecting surface phenomena dynamically to macroscopic results such as reaction rates and product formation (Hoffmann et al., 2014). Furthermore, developments in computational method have identified the key deficiencies in earlier modeling approaches based on thermochemical equilibrium and chemical vapor deposition (CVD) phase diagrams. These developments reflect the ability of simulations accounting for gas-phase mass transport and its impact on deposition rates. Successful modeling is dependent on the correct selection of boundary conditions, which have a significant influence in determining the outcomes of CVD processes. The emphasis on precise kinetic data for gas-phase and surface reactions indicates the true challenge facing scientists in terms of enhancing the precision of kinetic models. The emphasis lays the groundwork for greater efficiency and effectiveness in process design and optimization, marking a tremendous leap in the field (Gokoglu et al., 2025). The exchange between computational models and experimental data not only validates theoretical models but also informs future research in surface reactions and adsorption kinetics. By refining kMC calculations to represent the complexity of real-world situations, researchers can more precisely predict the behavior of catalytic systems under various operating conditions. The ongoing k_{mo} optimization combined with a combined effort towards the advancement of gas-phase and surface kinetic data is an evolutionary process toward eliminating existing limitations in reaction modeling. This research will definitely lead to the development of more efficient catalytic materials and processes in direct response to the industry's need for enhanced catalytic performance and sustainability in chemical manufacturing. All these developments therefore matter to the continued advancement of technological applications in heterogeneous catalysis and materials design.

REFERENCES

1. Alphonse, P., Tendero, C., & Varghese, A. M. (2010). Stable hydrosols for TiO_2 coatings. https://core.ac.uk/download/12042515.pdf (accessed on 10th September 2025).
2. Annesini, M. C., Cerbelli, S., & Murmura, M. A. (2018). Modeling fixed bed membrane reactors for hydrogen production through steam reforming reactions: A critical analysis. https://core.ac.uk/download/159268736.pdf (accessed on 10th September 2025).
3. Bailleul, S., De Wispelaere, K., & Van Speybroeck, V. (2016). Towards molecular control of elementary reactions in zeolite catalysis by advanced molecular simulations mimicking operating conditions. https://core.ac.uk/download/55855326.pdf (accessed on 10th September 2025).
4. Bandyopadhyay, S., Bhattacharjee, C., Datta, S., Dutta, A., et al. (2009). Kinetic study of adsorption and photo-decolorization of Reactive Red 198 on TiO_2 surface. https://core.ac.uk/download/139350.pdf (accessed on 10th September 2025).
5. Baranski, A., Boudard, M., Campbell, C. T., Campbell, C. T., Chorkendorff, I., Dumesic, J. A., Dumesic, J. A., et al. (2008). Examination of the concept of degree of rate control by first-principles kinetic Monte Carlo simulations. https://core.ac.uk/download/267951742.pdf (accessed on 10th September 2025).
6. Beenackers, A. A. C. M., & van Swaaij, W. P. M. (1993). Mass transfer in gas-liquid slurry reactors. https://core.ac.uk/download/pdf/11461337.pdf (accessed on 10th September 2025).
7. Bell, A. T., Head-Gordon, M., Janda, A., Lin, L. C., et al. (2018). Understanding Brønsted-acid catalyzed monomolecular reactions of alkanes in zeolite pores by combining insights from experiment and theory. https://core.ac.uk/download/158345582.pdf (accessed on 10th September 2025).
8. Bergkvist, P. (2003). Long-term fate of sewage-sludge derived cadmium in arable soils. https://core.ac.uk/download/11694621.pdf (accessed on 10th September 2025).
9. Boeckx, P., Cid Farina, I., Prévoteau, A., Rabaey, K., et al. (2016). The electron donating capacity of biochar is dramatically underestimated. https://core.ac.uk/download/55803407.pdf (accessed on 10th September 2025).
10. Bonrath, W., Borretto, E., Cherkasov, N., Cravotto, G., et al. (2015). Ultrasound- and microwave-assisted preparation of lead-free palladium catalysts: Effects on the kinetics of diphenylacetylene semi-hydrogenation.

https://core.ac.uk/download/151160887.pdf (accessed on 10th September 2025).

11. Borman, P. C., Borman, P. C., Schouten, E. P. S., Schouten, A., et al. (1996). Influence of reaction products on the selective oxidation of ethene. https://core.ac.uk/download/pdf/11452680.pdf (accessed on 10th September 2025).
12. Borman, P. C., Bos, A. N. R., Kuczynski, M., Westerterp, K. R., et al. (1989). The kinetics of the methanol synthesis on a copper catalyst: An experimental study. https://core.ac.uk/download/pdf/11473983.pdf (accessed on 10th September 2025).
13. Burggraaf, A. J., Gellings, P. J., & Koopmans, H. J. A. (1988). Electrocatalytic phenomena in gas phase reactions in solid electrolyte electrochemical cells. https://core.ac.uk/download/pdf/11473689.pdf (accessed on 10th September 2025).
14. Cadilhe, A., Araújo, N. A. M., Cadilhe, A., Privman, V., Barker, G. C., Burgos, E., Elimelech, M., Levich, V. G., Lewis, P. C., et al. (2007). Random sequential adsorption: From continuum to lattice and pre-patterned substrates. http://arxiv.org/abs/cond-mat/0611413 (accessed on 10th September 2025).
15. Cai, C. M., Li, M., Meng, X., Ozcan, S., et al. (2020). Synthesis, characterization, and utilization of a lignin-based adsorbent for effective removal of azo dye from aqueous solution. https://core.ac.uk/download/305124897.pdf (accessed on 10th September 2025).
16. Čechová, E. (2012). Thermooxidative stability of PMMA composites. https://core.ac.uk/download/30296004.pdf (accessed on 10th September 2025).
17. Chmiola, J., Gogotsi, Y., Huang, P., Largeot, C., et al. (2009). Solvent effect on the ion adsorption from ionic liquid electrolyte into sub-nanometer carbon pores. https://core.ac.uk/download/12041895.pdf (accessed on 10th September 2025).
18. Dengo, N., Gross, S., & Vittadini, A. (2017). Functionalization of colloidal transition metal sulphides nanocrystals: A fascinating and challenging playground for the chemist. https://core.ac.uk/download/141699569.pdf (accessed on 10th September 2025).
19. Di Palma, L., Verdone, N., & Vilardi, G. (2017). Competitive reaction modelling in aqueous systems: The case of contemporary reduction of dichromats and nitrates by nZVI. https://core.ac.uk/download/127587940.pdf (accessed on 10th September 2025).

20. Farra, R., Hevia, M., Knop-Gericke, A., López, N., et al. (2012). *In situ* surface coverage analysis of RuO_2-catalyzed HCl oxidation reveals the entropic origin of compensation in heterogeneous catalysis. https://core.ac.uk/download/210619064.pdf (accessed on 10th September 2025).
21. Gokoglu, S. A. (2025). Chemical vapor deposition modeling: An assessment of current status. https://core.ac.uk/download/pdf/42820573.pdf (accessed on 10th September 2025).
22. Gross, A., McQuarrie, D. A., Dumesic, J. A., Reuter, K., Reuter, K., Reuter, K., Reuter, K., et al. (2005). First-principles kinetic Monte Carlo simulations for heterogeneous catalysis, applied to the CO oxidation at $RuO_2(110)$. http://arxiv.org/abs/cond-mat/0510234 (accessed on 10th September 2025).
23. Güell, A. G., Zhang, G., Unwin, P. R., Patten, H. V., & Schmickler, W. (2016). Nanoscale electrochemistry of sp^2 carbon materials: From graphite and graphene to carbon nanotubes. https://core.ac.uk/download/293881802.pdf (accessed on 10th September 2025).
24. Hernández-Hernández, S. I., Ledesma-Durán, A., & Santamaría-Holek, I. (2016). Generalized Fick–Jacob's approach for describing adsorption–desorption kinetics in irregular pores under non-equilibrium conditions. http://arxiv.org/abs/1604.08549 (accessed on 10th September 2025).
25. Hoffmann, M. J., Matera, S., & Reuter, K. (2014). *kmos*: A lattice kinetic Monte Carlo framework. https://core.ac.uk/download/267951745.pdf (accessed on 10th September 2025).
26. Ibbotson, D. E., Taylor, J. L., & Weinberg, W. H. (1981). Carbon monoxide oxidation catalysis over *Ir(110)*. https://core.ac.uk/download/9124071.pdf (accessed on 10th September 2025).
27. Jones, D. L., Leitner, D., Oburger, E., Roose, T., et al. (2010). Adsorption and desorption dynamics of citric acid anions in soil. https://core.ac.uk/download/pdf/1633619.pdf (accessed on 10th September 2025).
28. Kaplani, E., Panagiotaras, D., Papoulis, D., Rapsomanikis, S., et al. (2014). Nanocrystalline TiO_2 and halloysite clay mineral composite films prepared by sol–gel method: Synergistic effect and the case of silver modification to the photocatalytic degradation of basic blue-41 azo dye in water. https://core.ac.uk/download/41990680.pdf (accessed on 10th September 2025).
29. Kratzer, P., Penev, E., & Scheffler, M. (2002). First-principles studies of kinetics in epitaxial growth of III–V semiconductors. http://arxiv.org/abs/cond-mat/0105430 (accessed on 10th September 2025).

30. O'Shaughnessy, B., Brandrup, J., Clegg, D. W., Clement, F., Cohen Addad, J. P., Cohen Addad, J. P., Cohen Stuart, M. A., et al. (2004). Non-equilibrium in adsorbed polymer layers. http://arxiv.org/abs/cond-mat/0407246 (accessed on 10th September 2025).

31. Perez, J. P. H. (2020). Green rust formation and reactivity with arsenic species. https://core.ac.uk/download/288114074.pdf (accessed on 10th September 2025).

32. Reuter, K., & Wang, T. (2015). Structure sensitivity in oxide catalysis: First-principles kinetic Monte Carlo simulations for CO oxidation at *RuO_2(111)*. https://core.ac.uk/download/532709023.pdf (accessed on 10th September 2025).

33. Rogal, J., Reuter, K., Scheffler, M., & Blaha, P. (2006). First-principles statistical mechanics study of the stability of a sub-nanometer thin surface oxide in reactive environments: CO oxidation at *Pd(100)*. https://core.ac.uk/download/210850442.pdf (accessed on 10th September 2025).

INDEX

A

Acrylated polyethylene glycol diacrylate 174
Activation energy 90
Active site geometric orientation 208
Addition polymerization 168, 169, 172
Adsorption 371, 373, 374, 377, 380, 381, 384, 385, 386, 390, 395, 400, 401, 403, 410
Adsorption kinetics 371, 373, 395, 401
Agricultural management 384
Agricultural practice 385
Ammonium persulfate-ammonium sulfite monohydrate (APS-AS) 177
Arene-ruthenium complexes 130, 131
Arrhenius effect 296
Arrhenius equation 6, 7, 9, 20, 21, 22, 23, 31
Astrophysics 264
Atmospheric pressures 127
Atmospheric science 333, 335
Atomic force microscopy (AFM) 99, 392
Atomistic molecular dynamics simulations 298
Atom transfer radical polymerization (ATRP) 167

B

Basal plane activity 377
Bimolecular kinetics 267
Biocatalysts 216
Biochemical pathways 42, 43
Biochemical reactions 296
Biochemistry 84, 97, 111
Biological systems 143, 144, 153, 291, 292, 294, 295, 296, 303, 306, 308, 309, 310, 326
Biology 86, 111
Biosphere 132
Butanol 4, 5, 37

C

Cadmium 385, 408
Carbon dioxide 298, 318
Carbon materials 374, 376, 377, 410
Carbon monoxide 86, 97, 102, 105, 121
Catalysis 207, 208, 209, 210, 211, 218, 227, 232, 237
Catalyst design 213
Catalysts 1, 3, 7, 8, 10, 14, 24, 25, 26, 27, 28, 29, 34, 35, 210, 211, 212, 216, 230, 233, 242
Catalyst surface 213, 221, 222
Catalytic performance 377
Catalytic reactions 130, 211, 218
Cavity-MicroElectrode (CME) 375
Cell signaling 259
Chain-growth polymerization 164, 167

Chain reaction dynamics 261
Chain reactions 128, 131, 260
Chemical behaviors 4
Chemical conversions 3
Chemical kinetics 1, 2, 3, 4, 14, 34
Chemical reaction networks (CRNs) 129
Chemical reaction rates 1
Chemical synthesis 46, 50, 69
Chemical vapor deposition 48, 56, 64
Chemisorption 381
Chlorofluorocarbons 333, 335, 357, 367
Classical thermodynamic principles 89
Climate change 127, 149
Cofactors 132
Collisions 83, 85, 86, 88, 89, 93, 94, 95, 97, 98, 103, 105, 109, 113, 114, 115, 116, 118
Collision theory 89, 90
Combustion dynamics 299
Complex-balancedness 308
Complex reactions 125, 128, 132
Computational fluid dynamics (CFD) modeling 52
Computational sciences 211
Computational Singular Perturbation (CSP) method 261
Computer modeling 109
Concentration 42, 43, 44, 45, 46, 47, 48, 49, 52, 53, 54, 55, 56, 57, 58, 59, 60, 61, 62, 63, 64, 65, 66, 69, 70, 71, 72, 73, 74, 76
Condensation polymerization 170, 171
Configurational-bias Monte Carlo (CBMC) simulations 304
Conservation laws 291, 292, 326
Controlled radical polymerization (CRP) 180
Conventional monomeric catalysts 223
Coordination polymers (CPs) 176
Copolymerization 171
Cyclic voltammetry (CV) 257

D

Decomposition reactions 299, 301
Defective sites 376
Degree of rate control (DRC) policy 382
Density functional theory (DFT) 391
Diacetylene (HCCCCH) 300
Diels-Alder cycloadditions 140
Differential equations 2
Differential Scanning Calorimetry (DSC) 169, 186
Dinitrosyl iron complex (DNICs) 275
DNA adsorption 384
DNA sequences 262
Drug discovery 2, 33, 34

E

Economic viability 254
Electroactive biofilms 218
Electrochemical imaging 374, 376, 392
Electron 336, 345, 346, 351, 354, 355, 369, 370
Electronic excitations 334
Electron transfer (ET) 257
Electron transfer (ET) reactions 345
Elementary reactions 259, 260
Energy 83, 84, 85, 86, 87, 88, 89, 90, 91, 92, 93, 94, 95, 96, 97, 98, 99, 100, 102, 103, 104, 105, 106, 107, 108, 109, 110, 111, 112, 113, 114, 115, 116, 117, 121
Energy profiles 135, 136
energy storage devices 126
Engineered nanomaterials (ENMs) 340, 345, 347
Engineering 163, 167, 172, 174, 181, 191, 197
Environmental chemistry 252, 263, 280, 281
Environmental degradation 166
Environmental evaluation 333
Environmental restoration 384
Environmental science 2, 3, 14, 44, 55, 62, 63, 67, 72
Enzyme catalysis 209
Enzyme-catalyzed processes 249
Enzyme-catalyzed reactions 220, 296, 308
Enzymes 209, 216, 236

Ethene selective oxidation 42
Ethynyl radical (CCH) 300
Excited state formation 343
Experimental evidence 92, 98

F

Fast reactions 252, 280, 281
Fatty acid methyl esters (FAME) 215, 233
Fick-Jacobs equation 372, 374
First-order kinetics 15
Food safety policy 385
Fourier transform infrared (FTIR) spectroscopy 344
Fourier-transform infrared spectroscopy (FTIR) 392
Free radical mechanism 174
Freundlich adsorption model 388

G

Gas chromatography 50, 79
Gas-liquid mass transfer reactions 376
Gibbs free energy 294, 309, 311, 312
Glioblastoma multiforme (GBM) 357
Graph theory 129
Greenhouse gases 127, 335

H

Heterogeneous catalysts 212, 215, 216
High-temperature molecules 95
Homogeneous catalysts 212, 214, 215
Hydrogen combustion 2, 3
Hydrogen cyanide (HCN) 334, 342, 366

I

Industrial manufacturing 1
Industrial processes 141, 142
Inorganic chemistry 300
Internal combustion engines 299
Inter-reactant interaction 12

J

Jacobian matrix 292, 294, 309, 310

K

Kinetic activity 11, 18
Kinetic analysis 41, 72, 75, 78
Kinetic control 3
Kinetic experiments 1, 7
Kinetic Molecular Theory (KMT) 93
Kinetic Monte Carlo (kMC) 373, 386, 401, 406
Kinetics 249, 250, 253, 278, 287

L

Langevin collision dynamics 93
Langmuir adsorption model 387
Laser-induced breakdown spectroscopy (LIBS) 137
Le Chatelier's principle 139
Light absorption 334, 336, 338, 339, 354, 362
Light intensity 46, 333, 337, 341, 347, 349, 364
Light-matter interaction 338
Linear programming 129
Linear regression 394
Liquid ammonia 352, 368

M

man-made materials 164
Markovian dynamics 290, 325
Markovian reaction networks 290
Mass action kinetics 42
Mass Action Law 13
Material interaction 373
Materials science 45
Materials technology 44
Material synthesis 125
Mathematical modeling 294
Mathematical models 1
Mechanical motions 92
Mechanical strength 164
Metabolic pathways 251, 252, 258, 259, 281
Methane 86, 97, 102, 115
Michaelis-Menten kinetics 250, 270, 284

Microbial catalysis 218
Microbial fuel cells (MFCs) 218
Microcapillary systems 257
Microscopic analyses 218
Modularity 124
Molecular weight 163, 164, 165, 167, 168, 170, 173, 178, 179, 180, 181, 182, 183, 184, 188, 189
Monte Carlo simulations 219, 226, 243, 246
Montreal Protocol 335
Morphology 391, 392
Multi-component systems 338
Multicomponent systems 43, 74

N

Nanoplastics 166, 198
Nanoscale 376, 392, 399, 400, 404
Network dynamics 44
Nitrogen oxides (NOx) 348
Nonextensive system 100
Nonlinearity 290
Nucleosynthesis 84, 87, 116, 120

O

Organic synthesis 211
Organocatalyst pathways 140
Oxygen reduction reaction (ORR) 218
Ozone depletion 333, 335, 357, 367

P

Perfluoroalkyl acids (PFAAs) 341, 345, 347
Perfluorooctane sulfonate (PFOS) 345
Perfluorooctanoic acid (PFOA) 345
Petrochemical 210, 211
Pharmaceutical field 44
Pharmacokinetics 290
Photochemical efficiency 336, 343, 348, 355
Photochemical kinetics 331, 333, 335, 357, 361, 367
Photochemical reactions 46
Photodynamic therapy (PDT) 357, 362
Photosynthesis 298, 302, 331, 333, 336
Photothermal oxidation 337, 341, 349, 369
Phototoxicity 331
Photovoltaic cell 331
Physisorption 377, 381
Phytochromes 332, 343, 344, 353, 354
Polar stratospheric clouds (PSCs) 312, 335
Polycyclic aromatic hydrocarbons (PAHs) 131
Polyelectrolyte gels 166
Polymerization 163, 164, 165, 166, 168, 169, 170, 171, 172, 174, 175, 177, 181, 185, 190, 191, 196, 198, 202
Polymerization dynamics 164, 187, 197
Polymerization kinetics 163, 164, 191, 196
Polytetrafluoroethylene (PTFE) 333
Power plants 299
Precision spectroscopy 93
Pressure 42, 43, 44, 45, 46, 47, 48, 51, 52, 59, 60, 67, 73, 76
Primary reference fuel (PRF) 87
Product synthesis 300, 319

Q

Quantum chemistry 93, 109
Quantum mechanics/molecular mechanics (QM/MM) 298
Quantum yield 340, 341

R

Ramifications 45
Rapid compression machines (RCMs) 87
Rapid reactions 249, 252
Rate laws 12, 13
Reactant concentration 179
Reaction dynamics 289, 291, 292, 317, 323
Reaction mechanisms 1, 2, 3, 4, 7, 12, 13, 15, 16, 17, 18, 20, 28, 29, 30
Reaction networks 42, 43, 45, 46, 75, 81
Reaction order 42, 43, 45, 46, 48, 49, 51,

57, 58, 59, 61, 62, 63, 64, 65, 67, 68, 69, 70, 72, 73, 74, 76
Reaction pathways 133
Reaction performance 44
Reaction rate calculation 43, 71
Reaction rates 1, 3, 4, 7, 8, 9, 10, 11, 17, 18, 19, 20, 21, 22, 23, 25, 27, 28, 30, 31, 32, 34, 35, 36
Reactive oxygen species (ROS) 174
Redox reactions 302
Reduction-oxidation reactions 302
Reflective high-energy electron diffraction (RHEED) 391
Rero-order kinetics 14
Reversible Addition-Fragmentation chain Transfer (RAFT) 170
Reversible reactions 129, 130
Risk assessment 385

S

Scanning electron microscopy (SEM) 392
Second-order reactions 15
Selective catalytic reduction (SCR) 51
Sensor development 384
Silicon-technology-based microreactors 263
Silver catalysts 378, 385, 390
Silver nanoparticles 353
Situation 2
Social sciences 290
Soil health 385
Solid catalysts 262
Solid oxide fuel cells (SOFC) 267
Solid particles 376
Solid-state interactions 130
Solid surfaces 376, 377
Sophisticated quantum mechanics 89
Spatial dynamics 258
Step-growth polymerization 167
Stopped-flow spectroscopy 274
Stratospheric chemistry 335, 367
Substrate orientation 208
Sudden responses 258
Sulfur photochemistry 46
Surface area 383, 399, 400
Surface reactions 371, 372, 375, 377, 401
Sustainable materials 166, 176
Synthesis reactions 298, 300
Systematic responses 136

T

Tangential Stretching Rate (TSR) 251, 256, 261
Temperature 1, 3, 5, 6, 7, 8, 9, 10, 14, 18, 19, 20, 21, 22, 23, 24, 25, 27, 29, 31, 32, 34, 35, 42, 43, 44, 46, 47, 48, 49, 50, 52, 54, 56, 57, 59, 60, 71, 72, 73, 76, 77, 83, 84, 86, 88, 91, 98, 105, 107, 111, 113, 114, 116
Temperature dependence analysis 88
Temperature differences 49
Temperature-programmed desorption (TPD) 391, 393
Temperature sensitivity 83, 109, 115, 117
Thermal stability 164, 167, 168, 178, 187
Thermodynamic information 292, 294, 310
Thermodynamic parameters 99
Thermodynamics 83, 92, 104, 111, 112, 114
Thermogravimetric Analysis (TGA) 169
Three-dimensional (3D) printing 166
Time-dependent Hartree-Fock (TDHF) 305
Time-resolved techniques 355
Transition states 134
Transition State Theory 86
Transition State Theory (TST) 305
Tropical analysis 251, 287
Tropical geometry principles 8
Tryptophan synthase (TRPS) 298

U

Ultrafast laser 3D lithography 166

Ultraviolet (UV) light 348
Ultraviolet-visible spectroscopy (UV-Vis) 353
UV-visible light exposure 338

V

Vinyl addition copolymerization 168
Vinylic addition polynorbornenes (VA-PNB) 130
Volatile organic compounds (VOCs) 348

W

Waste cooking oil (WCO) 215, 234
Waste treatment 384
Water 295, 298, 299, 307, 319, 321, 325
Wavelength 46, 333, 336, 337, 341, 342, 348, 349, 359, 362, 369

X

X-ray photoelectron spectroscopy (XPS) 392

Z

Zeolites 212, 218, 225
Zero-order reactions 14